Applied Statistics for Eco

This book introduces undergraduate students to commonly used statistical methods in economics. Using examples based on contemporary economic issues and readily available data, it not only explains the mechanics of the various methods, but also guides students to connect statistical results to detailed economic interpretations. Because the goal is for students to be able to apply the statistical methods presented, online sources for economic data and directions for performing each task in Excel are also included.

The book begins by demonstrating the role of empirical analysis in economics. It next discusses how economists use statistics to describe a single economic variable through frequency distributions, measures of central tendency, and measures of absolute and relative dispersion. It then considers methods for describing temporal changes in a variable. The remainder of the book focuses on basic statistical inference procedures used by economists. This includes statistical estimation and hypothesis testing of a single parameter, and correlation and simple linear regression analysis. It concludes with an introduction to multiple linear regression analysis, which serves as a springboard to a dedicated course in econometrics.

The text bridges the current gap between quantitative methods texts designed for business and economics that primarily focus on business applications, and dedicated econometric texts that give short shrift to descriptive statistics. It emphasizes the economic interpretation of statistical results and the appropriate economic applications of each method, features rarely found in existing texts.

Researchers with a general background in statistics who wish to better understand their application to economic questions may also find the clear explanations of each method and interpretation of economic results to be of value.

Margaret Lewis is Professor of Economics at the College of Saint Benedict/Saint John's University, Minnesota, USA.

Applied Statistics for Economists

Margaret Lewis

Routledge
Taylor & Francis Group

LONDON AND NEW YORK

First published 2012 by Routledge
2 Park Square, Milton Park, Abingdon, Oxon OX14 4RN

Simultaneously published in the USA and Canada
by Routledge
711 Third Avenue, New York, NY 10017

Routledge is an imprint of the Taylor & Francis Group, an informa business

British Library Cataloguing in Publication Data
A catalogue record for this book is available from the British Library

Library of Congress Cataloging-in-Publication Data
Lewis, Margaret, 1956–
    Applied statistics for economists/Margaret Lewis.
    p. cm.
    1. Economic–Statistical models. 2. Statistics. I. Title.
HB137.L49 2011
519.5–dc22

ISBN: 978-0-415-77798-8 (hbk)
ISBN: 978-0-415-55468-8 (pbk)
ISBN: 978-0-203-80845-0 (ebk)

Typeset in Times New Roman
by Sunrise Setting Ltd, Torquay, UK

Contents

List of figures

List of tables

List of charts

List of boxes

Preface

Economists have used empirical information to understand economic phenomena for many years. While economic data, and the statistical methods for evaluating it, are increasingly sophisticated, the basic reason for using such evidence remains the same – learning about an economy by systematically measuring economic activity.

The following text presents quantitative methods frequently used by economists to gain such understanding. The methods examined are divided into two categories – descriptive statistics used to measure individual economic variables, and statistical inference used to measure relationships between economic variables and to test hypotheses about them. In addition to demonstrating how to generate a numerical result with each method, the current text applies each method to specific and contemporary economic questions. The text also provides detailed economic interpretations of the statistical results, a feature missing from virtually every current text, and it explains the strengths and limitations associated with each method. The goal is for students to learn to "do" empirical economics as they work through the text.

The text was initially developed as class notes for a course in quantitative methods in economics and grew out of a dissatisfaction with the scope and emphasis in existing textbooks. Many currently available texts, designed for business and economics courses in statistics, focus primarily on business examples, rarely using actual data for economic variables and seldom discussing questions of interest to economists. A second group of texts focuses on econometric analysis, an important statistical tool in economics, but certainly not the only quantitative method useful to economists.

Most contemporary texts also emphasize the mechanics of generating statistical measures, but provide little on the important questions of economic interpretation and economic application of the methods. As a result, students may gain technical knowledge from such an emphasis, but, when it comes to conducting their own economic research, they may not fully understand how the methods can be applied to their own research, nor can they always accurately interpret the statistical results generated. The current text attempts to bridge this gap.

Textbooks for statistical methods in economics have not always been so focused. Several years ago, I found a used copy of Harold W. Guthrie's *Statistical Methods in Economics* (Richard D. Irwin, Homewood, IL, 1966). Guthrie's book, like contemporary texts, provided instruction on how to use a wide range of quantitative methods, but it did more than that: it used actual economic data in most examples and, most importantly, it provided economic interpretations of the empirical results generated. The text also emphasized the appropriate, and inappropriate, use of statistics and quantitative methods in economics, another feature sometimes given short shrift in contemporary textbooks.

Guthrie's book inspired the class notes that evolved into the current text. While substantially different, it is my hope that the current text retains the clarity, accessibility, and applicability of Guthrie's text. I also want today's students to better understand, as those who used Guthrie's text surely did, the empirical dimensions of contemporary economic problems and how quantitative methods can increase our understanding of them. Indeed, to put a twist on an old economic adage, theory without measurement can lead us to conclusions that have little basis in reality.

Margaret Lewis
College of Saint Benedict/Saint John's University
St. Joseph/Collegeville, Minnesota

December 2010

A note on the textbook's focus and scope

This textbook is designed for undergraduate students of economics who have completed an introductory course in statistics. While we review key findings from probability theory, sampling theory, and probability distribution functions in Part III, the purpose is to apply these concepts to hypothesis testing of sample means, correlation analysis, and regression analysis. Students without the prerequisite statistics course may find this discussion too brief and require additional study in basic statistics.

The economic questions raised throughout the text are based on the assumption that students have completed courses in micro- and macroeconomic theory, and that they have acquired some sophistication in applying economic theory to economic issues. Since it is the goal for students to learn these methods so they can conduct empirical analysis, this background is crucial. The text is thus pitched to students in the second half of their undergraduate education in economics.

The text also contains detailed instructions at the end of each chapter for using Microsoft Excel 2007/2010 to apply the quantitative methods described. While those familiar with dedicated statistical software may find Excel's tools less efficient, the widespread use and availability of Excel recommends its use here.

As economists know, economic statistics are governed by detailed definitions that can vary significantly across nations. Thus the text uses data primarily from the United States (reflecting the author's background), although the economic variables measured here are available for most economies. For those unfamiliar with a data series, the text will include information about who collects the data and how the data is constructed. Those wanting additional information can access the agency responsible, where detailed handbooks of methods and other information can be found.

A note on the "Where can I find...?" boxes

These boxes, found throughout the text, provide information about several popular locations for online economic data. Online accessibility has greatly facilitated empirical economic analysis, much as the development of computing power and the availability of statistical software has transformed such work.

The growing and widespread availability of economic data necessarily means much more information is available than can be recorded in this book. While each data source is current as of its writing, it is possible that some details, or even the web address itself, may have changed. (Readers are directed to use a search engine if forwarding links are not provided.)

Information is also increasingly available through the agencies that generate the data. It is recommended that students first visit the agency responsible for a particular data series to see if it is available and easily downloadable from that source.

1 The role of statistics in economics

1.1 Understanding the economy using empirical evidence

Economic questions are at the heart of our daily lives. Many of us, for example, wonder if we are materially better off than our parents or if our children will make economic gains beyond our own. We also have concerns about current economic conditions, such as the likelihood of a recession and whether economic actors are efficiently and responsibly utilizing resources when engaged in economic activities. And as public policies are enacted, we question if they are achieving the desired effects or if they are generating unintended consequences.

Society turns to economists for answers to such questions. Economists, in turn, rely on economic theory combined with statistical methods to arrive at their answers. You are no doubt familiar with many theoretical frameworks developed by economists for understanding economic relationships, as they are the focus of most undergraduate education in economics. However, to obtain concrete answers to society's questions, economists typically marry theory with empirical evidence obtained from the collection and statistical analysis of data. This latter approach is the focus of the current text.

What do economists mean when they talk about **empirical evidence**? In general, it is information that is quantified and analyzed in a systematic fashion for the purposes of understanding economic relationships and outcomes. For example, statistical descriptions of a country's income distribution can help us answer the first pair of questions posed above, while statistical hypothesis testing might be used to evaluate the effects of a public policy.

How, then, do economists know what empirical evidence to use for which questions? The answer is twofold and demonstrates the vital connection between economic theory and statistical methods. Once a question is posed, economists then identify appropriate theoretical frameworks that articulate the relationships between the key economic variables based on the theory's assumptions about economic behavior and institutions. Such theoretical frameworks not only provide economists with an orderly way of thinking about economic activities, but also guide our choices of effective empirical evidence for the questions at hand.

Suppose, for example, we wanted to assess if today's middle-aged adults are materially better off than their parents. We would begin by first defining what it means to be "materially better off" and then we would identify a quantifiable variable, such as household income, to measure it. After collecting data for household income, we would fully describe the statistical characteristics of the income variable, which might include constructing a frequency distribution, calculating the "average" and spread of the series, and perhaps its change, in real terms, over time. From these descriptive measures, we could then test if average household income, its distribution, or its growth rate are statistically different between the two generations. If such differences do obtain, we could then use economic theory along with the statistical method known as regression analysis to measure what factors might contribute to the variation in household income between the two groups.

Generating the appropriate empirical evidence to understand economic activity requires us to learn a variety of statistical methods. These methods, however, are worth very little if learned without consideration of the theoretical connections between economic variables. Thus we will use statistical methods to provide a link between abstract economic theory and the complex details of the actual economy.

For purposes of this text, we will consider methods that serve one of the two following functions:

- Describing characteristics of an economic variable for which we have a large number of observations. We will refer to such methods as **descriptive statistics**, and they are the focus of Parts I and II of the text.
- Drawing conclusions about economic variables in the context of incomplete information. These methods will be referred to as methods of **statistical inference**, and they are the focus of Parts III, IV, and V.

Before we turn to these methodological systems for generating empirical evidence beginning in Chapter 3, we will first provide a brief overview of several data series that measure important economic phenomena, raising questions about appropriate methods of statistical analysis for each. We will next introduce, in Chapter 2, key principles for effectively communicating our empirical results, examining in detail one popular approach, the use of charts to illustrate economic phenomena.

1.2   Measuring economic welfare

Perhaps the most important issue in economics is the material well-being of individuals and society. Not surprisingly, economists have developed a number of theoretical models to evaluate economic welfare. Microeconomists working in the neoclassical tradition, for example, may define welfare in terms of the marginal utility an individual obtains from different levels of consumption, while macroeconomists often define welfare in terms of aggregate economic growth. Although the theoretical bases for defining welfare are quite different, both groups of economists face similar problems when trying to measure their concepts of welfare. Neoclassical microeconomists, for example, face challenges in accurately measuring an individual's utility, while macroeconomists must account for all sources of aggregate economic activity. Despite these problems, economists have developed statistical measures to approximate economic "well-being."

1.3   Distribution of household income

One commonly used measure of economic welfare is income. As the primary means for acquiring the material necessities and comforts of contemporary life, it is an obvious choice. But how do we define income? Is it the actual dollar amount of money we have at hand, or does it include transfer and in-kind payments? Do we use pre-tax or post-tax amounts as the appropriate measure? In terms of measuring economic well-being, do we consider the absolute number of dollars or do we compare our income to that received by others? These are all questions we must answer prior to any statistical analysis of economic well-being.

Fortunately, we can often draw on the question we wish to answer and the relevant economic theories and practices to help make this determination.

Let us decide to measure income, and thus economic well-being, as the distribution of total money income across U.S. households. Total money income is defined by the Current Population Survey (CPS) as: "the [pre-tax] arithmetic sum of money wages and salaries, net income from self-employment, and income other than earnings. The total income of a household is the arithmetic sum of the amounts received by all income recipients in the household" (U.S. Census Bureau 2010b). We have chosen this particular measure because of its ready availability, a criterion we cannot underestimate when conducting empirical economic analysis.

Economists are fortunate that numerous surveys are conducted to gather economic data on variables we can use for economic analyses. Here we will utilize published statistics from the Annual Social and Economic Supplement to the CPS to describe the income distribution for households in the U.S. for 2008 (U.S. Census Bureau 2009a).

As we see in Table 1.1, there were (approximately) 171,181,000 U.S. households in 2008. Here we observe the percentage of households whose income falls into each specific income class, noting, for example, that 14 percent of U.S. households in 2008 had total money income between $35,000 and $49,999. This construction is known as a frequency distribution, a method for organizing many data observations into easily understood groups. The final set of statistics in Table 1.1 measures the central tendency (mean and median) of U.S. household income in 2008.

What questions might we be able to answer from these descriptive statistics? Suppose we have determined that a particular level of household income, say $17,330, is required to meet basic material needs for a two-adult, one-child household in 2008 (this was the 2008 poverty threshold for a three-person family, with one related child under 18 years of age). According to Table 1.1, at least 12.9 percent of U.S. households had total money income below that threshold. We further note that this income level is well below both the median and mean household incomes for 2008. Thus a household with total money income of only $17,330 is likely to consider itself less well off than most Americans.

Table 1.1 Percentage distribution of total money income across U.S. households, 2008

Total number of households (thousands)	*117,181*
Household income before taxes	*Percentage of all household units*
Under $5,000	3.0
$5,000 to $9,999	4.1
$10,000 to $14,999	5.8
$15,000 to $24,999	11.8
$25,000 to $34,999	10.9
$35,000 to $49,999	14.0
$50,000 to $74,999	17.9
$75,000 to $99,999	11.9
$100,000 and over	20.5
Total	100.0
Median household income	$50,303
Mean household income	$68,424

Data source: DeNavas-Walt *et al.* (2009, Table A-1).

We could also use the statistics presented in Table 1.1 to compare the 2008 distribution of household income with the distribution of household income from, say, 30 years ago. Assuming the data are in real (inflation-adjusted) units, this comparison will help answer the earlier question about generational well-being. We might also construct additional statistical measures that economists use to further assess economic well-being over time and across groups.

1.4   Assessing full employment

Most Americans receive the majority of their income from earnings obtained through employment in a labor market. Consequently, the extent to which an economy achieves full employment of its human resources is another important aspect of economic welfare. Economists have traditionally assessed this goal in terms of how many people in the civilian labor force are unemployed, or, more commonly, what percentage of the labor force is unemployed.

We begin our analysis by explaining how the U.S. unemployment rate is calculated to determine who is, and who is not, included in the statistic. According to the Current Population Survey of the U.S. Bureau of Labor Statistics (BLS), individuals who are under 16 years of age, serve in the Armed Forces, or live in institutions such as prisons are not included in the official statistics. In addition, the individual must be classified as a member of the labor force, which BLS defines as the sum of people who are employed and unemployed. Individuals who are not members of the labor force, which "includes retired persons, students, those taking care of children or other family members, and others who are neither working nor seeking work" (U.S. Bureau of Labor Statistics 2010d), will thus not counted in the unemployment rate.

The U.S. unemployment rate exhibits considerable movement between January 1992 and May 2010, and we can use those empirical patterns to comment on changes in economic welfare. As we observe in Chart 1.1 (p. 5), the U.S. unemployment rate initially increased until June 1992, after which it declined, for the most part, until January 2001. The unemployment rate reversed direction in 2001 through June 2003, during which time the U.S. economy was in recession (officially between March 2001 and November 2001[1]). After a period of declining unemployment between July 2003 and May 2007, the unemployment rate rose again, and after December 2007, it appears to have done so at an increasing rate, coinciding with the official start of the recent U.S. recession.

Increasing unemployment rates affect many individuals. Those who lose their jobs, of course, no longer have the money income required to meet basic needs, clearly lowering their economic well-being. Consumers also typically respond by cutting back on purchases of non-necessities due to fear of losing their jobs, while business owners may see profits drastically reduced or even be forced out of business. These widespread effects will certainly reduce many peoples' economic welfare.

We might also use the unemployment rate as one measure of labor under-utilization[2] and thus as an indicator of the economy's inefficient use of labor resources. We could, for example, calculate the average monthly change in the unemployment rate, citing periods of increasing rates as evidence of reduced economic efficiency. Alternatively we could calculate the average monthly rate of unemployment during recessions and compare it to the average non-recessionary rate using the statistical inference procedure of hypothesis testing discussed in Chapter 11. If that differential is determined to be statistically significant, we would have empirical evidence to support the argument that economic welfare falls during recessionary periods in the economy.

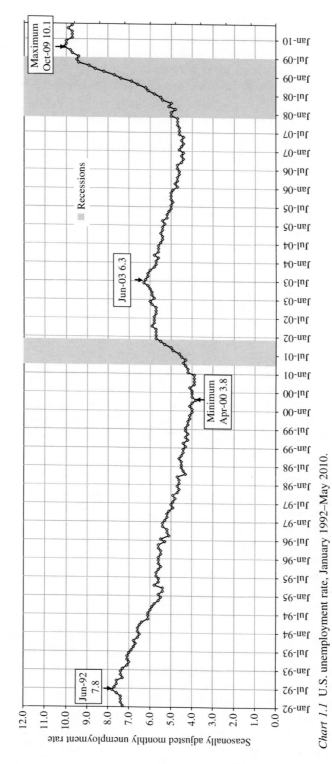

Chart 1.1 U.S. unemployment rate, January 1992–May 2010.

Data source: U.S. Bureau of Labor Statistics (2010b), Current Population Survey.

1.5   Calculating economic growth

Economic growth, measured as the change in national income or product, is closely connected to an economy's ability to fully utilize its resources. If growth is negative, for example, we expect to see some economic resources idled, but if the economy's national product is increasing over time, not only will more resources be employed, more income and more goods and services will be generated, improving overall economic well-being. This also makes economic growth a frequently used measure for assessing a nation's economic progress over time.

Economic growth is most commonly calculated as the average annual growth rate of a country's gross domestic product (GDP). Even though GDP is a frequently criticized measure for what it does and does not include, economists utilize GDP because "any estimate with known weaknesses is superior to ignorance and speculation as a basis for making a decision" (Guthrie 1966, p. 6). As long as we are explicit about such limitations, we can, with caution, employ GDP to assess how an economy changes over time.

Chart 1.2 (p. 7) illustrates how U.S. GDP changed between the first quarters of 1992 and 2010. Not only do we observe an overall upward trend in GDP, we can also calculate the amount by which GDP, on average, changed over the time period by fitting a linear trendline to the data. Applying this method to generate the trendline equation in Chart 1.2, we see that real GDP grew, on average, by $78,543,000,000 per quarter between the first quarter of 1992 and the first quarter in 2010.

We also note that GDP does not actually change by constant amounts over the time period. It appears, for example, that GDP grew more quickly during the 1990s than the 2000s. We can confirm this by calculating the average change for the two different decades and then testing if the difference between the two values is statistically different from zero. We could also perform a similar calculation and test to determine how GDP changes during recessionary periods as compared to expansion periods. Both sets of results will inform our assessment of economic well-being as measured by growth in the country's GDP.

1.6   Measuring inflation

Our discussion of economic growth referred to changes in *real* GDP. If domestic product increases over time but is measured in the current period's dollars, some or all of GDP's growth may be driven by increasing prices, thereby eroding purchasing power. Economists interested in measuring economic welfare typically use variables in which the effects of rising prices, or inflation, are factored out of the measured variable.

One method used to measure inflation is to construct a price index. Chart 1.3 (p. 8) illustrates how the most frequently used U.S. index, the Consumer Price Index for All Urban Consumers (CPI-U), changed between January 1992 and May 2010. We observe an overall upward trend for most months, indicating that inflation was indeed present in the U.S. economy. However, we also see that the U.S. experienced several periods of deflation during the 2000s, which some economists consider to be as serious a problem as inflation. Because uncertainty about future price trends destabilizes economic decision-making, economists and policy-makers prefer price stability in the economy.[3] Thus the patterns suggested in Chart 1.3 may challenge our ability to achieve economic well-being over time.

We also note that the CPI-U statistic is not measured in readily understood units. That is, what does "1982–84 = 100" on the Y-axis indicate? As we shall discuss in Chapter 6, index numbers are relative measures, which means that they are constructed using a base, or reference, period for comparison, with a value of 100 for that base period. To interpret

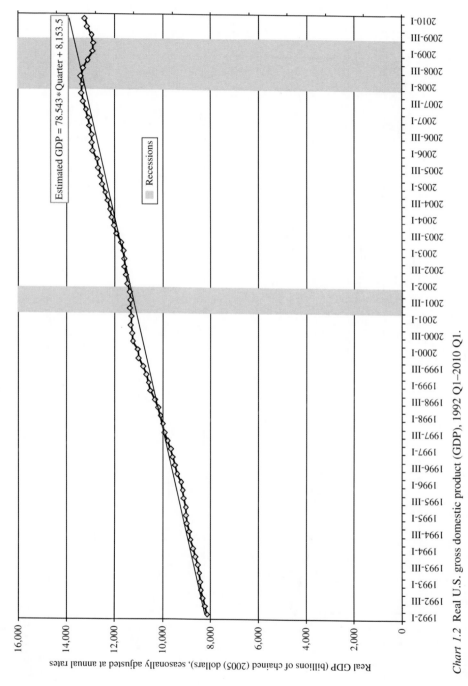

Chart 1.2 Real U.S. gross domestic product (GDP), 1992 Q1–2010 Q1.

Data source: Table 1.1.1, U.S. Bureau of Economic Analysis (2010c), National Economic Accounts.

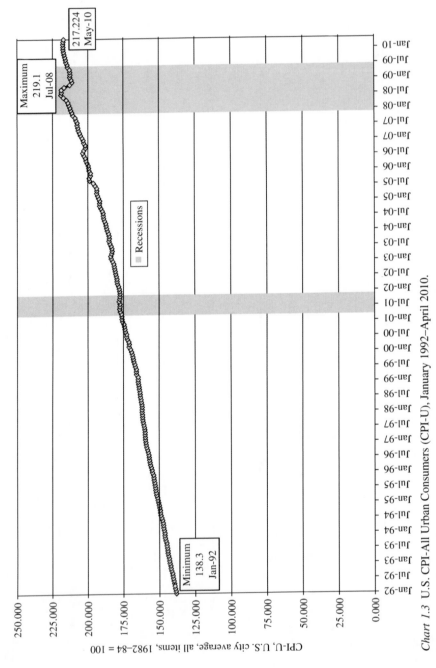

Chart 1.3 U.S. CPI-All Urban Consumers (CPI-U), January 1992–April 2010.

Data source: U.S. Bureau of Labor Statistics (2010b), Consumer Price Index.

an index correctly, we must know the base period because all price changes, as measured by the CPI, are measured relative to its reference period. Since U.S. CPI-U uses the August price level between 1982 and 1984 as its base, the July 2008 CPI-U's value of 218.1 tells us that the U.S. aggregate price level was 118.1 percent higher in July 2008 than the 1982–84 average price level in the U.S.

1.7 Theoretical relationships between unemployment and economic growth

Analyzing individual economic variables such as household income and the unemployment rate as we have seen can provide us with important empirical evidence for assessing questions about economic well-being. However, no economic variable changes in isolation, so to fully understand economic outcomes, we almost certainly need to consider how economic variables are related. That is, we want to explain in empirical terms the connections between two or more variables.

Let us reconsider the two macroeconomic goals of full employment and economic growth. As we have seen, we can assess whether either goal has been achieved by examining the descriptive statistics or empirical trends of the unemployment rate and the change in real GDP, respectively. But are the two variables related to one another? To answer this question, economists turn not only to theoretical explanations, but also to statistical methods to determine if the empirical evidence supports a relationship. Generating such evidence is not merely an academic exercise; it can also inform policy discussions. If economic growth, for example, corresponds to a *higher* unemployment rate, an undesirable social outcome, then policies promoting economic growth may not be supported by policy-makers. On the other hand, if positive economic growth coincides with *lower* unemployment rates, then growth policies are much more likely to be politically feasible.

Economists have identified important theoretical connections between GDP (the basis for assessing economic growth) and the unemployment rate (the basis for determining full employment) in both the short and long runs. In the short run, for example, we expect a widespread reduction in product demand, as manifested in declining GDP, may lead to firms reducing their labor force, which then increases unemployment.[4] This means we would expect to observe a negative relationship between a change in GDP and the unemployment rate in the short run. Given this hypothesized relationship, we can use statistical methods to provide us with empirical evidence for evaluating the theoretical relationship.

Let us consider the short-run relationship between changes in output and the unemployment rate by examining empirical evidence for the 50 U.S. states and the District of Columbia. In particular, let us first determine how gross state product (GSP) changed between 2007 and 2008 (defined as the variable, $\%\Delta\text{GSP}$) for the 51 geographical units and then plot that change for each state against the state's 2008 unemployment rate (STURT). Here we would expect, based on the theoretical underpinnings, that a negative change in a state's GSP would correspond to a higher-than-average state unemployment rate, while (positive) growth in GSP would correspond to a lower-than-average unemployment rate. In mathematical terms, we expect to observe a negative (or inverse) relationship between the two variables.

We now turn to the empirical evidence presented in Chart 1.4 (p. 10). This chart differs from the three previous charts in that we have now plotted paired information for two economic variables, gross state product and the unemployment rate. For each of the 51 geographical units (represented by the square markers in the chart), we observe the paired values of percentΔGSP for each state on the X-axis and its corresponding STURT on the Y-axis.

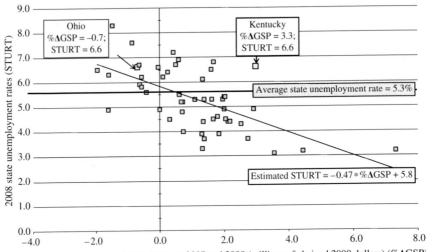

Chart 1.4 Percentage change in real gross state product (2007–2008) and state unemployment
rates (2008).

Data sources: Unemployment rates: U.S. Bureau of Labor Statistics (2010g), Local Area Unemployment
Statistics. Real gross state product: U.S. Bureau of Economic Analysis (2010d), Gross Domestic Product by
State.

Once we have empirically displayed the relationship between the annual percentage
change in U.S. gross state product between 2007 and 2008 and state unemployment rates
for 2008, we can generate a statistical estimation of that relationship by adding a line that
best fits the data. Here we see that a one percentage point increase in GSP corresponds, on
average, to an unemployment rate that is 0.47 percentage points lower, *ceteris paribus*.[5]

We can now consider how individual states compare with this overall relationship. We see,
for example, that Ohio's real GSP fell by 0.7 percentage points between 2007 and 2008. This
change corresponds to a higher-than-average 2008 unemployment rate in Ohio of 6.6 percent
and is consistent with the relationship predicted by economic theory.

The negative relationship does not, however, hold for all states. We see, for example, that
Kentucky's state unemployment rate equals the higher-than-average rate found in Ohio, but
Kentucky's GSP actually increased by 3.3 percent between 2007 and 2008. This result may
lead us to ask if our statistical estimates are incorrect.

The short answer is "no"; such examples do not negate the validity of our statistical
estimates. More importantly, such examples do not indicate that the predicted economic
relationship is incorrect. From a statistical perspective, we are estimating an average relationship between unemployment and GSP using a method that minimizes the sum of the
squared differences between each state's actual unemployment rate and that predicted by the
linear trendline. Such a method is not invalidated simply because some paired observations
deviate from the general tendency in the data.

Theoretically, we can well imagine many economic reasons for those cases that do
not exhibit the hypothesized relationship. One possible factor relates to differences across
industrial sectors. In Kentucky, for example, motor vehicle manufacturing, an industry
particularly hard hit in recent years, saw its employment levels decline by 13.2 percent

between 2007 and 2008. While not a perfect indicator of the unemployment rate, such a decline in an industry which accounted for 7.32 percent of Kentucky's GSP in 2007 surely contributed to its higher-than-average unemployment rate. However, Kentucky's economy also experienced both output and employment gains in other sectors, such as professional and technical services, educational and health services, and government, which may account for Kentucky's rising GSP over the same period.

The empirical evidence portrayed in Chart 1.4 invites further economic analysis. We know, for example, that the annual change in a state's gross product is not the only factor that contributes to the state's unemployment rate. Economists do, fortunately, have statistical methods for analyzing multiple factors, one of which we will examine in Part V.

1.8   The connection between economic theory and statistical evidence

The examples presented in Chapter 1 suggest that empirical evidence obtained through statistical methods can expand economic analysis from the merely theoretical to the measurement of actual economic activity. Applying statistical methods to economic questions can be quite exciting, but such work can also be the source of considerable frustration. Any economist engaging in empirical analysis quickly learns that the statistical evidence does not always line up with the underlying economic theory. While the discrepancy between our empirical results and the underlying economic theory may be the result of correctable statistical problems, we cannot always reconcile the two.

In such instances, we may be inclined to abandon economic theory in favor of our statistical results. But even if we were to assume (foolishly) that statistics do not lie, we must always remember that our statistical results are premised on relationships identified by economic theory. This is to say, our statistical results are *meaningless* if they are not derived from a theoretical relationship, since the underlying economic theory guides our empirical analysis, not vice versa. Consequently, it is critical we never forget that "measurement without theory" is of little use in economics.

Summary

In Chapter 1, we have introduced some examples of how economists use statistical methods to generate empirical evidence for answering economic questions. The remainder of the text explores these, and other, methods frequently used by economists. In the process, we will consider how economists identify which statistical methods are most appropriate for answering particular questions in empirical economics, and we will demonstrate the critical role that economic theory plays in the application of these methods. In addition, we will discuss some of the protocols used by economists to effectively communicate their empirical results to others, and we will learn how to use Microsoft Excel, beginning at the end of this chapter, to generate empirical evidence for our analysis of economic questions and outcomes.

Concepts introduced

Descriptive statistics
Empirical evidence
Statistical inference

Box 1.1 Where can I find…? *The Economic Report of the President*

In the 1946 Employment Act, the U.S. Congress established a Council of Economic Advisors (CEA) to the President, whose job is to "analyze and interpret economic developments, to appraise programs and activities of the Government and to formulate and recommend national economic policy" (*Economic Report of the President* 1947, p. vii).

A key element of CEA's work is to submit an annual report no later than ten days after Congress receives the President's annual *Budget of the United States Government*. These annual *Economic Reports of the President* include the following:

- Current and foreseeable trends and annual numeric goals concerning topics such as employment, production, real income, prices, and Federal budget outlays
- Employment objectives for significant groups in the labor force
- Plans for carrying out program objectives

A large component of the annual *Economic Report* is Appendix B: Statistical Tables Relating to Income, Employment, and Production. The appendix is a rich source for national economic data in the following categories:

- National income or expenditure
- Population, employment, wages, and productivity
- Production and business activity
- Prices
- Money stock, credit, and finance
- Government finance
- Corporate profits and finance
- Agriculture
- International statistics

Each published table also includes the source of the statistics, which makes it easy for researchers to locate historical, or more extended, data series for the particular variables.

As of this writing, full *Economic Reports of the President* can be downloaded (in PDF format) from the Economic Report of the President (EOP) home page <http://www.gpoaccess.gov/eop/index.html> for 1995 forward, and the statistical tables in Appendix B are also downloadable (in XLS format) from 1997.

The EOP site also provides a link to FRASER®, the Federal Reserve Archival System for Economic Research, of the Federal Reserve Bank of St. Louis. FRASER contains facsimiles of every *Economic Report* (in PDF format). Some statistical tables from more recent *Economic Reports* are also available in a single PDF file for a group of years. (For example, you can download a PDF file containing tables on farm income from 1981 through 2010.) Note that you must first identify a particular table and then follow the link to see what years are available.

Box 1.2 Economic statistics and Microsoft Excel

Conducting statistical analyses of economic variables has become increasingly feasible for economists due to the advent of the personal computer and its dedicated software. This text will take advantage of both technologies by using Microsoft Excel 2007/2010 for all calculations, and it will include detailed directions for using Excel to perform the statistical methods discussed in each chapter.

While Excel has considerable limitations for advanced statistical analysis, it will serve our purposes here quite well. In addition, because Excel is both widely available and the statistical software of choice throughout the business world and public policy arena, mastering its statistical capabilities not only will enhance one's undergraduate education in economics but also will provide valuable skills for conducting analytical work after college.

Box 1.3 Working with Excel: some basics

Creating a new worksheet

To open a new worksheet in Excel 2007, simply open the program. If a blank worksheet does not appear, click on the **Microsoft Office Button** and **select New**. Highlight **Blank Workbook** and click **Create.**

To open a new worksheet in Excel 2010, click on the **File** tab. Under **New**, select **Blank workbook**.

Entering information into a worksheet

An Excel worksheet consists of rows, columns and cells. Each individual rectangle is a **cell**, which is identified by its placement in both the **column** (A, B, C, ...) and the **row** (1, 2, 3, ...). For example, the cell B3 is in the 2nd column and the 3rd row.

Both numerical information and text can be entered into an Excel cell. To do so, the cell must first be *activated*, which is done by clicking on it. This will place a dark border around the cell with a dark square in the lower right-hand corner (Excel calls this the **fill handle**), as in cell A1 below.

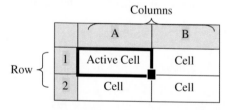

Enter the information either by typing directly in the cell or by typing the information into the formula bar, denoted as f_x, found above the worksheet.

Once the information has been typed in, press the <**Enter**> key. This will move the cursor to the cell immediately below the previously active cell. If that is not the correct location, move to the desired location either by using the arrow keys (up, down, right, left) or the mouse.

To **edit a cell**, double click on it, or use the **F2 key**. You may then edit the cell without re-entering the contents.

To **select multiple cells at once** (what Excel refers to as a **range**), place the mouse on the upper left cell of the range to be highlighted. Click and then hold and drag the mouse pointer to the lower right cell of the range. Once the desired range is highlighted, release the mouse button. The first cell shows a white background, while all other cells in the selected range show a shaded background.

Exercise

- Open a new Excel worksheet.
- Activate cell A2 by clicking on it. Key in "Gross domestic product"; press <Enter>.
- Cell A3 will now be the active cell. Enter a label for "Personal consumption expenditures".

Because personal consumption expenditures are *part* of GDP, we want to indent our entry to indicate that relationship. To indent the cell information, go to **Paragraph** on the **Home** tab, and click the **Increase Indent** icon. Type in the text, and hit <Enter>.

- Continue entering information until your worksheet looks like the one below.

	A	B
1		2007 II
2	Gross domestic product	13,755.9
3	Personal consumption expenditures	9,671.9
4	Gross private domestic investment	2,135.0
5	Net exports of goods and services	-721.6
6	Gover nment consumption expenditures and gross investment	2,670.5
7	Data source : Table 1.1.5, U.S. National Income and Product Accounts. U.S. Bureau of Economic Analysis. Online. Available from <http://www.bea.gov/national/nipaweb/>	

Moving or copying text to a new position

Select the range of cells B1:B4. Right-click and select the **Cut** button on the **Clipboard** on the **Home** tab, or use **Ctrl-X** to cut the selection.

Next activate the upper-left cell into which you want the cut information. Click the **Paste** button on the **Clipboard** on the **Home** tab, or use **Ctrl-V** to paste the selection.

If you wish to **copy text**, select the **Copy** button on the **Clipboard** on the **Home** tab, or use **Ctrl-C** to copy the selection.

Saving a worksheet

To **save a worksheet (file)** in Excel 2007 for the first time:

- Click on the **Microsoft Office Button**, and select **Save**.
- Key in the desired file name in the File name text box. *Word of advice*: use names that will make it easy to locate and identify the file in the future.
- Select an appropriate location for the file.
- Make sure the file type is "Excel Workbook" (with the *.xlsx extension).
- Click on **Save**.

To save a previously saved worksheet:

- Click on the **Save** icon on the **Quick Access Toolbar**, or
- Use the keyboard shortcut, **Crtl-S**.

To save a new worksheet in Excel 2010 for the first time:

- Click on the **File** tab, and select **Save**.
- Key in the desired file name in the File name text box. *Word of advice*: use names that will make it easy to locate and identify the file in the future.
- Select an appropriate location for the file.
- Make sure the file type is "Excel Workbook" (with the *.xlsx extension).
- Click on **Save**.

Box 1.4 Using Excel to perform basic mathematical operations

Excel can be invaluable for performing repeated mathematical operations. The following exercise demonstrates how to use Excel to sum columns of values and to calculate percentage changes. It also introduces several formatting procedures useful for managing large numbers of observations.

Exercise

Suppose we want to know how the number of new passenger cars imported into the U.S. by country of origin changed between 1988 and 2007. We will use the following data from the *World Almanac and Book of Facts* (2009) to determine the answer.

New and used passenger cars imported into the U.S., by country of origin (1), 1988–2007
Source: Foreign Trade Division, U.S. Census Bureau
(in number of units)

Year	Japan	Germany (2)	Italy	United Kingdom	Sweden	France	South Korea	Mexico	Canada
1988	2,123,051	264,249	6,053	31,636	108,006	15,990	455,741	148,065	1,191,357
1989	2,051,525	216,881	9,319	29,378	101,571	4,885	270,609	133,049	1,151,122
1990	1,867,794	245,286	11,045	27,271	93,084	1,976	201,475	215,986	1,220,221
1995	1,114,360	204,932	1,031	42,450	82,593	14	131,718	462,800	1,552,691
1996	1,190,896	234,909	1,365	44,373	86,619	27	225,623	550,867	1,690,733
1997	1,387,812	300,489	1,912	43,691	79,780	67	222,568	544,075	1,731,209
1998	1,456,081	373,330	2,104	49,891	84,543	56	211,650	584,795	1,837,615
1999	1,707,277	461,061	1,697	68,394	83,399	186	372,965	639,878	2,170,427
2000	1,839,093	488,323	3,125	81,196	86,707	134	568,121	934,000	2,138,811
2001	1,790,346	494,131	2,580	82,487	92,439	92	633,769	861,853	1,855,789
2002	2,046,902	574,455	3,504	157,633	87,709	150	627,881	845,181	1,882,660
2003	1,770,355	561,482	2,943	207,158	119,773	298	692,863	680,214	1,811,892
2004	1,727,065	547,008	3,373	185,621	98,131	2,417	860,424	652,509	2,035,345
2005	1,832,534	547,191	5,377	184,716	93,736	412	730,500	693,149	1,967,985
2006	2,347,532	532,022	5,469	148,014	81,008	567	697,061	947,824	1,963,922
2007	2,300,913	466,458	5,650	108,576	92,600	1,746	676,594	889,474	1,912,744

(1) Excludes cars assembled in U.S. foreign trade zones.
(2) Figures prior to 1991 are for West Germany.

Source: *World Almanac and Book of Facts* (2009). Accessed through EBSCOhost Academic Search Premier.

Step 1

- Open a new worksheet, and enter the nine countries in column A. Notice the names of some countries extend into column B:

	A	B	C
1	**Country**	**1988**	**2007**
2	Japan		
3	Germany (2)		

- To **automatically fit the column width** to the cell's contents, place the mouse on the column heading row between columns A and B. The pointer changes to a thick, black sign (✛). Double click the left mouse button. The column automatically widens so that the text in each row fits into the column.

	A	B	C
1	**Country**	**1988**	**2007**
2	Japan		
3	Germany (2)		

(You can alternatively go to Excel's **Home** tab, select **Cells, Format, AutoFit Column Width**.)

Step 2

- Enter the two dates (1988 and 2007) in cells B1 and C1, respectively.
- Fill in the amounts for 1988 in cells B2 to B10 and the amounts for 2007 in cells C2 to C10.
- In cell A11, type in Total, and use **Alignment** on the **Home** tab to **right-justify the label**.

	A	B	C
1	**Country**	**1988**	**2007**
2	Japan	2,123,051	2,300,913
3	Germany (2)	264,249	466,458
4	Italy	6,053	5,650
5	United Kingdom	31,636	108,576
6	Sweden	108,006	92,600
7	France	15,990	1,746
8	South Korea	455,741	676,594
9	Mexico	148,065	889,474
10	Canada	1,191,357	1,912,744
11	**Total**		

Step 3

- Sum the number of imported passenger car values for both years. As with many features in Excel, there are multiple ways to generate the result:
  - Click on the **Formulas** tab. From the **Formula Library,** select the **AutoSum** function button (**AutoSum Σ**) and then **Sum**.
  - If you do not know Excel's name for the formula or if you are not sure which formula to use, click on the **Formula Bar,** and the **Insert Function** dialog box will appear. Type a brief description, and click **Go**; Excel will generate a list of functions from which to choose.
    To add the values together, select **SUM** and click **OK**. You will then be taken to a new dialog box. Click on the **Number1** button (which looks like a miniature Excel worksheet), and then highlight the data series. Click **OK**. Excel will generate the sum of the values in the data series.
  - Type "**=SUM(B2:B10)**" into cell B11, and hit <Enter>.
    Note: Excel will guess which values you want to sum. In the above example, it will highlight all numbers in a column, including the year. Since we do not want the year added to the total, highlight *only* those values you want to include. (Ignore the error message, "Formula Omits Adjacent Cells".)

Note: To **enter a formula** in Excel, you must begin with an equal sign, =. Otherwise, Excel will not know to perform the desired calculation.

You could follow any of the above procedures to sum 2007 imports, but it is much more efficient to **copy the formula** from cell B11 into C11.

- Activate cell B11.
- Drag the fill handle to cell C11. Excel will copy the contents of cell B11, the summation formula, into cell C11.

	A	B	C
1	**Country**	**1988**	**2007**
2	Japan	2,123,051	2,300,913
3	Germany (2)	264,249	466,458
4	Italy	6,053	5,650
5	United Kingdom	31,636	108,576
6	Sweden	108,006	92,600
7	France	15,990	1,746
8	South Korea	455,741	676,594
9	Mexico	148,065	889,474
10	Canada	1,191,357	1,912,744
11	**Total**	4,344,148	6,454,755

According to the calculations, the number of new and used passenger cars imported into the U.S. from the nine countries increased from 4,344,148 units in 1988 to 6,454,755 units in 2007.

he difference between the number of imported cars between 1988 and 2007. In cell D11, **1-B11**". Hit <Enter>.

While absolute amounts of change may be of interest, economists and policy-makers are often more interested in the percentage change in an economic variable. Excel can quickly calculate the percentage change in the number of new passenger cars imported into the U.S. from 1988 to 2007.

Recall the **percentage change formula**:

$$\text{Percentage change} = \left(\frac{\text{final period value} - \text{initial period value}}{\text{initial period value}} \right) * 100$$

Note we **multiply the percentage change calculation by 100**. This follows the convention used by economists, in which the symbol "percent" is not reported. **Do** not **use** Excel's **Percentage** format, as it will unnecessarily complicate subsequent calculations.

Step 5

Calculate the percentage change in auto imports for Japan, the first country in the list.

- Go to cell D1, and Type "**Percentage change**". (Always use labels so it is clear what data is being analyzed.)
- Go to cell D2, and enter the percentage change formula for Japan.
    - Type in the actual formula using the cell addresses "=((C2−B2)/B2)∗100", or
    - Type the operators but click on the cells to be used in the calculations. Excel will insert the cell address. (You can toggle between typing in characters and clicking on the desired cells.)

You will obtain the following result, indicating that passenger car imports from Japan increased by 8.37766026 percent between 1988 and 2007.

	A	B	C	D
1	**Country**	**1988**	**2007**	**Percentage change**
2	Japan	2,123,051	2,300,913	8.37766026

Rarely are eight decimal places reported for a percentage change. To **reduce the number of decimal places**, highlight cell D2. On the **Home** tab, go to **Number** and select the **Decrease Decimal** button. Reduce the decimal places to two:

	A	B	C	D
1	**Country**	**1988**	**2007**	**Percentage change**
2	Japan	2,123,051	2,300,913	8.38

Step 6

Use the drag-and-drop approach to calculate the percentage changes in car imports for the remaining countries as well as for the total.

	A	B	C	D
	Country	**1988**	**2007**	**Percentage change**
1	**Country**	**1988**	**2007**	**Percentage change**
2	Japan	2,123,051	2,300,913	8.38
3	Germany (2)	264,249	466,458	76.52
4	Italy	6,053	5,650	-6.66
5	United Kingdom	31,636	108,576	243.20
6	Sweden	108,006	92,600	-14.26
7	France	15,990	1,746	-89.08
8	South Korea	455,741	676,594	48.46
9	Mexico	148,065	889,474	500.73
10	Canada	1,191,357	1,912,744	60.55
11	**Total**	4,344,148	6,454,755	48.59

Step 7

Include the following information in all Excel-generated tables *and* when presenting empirical results:

- *Descriptive title*. Here, "**New and used passenger cars imported into the U.S., by country of origin (1), 1988 and 2007**".
- *Units of measurement*. Here, "(in number of units)".
- *All relevant footnotes*. Here, include footnotes 1 and 2.
- *The source of the data*. While the original source is the Foreign Trade Division of the U.S. Census Bureau, we actually obtained the data from the *World Almanac and Book of Facts* (2009). Thus the *World Almanac* is the appropriate source to use.

To add the title and units of measurement, insert two rows above row 1:

- Activate cells A1 and A2.
- Right-click on the activated cells, and select **Insert**. . . .
- From the drop-down menu, select **Entire row**.

Add the footnotes and then the data source at the bottom of the table.

The completed data table should look like the following:

	A	B	C	D
1	**New and used passenger cars imported into the U.S., by country of origin (1), 1988 and 2007**			
2	(in number of units)			
3	**Country**	**1988**	**2007**	**Percentage change**
4	Japan	2,123,051	2,300,913	8.38
5	Germany (2)	264,249	466,458	76.52
6	Italy	6,053	5,650	-6.66
7	United Kingdom	31,636	108,576	243.20
8	Sweden	108,006	92,600	−14.26
9	France	15,990	1,746	−89.08
10	South Korea	455,741	676,594	48.46
11	Mexico	148,065	889,474	500.73
12	Canada	1,191,357	1,912,744	60.55
13	**Total**	4,344,148	6,454,755	48.59
14	(1) Excludes cars assembled in U.S. foreign trade zones.			
15	(2) Figures prior to 1991 are for West Germany.			
16	Source: *World Almanac and Book of Facts* (2009). Accessed through EBSCOhost Academic Search Premier.			

Exercises

1. Many government agencies and international organizations establish specific economic goals related to their missions and functions and regularly evaluate those goals using empirical evidence. A case in point is the World Bank, whose overarching goal is to aid governments in developing countries, through financial and technical assistance, in reducing poverty. Each year, the Bank focuses on a specific topic in their *World Development Report* within the context of reducing poverty; the 2007 report, for example, examines development policies and institutions that might enhance the capabilities of young people between the ages of 12 to 24, which have long-run ramifications for poverty alleviation. For this exercise, go to the World Bank's *World Development Report* website (<http://go.worldbank.org/LOTTGBE9I0>), select a specific year, and click on the Full Text link. Check the Overview for that year's selected focus, and describe how its characteristics may contribute to poverty. Next, discuss how empirical evidence is used to demonstrate the current conditions in developing countries.

2. Economic charts and tables are frequently included in news articles on various aspects of the economy, such as changes in GDP, price levels, the unemployment rate, and the federal deficit and/or debt.

Select one of these variables, and identify which federal agency collects the data used to measure the change. In addition, download a recent agency news release that contains charts and tables for the variable.

Next, go to a well-respected news source, such as the *Wall Street Journal*, the *New York Times*, *The Times* (U.K.), or *The Economist*, and download a news story based on a news release that is accompanied by data tables or charts. Finally, go to a blog related to economic issues, and download a related story that uses empirical evidence. Discuss both the quality of evidence offered from each source and how effectively the source uses empirical evidence in its analysis.

3. The 1946 Employment Act established three macroeconomic goals for the U.S. economy: full employment, price stability, and economic growth. Since then, additional macroeconomic goals have been identified by U.S. officials as circumstances and political priorities have changed. A good source for identifying such goals is the annual *Economic Report of the President* (see above), in which an administration's goals are articulated and then evaluated using empirical evidence. These reports also contain extensive tables of economic data, which provide overviews of the U.S. economy, key domestic sectors, and the international economy.

   (a) Macroeconomic conditions typically influence an administration's economic priorities. During a recession, for example, an administration is likely to be very concerned about unemployment. In the *Economic Report of the President* (1983), for example, Ronald Reagan's Council of Economic Advisors (CEA) argued that "Unemployment is the most serious economic problem now facing the United States." To recommend appropriate policies for combating this problem, the CEA first distinguished cyclical and structural unemployment. Go to Chapter 2, and discuss how the CEA used empirical evidence (Charts 2.1 through 2.5) to explain the differences between these two major types of unemployment.

   (b) Health-care reform has been the focus of U.S. policy-makers for many years because of rising health-care costs and increasing numbers of Americans being inadequately insured. The Clinton administration famously (and unsuccessfully) targeted health care as a major policy priority, making its case for reform in Chapter 4 of its *Economic Report of the President* (1994). Discuss the empirical evidence used by the CEA to argue for the inadequacy of health security for all Americans, the limitations of private health insurance, the lack of competition in health-care markets, and the growing burden on government finances.

   (c) The U.S. economy was plunged into recession in December 2007, fueled, in part, by the collapse of the U.S. housing market. Go to Chapter 2 of the *Economic Report of the President* (2009a), and discuss how George W. Bush's Council of Economic Advisors utilized empirical evidence to explain the origins and evolution of this crisis.

2 Visual presentations of economic data

2.1 Economic graphs and charts

All data presented in Chapter 1 was readily obtained from U.S. federal government sources via the internet. Access to such economic data continues to expand, facilitating empirical economic analysis through the application of statistical methods. We begin our examination of these methods by discussing how visual representations of economic data, or charts, can be used to identify overall patterns and trends in economic variables.

Let us first distinguish the charts developed in this chapter from the graphs typically encountered in economic theory classes. The best-known example of the latter is the supply-and-demand framework for a product market such as coffee. As shown in Figure 2.1, the supply schedule for coffee is depicted graphically as an upward-sloping curve that reflects the positive relationship between the price of coffee and the quantity of coffee that sellers are willing and able to supply, while demand is a downward-sloping schedule that illustrates the negative relationship between the price of coffee and the quantity that buyers are willing and able to pay for coffee. The point at which the two schedules intersect (E^*) determines the equilibrium price (P^*) and quantity (Q^*) that will obtain in the market. In this theoretical construction, we do not use data to describe either relationship; instead, the relationships are deduced from sets of assumptions about how sellers and buyers are expected to respond to changes in the product's price, and the point at which the two groups agree on price establishes market equilibrium.

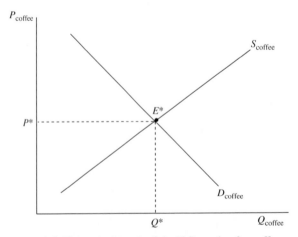

Figure 2.1 Theoretical graph of the U.S. market for coffee.

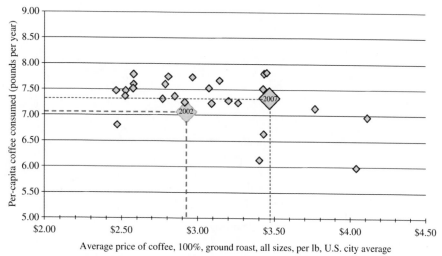

Chart 2.1 U.S. per-capita coffee consumption, 1980–2007.

Data sources: Coffee prices: U.S. Bureau of Labor Statistics (2010b), Consumer Price Index Per capita coffee consumption: Economic Research Service (2010a), Food Availability (Per Capita) Data System.

Note the rescaling of both axes.

Such graphical representations of abstract theoretical economic models may, or may not, correspond to actual empirical observations of the relationship between price and quantity. Let us now create a **scatter diagram** in which we plot paired data for average U.S. coffee prices and per-capita coffee consumption for each year between 1980 and 2007. Here we observe, for example, that the average price of coffee of $3.47 in 2007 corresponded to per-capita coffee consumption of 7.33 pounds, while in 2002 the average price of coffee was $2.92 and per-capita coffee consumption was 7.07 pounds.

The paired data points in Chart 2.1 (p. 23) do not obviously resemble the relationships depicted in Figure 2.1 which might, erroneously, suggest the economic theory of product markets does not obtain empirically. But what if each of the chart's data points reflects the equilibrium coffee price and quantity in a particular year? Such an interpretation is not unreasonable since we are using average annual data for both series. In addition, this interpretation reinforces the very important connection between the theoretical graph in Figure 2.1 and the empirical pattern in Chart 2.1.

For now, let us distinguish between abstract **graphs** that portray theoretical relationships between economic variables and data-based **charts** that empirically depict those variables. We begin our discussion of charts by noting that economists distinguish between three categories of empirical evidence: time-series, cross-section, and panel data. These distinctions will help inform our choices of the most appropriate chart types for illustrating patterns and trends in economic variables.

2.2   Time-series data and charts

The term **time-series** data refers to the movement of an economic variable over multiple time periods. Visual representations of time-series data allow economists and policy-makers to assess, for example, how an economy has grown (as we saw in Chart 1.2) or how the unemployment rate has changed over time (Chart 1.1), and they can illustrate trends in economic variables such as inflation (Chart 1.3).

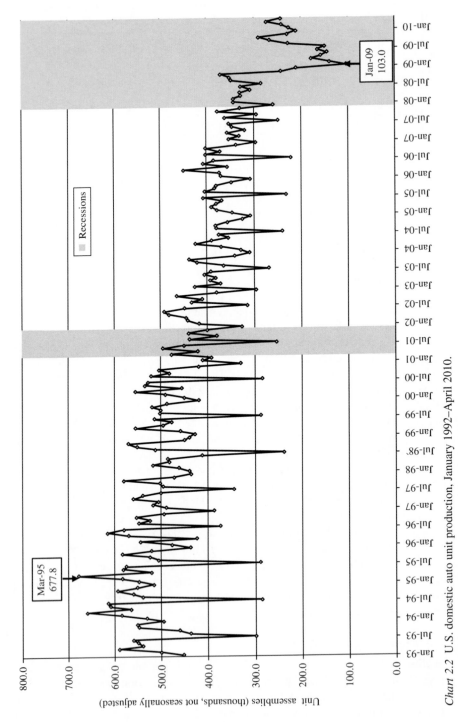

Chart 2.2 U.S. domestic auto unit production, January 1992–April 2010.

Data source: Table 7.2.5S, U.S. Bureau of Economic Analysis (2010c), National Economic Accounts.

One economic variable of interest to U.S. economists is domestic automobile production. The U.S. auto industry was for many decades the world leader in automobile production, but it began to lose market share to foreign competitors during the 1970s. Since then, the industry's production levels have exhibited a long-run downward trend, as suggested by Chart 2.2, declining from a monthly high of 677,800 units in March 1995 to a low of 103,000 units in January 2009. We further observe that domestic production has increased, on average, since January 2009, although it still appears to fall below the long-run trend.

We also notice that domestic production declines during recessionary periods, which are denoted by the shaded areas in Chart 2.2 (p. 24). (Directions for including recessionary periods in an Excel line chart may be found at the end of this chapter.) Such a decline is not unexpected; as a durable good, the purchase of a new car is something that consumers can defer until better economic conditions obtain.

We further observe noticeable fluctuations in automobile production for any given year. The regular annual dip in production corresponds to the traditional summer shutdown of U.S. auto plants to retool for the upcoming model year. Such cyclical and seasonal fluctuations in the domestic auto production data are readily apparent in the chart and provide economists with important information about how production in the auto industry changes over time.

Chart 2.3 (p. 26) offers another example of how time-series data for an economic variable might be presented. Suppose we are interested in examining the amount of U.S. agricultural output per unit of total factor inputs, known as total factor productivity (TFP), over the last 60 years. In the context of scarce resources, society hopes that TFP will increase over time as more efficient processes in agricultural production are introduced. We also recognize that such gains are not likely to be steady over the years. Technological advances, political actions, and factors beyond producers' control – such as weather – may all disrupt the long-run trend in agricultural productivity. If we can identify when TFP deviates from its long-term trend, we may gain a more complete understanding of changes in agricultural efficiency.

The first step in such an analysis is to evaluate how TFP has changed over the time period. Here we have plotted the index of U.S. agricultural total factor productivity between 1948 and 2006.[1] We have also added a line that estimates the long-run secular trend[2] for U.S. agricultural TFP, and a series that measures the deviation between the long-run secular trend and the actual value of the TFP index. The latter series allows us to quickly observe for which years total factor productivity diverged from its long-run trend so we might identify possible reasons for the deviations.

As indicated in Chart 2.3, some factors contributing to below-trend agricultural TFP since 1974 can be attributed to specific events, such as oil price shocks, weather conditions, and government programs. Higher oil prices, for example, raise the costs of petroleum-based agricultural chemicals, machinery operation, and overall energy costs, leading farmers to let land stand idle or to leave farming altogether, thereby reducing total agricultural output and thus total factor productivity. Droughts also shrink output, either by reducing yield or by destroying crops or livestock. While both factors are typically considered beyond the control of producers or the government, the deviation from trend in 1983 coincides with the U.S. federal government's Payment-in-Kind (PIK) program. (PIK essentially paid farmers to remove land from agricultural use and into conservation. By reducing surplus agricultural output, the program's goal was to raise farm incomes.) While we will leave it to policy-makers to debate the efficacy of such programs, knowing the reason behind the 1983 deviation (and the others) provides invaluable information to economists interested in understanding patterns in agricultural TFP. Thus creating an empirical portrait of the trend in the TFP variable not

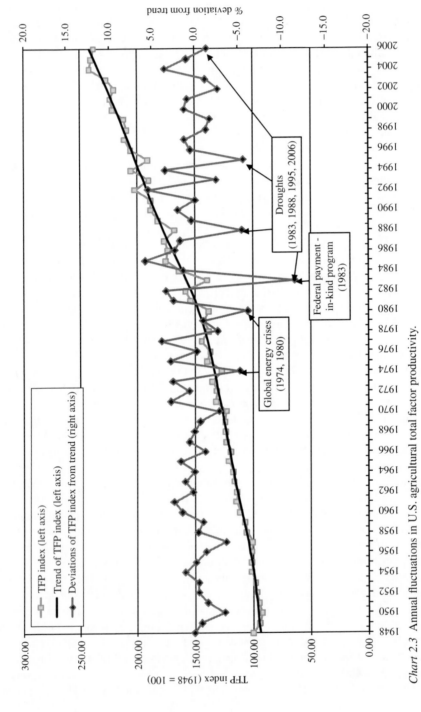

Chart 2.3 Annual fluctuations in U.S. agricultural total factor productivity.

Data source: Economic Research Service (2009), Agricultural Productivity in the United States.

only provides a starting point for analysis, but also can help direct subsequent research on agricultural productivity.

Chart 2.3 uses a dual Y-axis construction which allows us to compare time-series data for economic variables that are measured in different units. Suppose, for example, we wanted to compare how new home sales (measured in thousands of units sold) and fixed-rate mortgage rates (measured in percentage points) changed between January 1990 and March 2009. By scaling the left-hand axis in units sold and the right-hand axis in percent, as in Chart 2.4 (p. 28), we can readily make that comparison.

The patterns depicted in Chart 2.4 suggest that, until summer 2003, the average contract rate for 30-year fixed-rate first mortgages and the number of new homes sold in the U.S. were negatively related, the relationship economists expect since mortgage rates contribute to the costs of purchasing a new home. Over the following two years, however, home sales rose on average while mortgage rates started to creep upward. By the end of 2006, new home sales started a precipitous decline even though mortgage rates were relatively stable, and, by 2008, both home sales and mortgage rates were declining. Thus after mid-2003, the U.S. housing market no longer exhibited the behavior predicted by economists.

We now know that the U.S. housing market experienced an unsustainable bubble followed by a dramatic collapse,[3] and Chart 2.4 clearly presents evidence of both. It also demonstrates how a basic **line chart** can illustrate the relationship between economic variables, showing us where the expected relationship obtains and where we see anomalies according to economic theory. Such information can then inform further economic analysis of the relationship.

Constructing charts of time-series economic data serves several purposes. First, it allows us to recognize the pattern of change in an economic variable over a given time period, as shown for U.S. domestic auto production in Chart 2.2. Charts of time-series data also provide us with a basis for estimating the long-run trend of an economic variable and how the variable's actual values might deviate from that trend, as illustrated in Chart 2.3. We can also use charts of time-series data to provide us with a first approximation of the relationship between two (or more) economic variables, as with Chart 2.4, which then informs our subsequent economic analysis.

2.3 Cross-section data and charts

The second type used by economists, **cross-section** data measures how an economic variable differs across categories at a given point in time. For example, we used cross-section data in Table 1.1 to show how U.S. household income differed across income classes in 2008. We could also examine how 2008 household income compared across such demographic categories as race-ethnicity, family type, and the presence of children, or across geographical locations such as the 50 U.S. states or between the U.S. and other nations (assuming comparable currencies).

Cross-section data is most frequently illustrated using a **bar chart**. Chart 2.5 (p. 29), for example, uses a vertical bar chart to depict the income data from Table 1.1. Here we observe that the "$100,000 and over" income class represents the largest percentage (20.5 percent) of U.S. households in 2008, followed by 11.9 percent of households in the $50,000–$74,999 income class. We can also determine that almost one-third (32.4 percent) of U.S. households had income of more than $75,000, while just under one-fourth (24.7 percent) had total money income less than $25,000.

Vertical bar charts can also be used to compare cross-section data for two different time periods. Suppose, for example, we want to examine how the 2008 distribution of U.S.

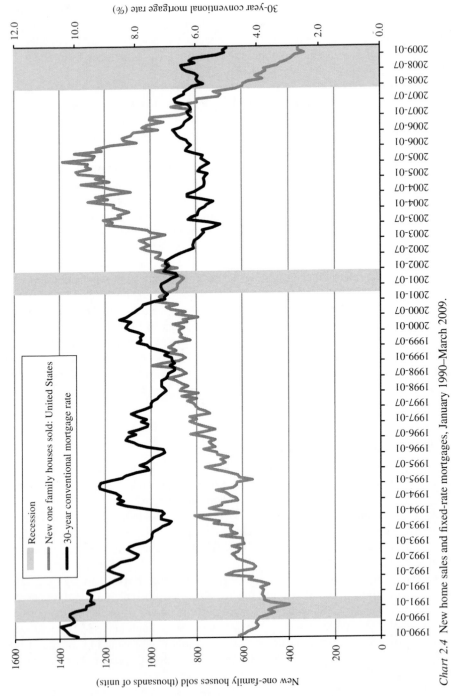

Chart 2.4 New home sales and fixed-rate mortgages, January 1990–March 2009.

Data source: FRED (20010b, c), Federal Reserve Economic Data, HSN1F and MORTG.

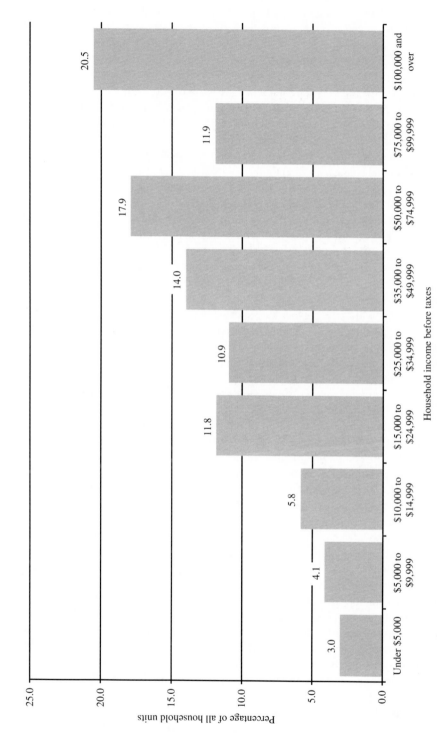

Chart 2.5 Percentage distribution of total money income across U.S. households, 2008.

Data source: DeNavas-Walt *et al.* (2009, Table A-1).

household income differs from the distribution of household income in, say, 1974 to see if the oft-repeated claim "the rich are getting richer, and the poor are getting poorer" has any empirical basis. Because we have inflation-adjusted income data for 1974 and 2008 that uses the same income classes, we can plot the 1974 and 2008 series in the same bar chart to facilitate the comparison. In Chart 2.6, we observe that the distribution of household income has indeed shifted over the 24-year period. Specifically, we observe that the percentage of households in the top two upper income classes increased from 19.0 percent in 1974 to 32.4 percent in 2008. We also see that the percentage of U.S. households in the bottom income class increased from 2.3 percent in 1974 to 3.0 percent in 2008. Based on this evidence alone, it appears that the above statement has empirical merit, leading us to conduct further empirical investigation. (We will develop more sophisticated statistical methods for assessing distributional changes in Chapter 5.)

Bar charts that present cross-section data may be vertically aligned as above, or horizontally aligned. A **vertically aligned bar chart** is more effective when the classes, or categories, into which the variable is sorted, exhibit a recognizable order such as increasing dollar amounts. A **horizontally aligned bar chart** may be more effective when the variable is sorted into categories that exhibit no logical order.

Chart 2.7 (p. 32) is an example of a horizontally aligned bar chart. It presents the total number of U.S. workers employed in the three U.S. supersectors (private goods-producing, private service-providing, and public government) by major industry sectors for 2009. Because the order in which we present the supersectors has no economic significance, we can horizontally align the data to emphasize key differences. We see, for example, that the private service-providing supersector employed almost three times the number of workers as the public supersector, and three and a half times those employed in the private goods-producing supersector. We also observe that more U.S. workers (24,949,000) were employed in the private service-providing trade, transportation, and utilities major industry sector than in any other major industry sector.

Economic variables measured in millions or billions of units are difficult for most of us to grasp, so economists typically prefer to use percentages rather than actual numbers. Chart 2.8 (p. 33) presents the same employment information as Chart 2.7, but now the categories reflect the percentage contribution of each major industry sector to total employment. Here we can more readily see, for example, that employment in trade, transportation, and utilities industries accounted for the largest percentage (19.1 percent) of the U.S. workforce. We can also determine that, when combined with the next three largest employers – public government (17.2 percent), private education and health services industries (14.7 percent), and professional and business services industries (12.7 percent) – these four sectors comprised almost two-thirds (63.7 percent) of all U.S. employees in 2009, while the private goods-producing supersector accounted for only 14.2 percent of the workforce.

As with the distribution of household income, we might also be interested in how the distribution of employment across major industry sectors changed in the U.S. between a year of economic expansion, say 2006, and 2009, a year of transition from recession to expansion. To make comparisons, we will return to a vertically aligned bar chart in Chart 2.9 (p. 34) so we can more clearly distinguish the changes in the 11 major industry sectors between the two years. (Data for the supersectors were omitted to streamline the chart, but two dark vertical lines were added to separate the industries by supersector.)

We observe that the distribution of workers shifted across most major industry sectors between 2006 and 2009. In particular, construction and manufacturing's shares of total

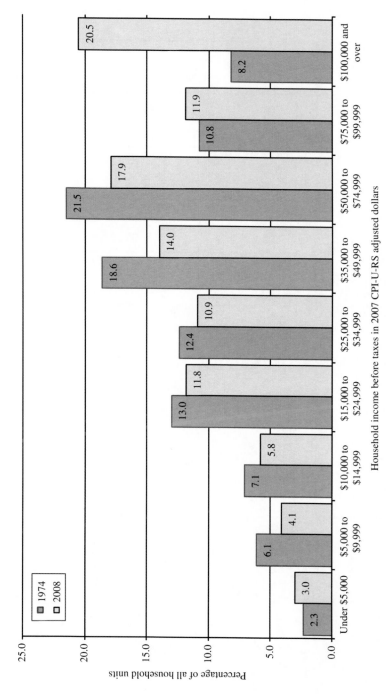

Chart 2.6 Percentage distribution of U.S. household total money income, 1974 and 2008.

Data source: DeNavas-Walt *et al.* (2009, Table A-1).

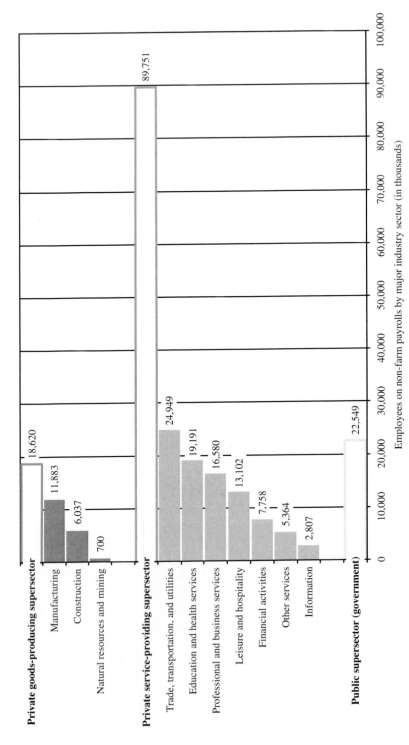

Private goods-producing supersector

Manufacturing — 11,883
Construction — 6,037
Natural resources and mining — 700

18,620

Private service-providing supersector

Trade, transportation, and utilities — 24,949
Education and health services — 19,191
Professional and business services — 16,580
Leisure and hospitality — 13,102
Financial activities — 7,758
Other services — 5,364
Information — 2,807

89,751

Public supersector (government) — 22,549

Employees on non-farm payrolls by major industry sector (in thousands)

Chart 2.7 U.S. employment by supersector and major industry sector, 2009 (total U.S. employees: 130,920,000).

Data source: Table B-1, U.S. Bureau of Labor Statistics (2010c), Current Employment Survey.

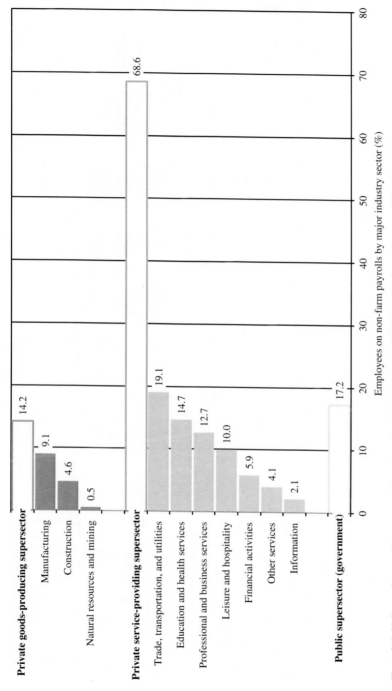

Chart 2.8 U.S. employment by supersector and major industry sector, 2009 (total U.S. employees: 100.0 percent).

Data source: Table B-1, U.S. Bureau of Labor Statistics (2010c), Current Employment Survey.

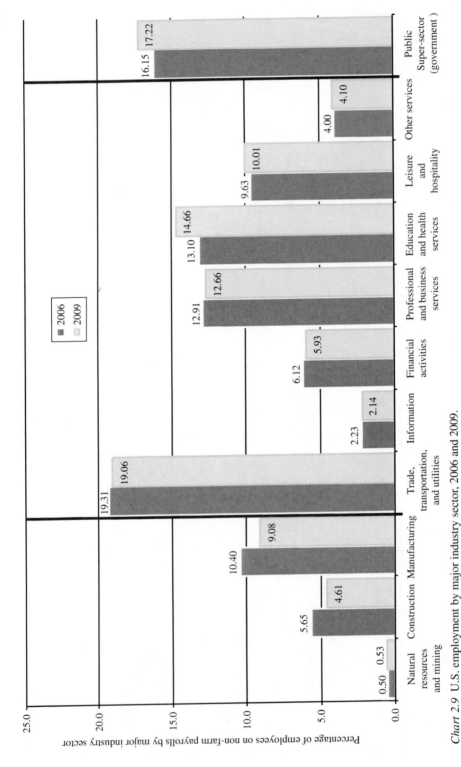

Chart 2.9 U.S. employment by major industry sector, 2006 and 2009.

Data source: Table B-1, U.S. Bureau of Labor Statistics (2010c), Current Employment Survey.

employment visibly dropped, while public government, and leisure and hospitality's shares noticeably increased.

Chart 2.9 is, however, of limited value for understanding employment changes, since each year's shares must sum to one (100 percent). Consequently, we do not know how the number of workers in each major industry sector changed, information of particular interest when comparing employment at different points in the business cycle. How, then, might we more effectively visualize those changes in a chart? One solution is to construct a new statistic that measures the percentage change in employment by supersector and major industry sector between the two years.

Chart 2.10 (p. 36) illustrates the percentage change in the number of non-farm payroll employees between 2006 and 2009. We immediately observe that employment in both private supersectors and all but three major industry sectors declined between 2006 and 2009. The number of construction and manufacturing workers took the hardest hit, declining by 21.5 percent and 16.1 percent, respectively. While we might expect employment in both sectors to decline during a recession, these sectors were probably further affected by the collapse in the U.S. housing market and the U.S. auto industry's difficulties.

Other sectors experiencing declines include the information and financial activities sectors, where employment fell by 7.6 percent and 6.8 percent, respectively, which may be attributable to the crisis in the U.S. financial sector. On the other hand, we see that employment in the education and health services industry sector bucked the overall trend, growing by 7.7 percent over the three-year period. Government employment also increased, by 2.6 percent, a likely result of the services it provides and economic stimulus policies. Such initial speculation, of course, suggests further analysis.

The growth in educational and health services employment between 2006 and 2009 might suggest that this major industry sector is recession-proof, particularly when we consider that such services are necessary in all phases of an economy's business cycle. To investigate this supposition further, we might first want to determine how employment within the educational services sector was distributed among industry groups at a given point in time.

Table 2.1 presents employment data for the educational services subsector by industry group. According to the U.S. Bureau of Labor Statistics, the education services economic sector is divided into six industry groups that are demarcated by the specialized instruction or training offered by institutions, plus a seventh industry group of workers who provide non-instructional support services to educational institutions. Two of the industry groups (Business, computer, and management training; and Other schools and instruction) are further divided into NAICS industries[4] to differentiate the varied types of instruction offered.

Let us now note a key feature of Table 2.1 – the use of **indentation** to identify which categories belong to each level of classification within a hierarchical structure. We see, for example, that the categories of elementary and secondary schools (NAICS 61111) and junior colleges (NAICS 61112) are indented by the same amount, which indicates that they are both major industry groups, while the additional indentation for business and secretarial schools and computer training (611412) and management training (611413) signifies that these establishments offer more specialized educational services under the industry group of "Business, computer, and management training" (61114). Recognizing the economic meaning of indentation is crucial for accurately comparing categories, and it must be retained to ensure we only compare like categories.

Let us now illustrate how employment was distributed across the seven education services industry groups in 2009. Since we want to compare the relative size of the parts of a whole, we will use a new chart type, a **pie chart**. As Chart 2.11 (p. 37) illustrates, half (50.3 percent)

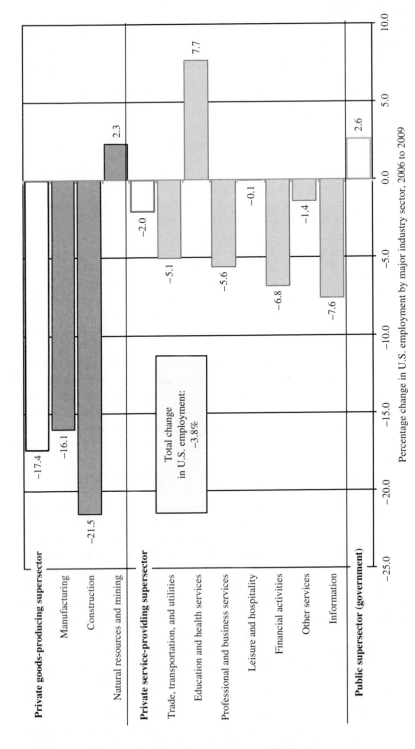

Private goods-producing supersector

Manufacturing −16.1

Construction −21.5

Natural resources and mining 2.3

−17.4

Private service-providing supersector

Trade, transportation, and utilities −2.0

Education and health services 7.7

Professional and business services −5.6

Leisure and hospitality −0.1

Financial activities −6.8

Other services −1.4

Information −7.6

−5.1

Total change
in U.S. employment:
−3.8%

Public supersector (government) 2.6

−25.0   −20.0   −15.0   −10.0   −5.0   0.0   5.0   10.0

Percentage change in U.S. employment by major industry sector, 2006 to 2009

Chart 2.10 Percentage change in U.S. employment by supersector and major industry sector between 2006 and 2009.

Data source: Table B-1, U.S. Bureau of Labor Statistics (2010c), Current Employment Survey.

Table 2.1 Distribution of U.S. education services* employees, 2009

	All employees (in thousands)
Education services (61)	3,089.9
Elementary and secondary schools (61111)	861.5
Junior colleges (61112)	80.5
Colleges and universities (61113)	1,553.7
Business, computer, and management training (61114)	75.8
Business and secretarial schools and computer training (611412)	30.1
Management training (611413)	45.7
Technical and trade schools (61115)	119.5
Other schools and instruction (61116)	300.3
Fine arts schools (61161)	69.9
Sports and recreation instruction (61162)	72.6
Miscellaneous schools and instruction (611439)	157.9
Educational support services (61117)	98.5

*Parenthetical numbers refer to North American Industry Classification System (NAICS) codes.

Data source: Table B-1, U.S. Bureau of Labor Statistics (2010c), Current Employment Survey.

of all workers in educational services in 2009 were instructors at "Colleges and universities" while more than a quarter (27.9 percent) worked as instructors in "Elementary and secondary schools". The remaining 21.8 percent of employees were distributed among "Other schools and instruction" (which includes instruction in fine arts, and sports and recreation), other types of post-secondary instruction, and the all-important "Educational support services" group.

Chart 2.11 clearly shows that almost 80 percent of education services employment in 2009 occurred in traditional locations for education. We might now ask if a similar distribution

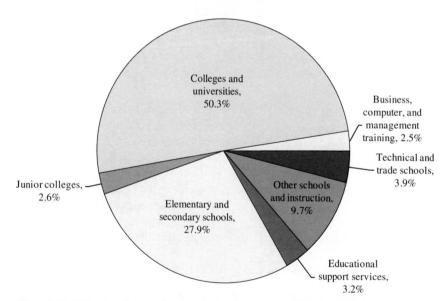

Chart 2.11 U.S. education services employees by industry group, 2009.

Data source: Table B-12, U.S. Bureau of Labor Statistics (2010c), Current Employment Survey.

obtained in 2006, a period of economic expansion. Unfortunately, such comparisons are not feasible using pie charts. While we might consider constructing a separate pie chart for the other year and doing a side-by-side comparison, we need to recall the adage about looks being deceiving. Thus to effectively compare two years, we would be better served by constructing a vertical bar chart that includes data for both years or a horizontal bar chart that depicts the change in distributional shares between the two years.

Some have also criticized the use of pie charts for other reasons. Edward Tufte, for example, claims that, because of "their low data-density and failure to order numbers along a visual dimension," pie charts "should never be used" (Tufte 2007). As Tufte's criticism suggests, we can develop certain principles of design for the effective visual presentation of economic data, which we will do in Section 2.5. For now, we shall simply note that, despite such criticisms, pie charts remain popular in the press and among policy-makers and economists, so understanding when it is appropriate to employ this chart type is useful to know.

2.4   Panel data and charts

The third category of data, **panel** data, also called **longitudinal** data, contains both cross-section and time-series components. Specifically, panel data tracks the *same* cross-section individuals over multiple time periods. These individuals may be people, families, firms, cities, or nations – the key is that the same characteristics are collected for the same individuals over the time frame.

The income and employment data presented in Charts 2.6 (p. 31) and 2.9, respectively, are not panel data because neither represents the same households or firms for the two years presented. This data, called **pooled** data, combines cross-section data over multiple time periods but does not necessarily refer to the same cross-section groups. Thus panel data is one type of pooled data.

Creating a visual representation of panel data is rarely feasible, in part because the two-dimensional nature of charts limits the number of individual units and the number of time periods we can depict. Second, if we can construct a chart using all observations in a panel data series, we can probably just as effectively present this information in tabular form. Consequently, we will not try to render charts of panel data, as they are unlikely to increase our understanding of the empirical features of economic variables.

2.5   Creating effective charts

What, then, makes a chart effective? Remember, we construct charts to illustrate patterns or trends in economic variables in ways that communicate clearly and completely such information to our audience. Thus the purpose of our chart-making is to effectively convey economically meaningful empirical evidence so others can better understand economic phenomena.

The **effective communication** of empirical results is both efficient and honest. Efficient communication requires that we completely and unambiguously specify the important ideas we wish to express. In other words, we want to provide the audience with complete information for interpreting the empirical results without overwhelming them with extraneous details.

Conventions for **efficient communication** with a chart include the following:

- Providing a succinct descriptive title for the chart.
- Specifying fully the titles of both axes and, where appropriate, the associated tick marks on each axis.

- Specifying precisely all units of measurement for the data presented.
- Including, where appropriate, labels on individual data points and a legend.
- Providing descriptive captions to highlight key information.
- Indicating clearly which subclasses are members of the same group.
- Excluding superfluous or distracting information. This includes avoiding elaborate options such as 3D charts and the distracting colors and patterns offered by many software packages.

Honest communication requires that we provide the audience with complete information about our terminology, methods, and data sources. Perhaps more importantly, honest communication also means that we do not (mis)lead the audience to predetermined conclusions. In other words, we do not want to be guilty of "lying with statistics" to make a point.

Conventions for **honest communication** with a chart include the following:

- Always indicating your data sources and methods so researchers can replicate the results.
- Clearly defining all variables and the data used to measure them, acknowledging any limitations of that data for the question at hand.
- Not misleading the reader with charts that give false visual impressions of magnitudes or changes.
- Scaling the chart's axes using contiguous values. Generally we include the origin in a chart, but for some series (such as the coffee price and quantity data in Chart 2.1), doing so may obscures the trend of interest. As shown in Chart 2.1, if we omit the origin, we must clearly indicate that omission on the chart so as not to mislead the reader.
- In economics, most data is more effectively presented in charts where the horizontal axis is longer than the vertical axis. Such construction is less likely, in particular, to distort the changes in an economic variable over time.[5]

To demonstrate these principles, let us turn to Chart 2.12 (p. 40), a less effective version of Chart 2.2, which depicted the distribution of U.S. household income in 2008.

We can clearly see that Chart 2.12 does not efficiently communicate information about the distribution of U.S. household income. First of all, the title of Chart 2.12 does not indicate what economic variable is being presented, which is the primary function of a title. In addition, while the X-axis title indicates that the chart depicts household income before taxes, we do not know what each bar in the chart represents, nor do we know what units of measurement are used. We also have no data labels for the chart's individual bars, making it difficult to know precisely how many households fall into each class.

Chart 2.12 is also an example of software run amok. The bars in the chart are unnecessarily patterned. The shadows serve no purpose and actually make it difficult to determine the value associated with each bar. The patterned backgrounds for the plot and chart areas are also distracting, again serving no clear purpose for understanding how household income was distributed for whichever country and whatever year this data represents.

Chart 2.12 includes other misleading features. There is no indication that the Y-axis begins at a value of 2 percent rather than the origin, and the chart's vertical orientation further distorts the relative heights of the chart's bars. More importantly, we do not know the source of the data presented, leading some readers to question if the data was simply made up. Since being able to replicate empirical results is an important principle in the social sciences, this last omission is particularly problematic.

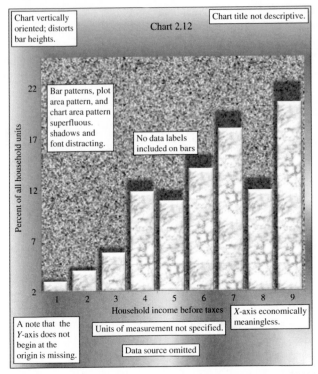

Chart 2.12 Ineffective chart of the percentage distribution of 2008 U.S. household income.

While Chart 2.12 was constructed for the purpose of illustrating an ineffective chart, examples of poor chart-making abound, particularly in the popular press, as an internet search for "bad charts" will quickly reveal. In addition, while charts often facilitate our understanding of economic data, they may not always be the most effective means for presenting such information. In cases where we have only a few categories, for example, constructing a chart of cross-section data may represent unnecessary effort when we can present the same information just as effectively in tabular form. At the other end, too many categories may result in a jumble of lines or bars that gives the audience no insight into the underlying patterns or trends. Constructing an effective chart thus also requires us to consider if the chart is the method most helpful for the intended audience to understand the empirically interesting features of the variable.

2.6 Constructing charts using Excel

All charts presented in Chapters 1 and 2 were generated using Microsoft Excel. Those of us born after 1982 may not realize how much the introduction of personal computers and spreadsheet software in the 1980s revolutionized the statistical methods used by economists. Now, for example, instead of investing a considerable number of hours hand-drawing charts, popular software programs such as Microsoft Excel make constructing a chart quick work.

Let us identify Excel's chart types in the context of the chart types discussed above. They are listed in Table 2.2. We will discuss only those Excel chart types frequently used by economists, and we will follow the order in which they appear on Excel's **Insert** tab.

Table 2.2 Excel chart types and economic charts

Excel chart types	Economic charts discussed in Chapter 2	Examples in Chapters 1 and 2
Column	Vertical bar chart	Charts 2.5, 2.6, 2.9
Line	Line chart	Charts 1.1, 1.2, 1.3, 2.2, 2.3, 2.4
Pie	Pie chart	Chart 2.11
Bar	Horizontal bar chart	Charts 2.7, 2.8, 2.10
Scatter	Scatter chart	Charts 1.4, 2.1

Detailed instructions for creating charts in Excel may be found below following this chapter's exercises.

Excel's **Column** chart type corresponds to the vertical bar charts presented in Section 2.3. As we have discussed, this **column chart** type can be used to depict, in either absolute or relative terms, how an economic variable differs across multiple categories at a given point in time. Vertical bar charts can also be used to illustrate how a variable changes over short periods of time.

Excel's next chart type is the **Line** chart, which we would use to depict time-series data when that data spans more than a few time periods. We can also include trend-lines in Excel's line charts, which estimate the variable's secular trend, as demonstrated in Chart 2.3, and we can plot differently scaled variables by using a secondary Y-axis, as illustrated in Charts 2.3 and 2.4.

The **Pie** chart is the third chart type presented in Excel. As noted in Section 2.3, pie charts are used to highlight the relative sizes of the components of a whole, but they can only be constructed for one data series at a time. When constructing an effective pie chart, we do not want to include so many slices as to clutter the image, which is a particular problem when a number of the slices are small relative to the entire pie. In such cases, we might more effectively communicate the same information by constructing a bar chart or using a data table.

Excel's fourth chart option, **Bar** charts correspond to the horizontal bar charts discussed in Section 2.3. The horizontal bar chart is effective for demonstrating differences in a variable across geographical units or demographic groups, particularly when we want to depict classes and subclasses in the same chart as we did for U.S. employment in the three economic supersectors and major industry groups.

The final Excel chart type used by economists is the **Scatter** (or X–Y) chart, which we use to construct a **scatter chart**, X–Y chart, scatter diagram or scatterplot. While our focus thus far has been on charts depicting a single economic variable, economists are often interested in the relationship between two economic variables. As we saw in Charts 1.4 and 2.1, we can visually illustrate such relationships using a scatter diagram. We shall return to this type of chart in Part IV when we discuss correlation and regression analysis.

Summary

The growing availability of ready-to-use economic data has increased significantly the ability of economists to examine empirical patterns and trends in economic variables. One popular method for identifying regularities in economic variables is the construction of charts that depict how an economic variable differs across categories, compares with another variable, or changes over time. These charts may also help economists to identify key disruptions in overall patterns and trends, which are then considered in subsequent statistical analysis.

While economists use several different types of charts, depending on the variable and the purpose of their analysis, each chart must adhere to economic conventions for effective communication. Such considerations include the efficient presentation of the data so the audience can quickly see and understand the key elements of the variable's pattern or trend. It is also imperative that our charts communicate empirical evidence honestly. We want readers to be able to reproduce our results if they choose, and we do not want to mislead them to an erroneous conclusion about the variable. As we shall see in subsequent chapters, such concerns about effective communication of empirical results not only inform the charts we construct, but also help us to define the desirable properties of the various economic statistics we will examine.

Concepts introduced

Bar chart
Chart
Column chart
Cross-section
Effective communication
Efficient communication
Graph
Honest communication
Horizontally aligned bar chart
Indentation
Line chart
Longitudinal data
Panel data
Pie chart
Pooled data
Scatter chart (X–Y chart)
Scatter diagram
Time-series
Vertically aligned bar chart

Box 2.1 Where can I find. . .? *The Statistical Abstract of the United States*

The *Statistical Abstract of the United States* is an annual compendium of national statistics, published by the United States Census Bureau. This "national data book" (U.S. Census Bureau 2011):

> . . . is the standard summary of statistics on the social, political, and economic organization of the United States. It is also designed to serve as a guide to other statistical publications and sources. The latter function is served by the introductory text to each section, the source note appearing below each table, and Appendix I, which comprises the Guide to Sources of Statistics, the Guide to State Statistical Abstracts, and the Guide to Foreign Statistical Abstracts.
>
> This volume includes a selection of data from many statistical sources, both government and private. Publications cited as sources usually contain additional statistical detail and more comprehensive discussions of definitions and concepts. Data not available in publications issued by the contributing agency but obtained from the internet or unpublished records are identified in the source notes. More information on the subjects covered in the tables so noted may generally be obtained from the source.

All earlier editions of the *Abstract* are available from the "The 2010 Statistical Abstrac tions" webpage (<http://www.census.gov/compendia/statab/past_years.html>). This link and ZIP versions of the *Bicentennial Edition: Historical Statistics of the United States, to 1970*, and statistical tables from the 2006 edition forward may be downloaded in XL The *Abstract* also includes the following auxiliary information.

- **Appendix I** contains references to the primary sources of statistical information in the *Abstract*; internet links to U.S. federal agency reports; links to State Abstracts; and a guide to Foreign Statistical Abstracts.
- **Appendix II** contains information about the concepts, components, and population for Metropolitan and Micropolitan Statistical Areas.
- **Appendix III** describes limitations of the data presented in the *Abstract*.
- **Appendix IV** provides metric conversion rubrics.
- **Appendix V** contains those tables deleted from the 2009 edition of the *Statistical Abstract*.
- **Index**.

Box 2.2 Using Excel to create charts

Constructing a basic chart with Excel is quite straightforward. Once created, we can format the basic chart to meet the principles for effective visual communication discussed in Section 2.5.

We will use the following data on U.S. real personal consumption expenditures (PCE) to create column (vertical bar), bar (horizontal bar), and pie charts in Excel:

U.S. Real Personal Consumption Expenditures by Major Type, 4th quarter, 2009

2009 IV	Billions of chained (2000) dollars
Personal consumption expenditures	9,289.5
Durable goods	1,123.7
Nondurable goods	2,053.4
Services	6,105.9

Data source: Table 3: Gross domestic product and related measures: Level and change from pre- ceding period. U.S. Bureau of Economic Analysis. *News Release on Gross Domestic Product.* Online. Available from <http://www.bea.gov/newsreleases/national/gdp/2010/pdf/gdp1q10_ 3rd.pdf> (accessed 1 July 2010).

Entering data for chart

Enter the above data into an Excel worksheet.

1. Type the title in cell A1.
2. Type the units of measurement "Billions of chained (2000) dollars" in cell B2.
3. Type the data labels in cells A3 through A6, and the actual values in cells B3 through B6.
4. In cell A7, enter the source information.

Formatting
1. To *run the title across both columns*, highlight cells A1 and B1. Under **Alignment** on the **Home** tab, click **Merge & Center**.
2. To *spread the data source information across cells* A7 and B7, highlight both cells and select **Merge Across** under **Merge & Center**. Next click the **Wrap Text** button immediately above **Merge & Center**.
3. To *indent the labels* for the three major types of PCE, highlight cells A4 through A6, and click the **Increase Indent** button under **Alignment** on the **Home** tab. This presents the data according to the indentation conventions discussed in Section 2.3.

The worksheet should now look like the following:

	A	B
1	**U.S. Real Personal Consumption Expenditures by Major Type, 4th quarter, 2009**	
2		**Billions of chained (2000) dollars**
3	Personal consumption expenditures	9,289.5
4	Durable goods	1,123.7
5	Nondurable goods	2,053.4
6	Services	6,105.9
7	Data source: Table 3: Gross domestic product and related measures: Level and change from preceding period. U.S. Bureau of Economic Analysis. *News Release on Gross Domestic Product*. Online. Available from <http://www.bea.gov/newsreleases/national/gdp/2010/pdf/gdp1q10_3rd.pdf> (accessed 1 July 2010).	

Column charts

To *create a column chart* of the three major types of personal consumption expenditures:

1. Highlight the information in cells **A4 through B6** (do *not* include A3 or B3).
2. Under **Charts** on the **Insert** tab, go to **Column** and select the **2-D Clustered Column** chart type. Excel will generate the following chart:

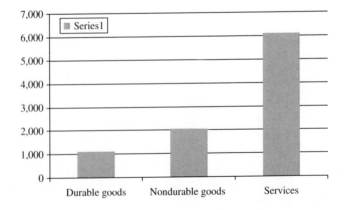

Formatting

The above chart obviously does not meet our standards for effective graphical representation of data. Specifically, we want to:

1. add a descriptive chart title,
2. add a label to the *Y*-axis label,
3. add informative labels to each data point,
4. remove the unnecessary legend, and
5. include the data source.

---

1. **Add** a meaningful **chart title**

- Activate the chart.
- Select **Layout** and then Labels.
- Select **Chart Title**.
- From the drop-down menu, select the desired option. (The following chart below places the title above the chart.)
- In the **Insert Function** (f_x) bar, type in the information you want to add.

---

2. **Add** a meaningful *Y*-axis label

- Activate the chart.
- Select **Layout** and then **Labels**.
- Select **Vertical Axis Titles**.
- From the drop-down menu, select the desired option. (The chart below uses a Rotated Title.)
- In the **Insert Function** (f_x) bar, type in the information you want to add.

---

3. **Add Data Labels** to columns, bars or pie slices

From Chart Tools	*Directly on the Chart*
- Activate the chart.	- Activate the chart.
- Select **Layout**, and click on **Data Labels**.	- Right-click on any column, bar, or slice.
- From the drop-down menu, select the desired option. (The chart below uses Outside End.)	- From the drop-down menu, select **Add Data Labels**.

---

4. **Remove** the **Legend**

- Activate the chart.
- Select **Layout** and then **Labels**.
- Select **Legend**.
- Select **None**.
  - Alternatively, simply left-click on the legend and hit the **Delete** key.

---

5. **Add** a source or other information – **Text Boxes**

- Activate the chart.
- Click on the **Insert** tab, and under the **Text** group, select **Text Box**.
- When you move your cursor over the chart, it will be an upside-down dagger. Click where you want the Text Box to appear. A box with a dashed border will appear on the chart, with the cursor blinking inside.
- Type the source information.
- Click outside the Text Box to close it.

These formatting changes will generate the following column chart:

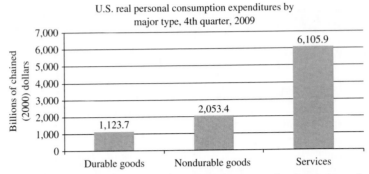

Data source: Table 3: Gross domestic product and related measures: Level and change from preceding period. U.S. Bureau of Economic Analysis. News Release on Gross Domestic Product, Online. Available from <http://www.bea.gov/newsreleases/national/gdp/2010/pdf/gdp1q10_3rd.pdf> (accessed 1 July 2010).

Bar charts

The PCE data can also be displayed as a bar chart by following largely the same procedures. Instead of selecting **Column** as the chart type, under the **Insert** tab, select **Bar**. Next, select the first option, the **2-D Clustered Bar chart** type. Click **OK**. The resulting chart looks like the following:

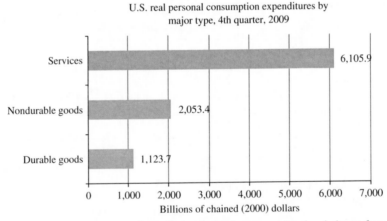

Data source: Table 3: Gross domestic product and related measures: Level and change from preceding period. U.S. Bureau of Economic Analysis. News Release on Gross Domestic Product, Online. Available from <http://www.bea.gov/newsreleases/national/gdp/2010/pdf/gdp1q10_3rd.pdf> (accessed 1 July 2010).

Formatting

The vertical grid lines above make it difficult to read the data labels, so we want to change the label's background.

Change the Background on Data Labels to Columns, Bars or Pie Slices

From Chart Tools	*Directly on the Chart*
o Activate the chart. o Select **Layout**, and click on **Data Labels**. o From the drop-down menu, select **Fill**. o Under **Fill**, select **Solid fill**, and choose the color that corresponds to the chart's plot area.	o Right-click on any data label. o From the drop-down menu, select **Format Data Labels**. o From the drop-down menu, select **Fill**. o Under **Fill**, select **Solid fill**, and choose the color that corresponds to the chart's plot area.

The chart will now look like:

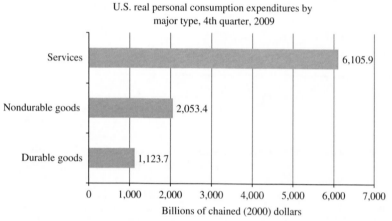

Data source: Table 3: Gross domestic product and related measures: Level and change from pre-ceding period. U.S. Bureau of Economic Analysis. News Release on Gross Domestic Product, Online. Available from <http://www.bea.gov/newsreleases/national/gdp/2010/pdf/gdp1q10_3rd.pdf> (accessed 1 July 2010).

Pie charts

We can also display the PCE data in a **Pie chart**. While we could begin from scratch, let us simply modify the bar chart just created.

1. Activate the bar chart.
2. Under **Chart Tools**, click on **Design** and then on **Change Chart Type**.
3. Select the first option under **2-D Pie**, and click **OK**.

U.S. real personal consumption
expenditures by major type, 4th quarter, 2009

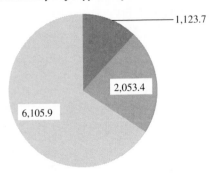

Data source: Table 3: Gross domestic product and related measures: Level and change from pre-
ceding period. U.S. Bureau of Economic Analysis. News Release on Gross Domestic Product,
Online. Available from <http://www.bea.gov/newsreleases/national/gdp/2010/pdf/gdp1q10_3rd.pdf>
(accessed 1 July 2010).

Formatting

The pie chart omits what each pie slice signifies. We can add this information to the pie slices in one
of the following ways:

Formatting the Pie Slices

Add a Legend	*Adding Information in Data Label*
1. Activate the chart.	1. Right-click on any data label.
2. Select **Layout**, and click on **Legend**.	2. Select **Format Data Labels**. Under **Label Contains** in **Label Options**, we can choose several pieces of information to include in each slice's data label:
3. Select the location for the legend to appear. Here the **Show Legend at Left** option was chosen.	a. Check the **Category Name** box to add the descriptive labels entered in cells A4 through A6.
4. Click **OK**.	b. Check the **Percentage** box to display the percentage contribution of each category. To add a decimal place to these values, go to **Number**, select **Percentage**, and enter 1 under **Decimal places**.
	3. If *dissatisfied with the location of the data labels*, select the desired option under **Label Position**. (The **Outside End** option was chosen in this example.)
	4. To *rotate the pie slices*, click on any slice and from the drop-down menu, select **Format Data Series**. Under **Series Options**, use the slider in **Angle of first slice** to rotate the slices so their labels are easy to read. In the chart below, 45 was chosen to reorient the pie chart.

The following two charts illustrate both methods for formatting the pie slices:

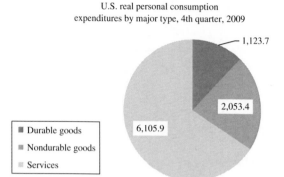

Data in billions of chained (2000) dollars.
Data source: Table 3: Gross domestic product and related measures: Level and change from pre-ceding period. U.S. Bureau of Economic Analysis. News Release on Gross Domestic Product, Online. Available from <http://www.bea.gov/newsreleases/national/gdp/2010/pdf/gdp1q10_3rd.pdf> (accessed 1 July 2010).

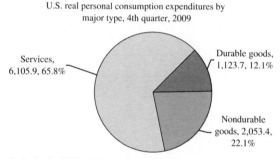

Data in billions of chained (2000) dollars and percentage contribution to total real personal consumption expenditures.
Data source: Table 3: Gross domestic product and related measures: Level and change from pre-ceding period. U.S. Bureau of Economic Analysis. News Release on Gross Domestic Product, Online. Available from <http://www.bea.gov/newsreleases/national/gdp/2010/pdf/gdp1q10_3rd.pdf> (accessed 1 July 2010).

Formatting

The chart also does not indicate the data's units of measurement. To *include the units of measurement*, a line was added to the text box containing the source of the data.

Which chart type is most effective?

Which of the three chart types is most effective for displaying the differences in the relative size of the three major categories of personal consumption expenditures?

The column (vertical bar) chart is least effective because there is no economic meaning in the order of the three categories.

Both the bar (horizontal bar) and pie charts clearly illustrate the relative sizes of the three expenditure categories. Both have the added advantage of clearly indicating that the data is for one time period, whereas the column chart might suggest a progression over time.

Excel's **Label Contains** options for including the category name (also available for the bar chart) and the percentage contribution (unique to the pie chart) suggest the pie chart as the chart type that most effectively conveys the relative magnitude of the three major types of real personal consumption expenditures.

Box 2.3 Using Excel to format a line chart

Adding secondary axes (Charts 2.3 and 2.4)

Economists often want to compare data series that are measured in different units. One example was Chart 2.4, in which we compared "thousands of units" and "percentages." If we were to plot both series on the same Y-axis, our chart would look like this:

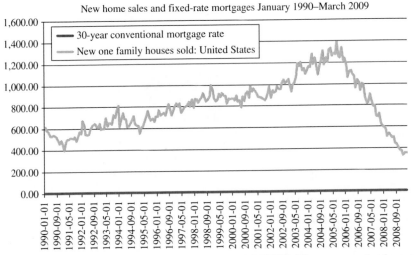

New home sales and fixed-rate mortgages January 1990–March 2009

Data source: HSNIF (New one family houses sold) and MORTG (30-year conventional mortgage rate. Economic Data – FRED®) (Federal Reserve Economic Data). Online. Available from <http://research.stlouisfed.org/fred2/> (accessed 15 May 2009).

As we see, all variation in the fixed-rate mortgages data is obscured at the bottom of the plot area. We can remedy this by plotting the mortage data on a secondary Y-axis.

Add a Secondary Axis

1. Right-click the data series that you want to plot on the secondary axis.
2. Select **Format Data Series**.
3. Under **Series Options**, select the **Secondary Axis** box; click **Close**.

Your chart will now look like this:

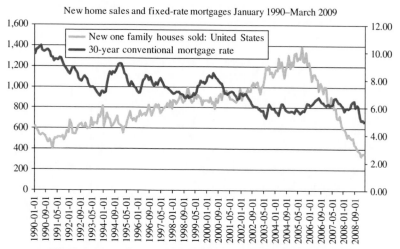

New home sales and fixed-rate mortgages January 1990–March 2009

Data source: HSNIF (New one family houses sold) and MORTG (30-year conventional mort-gage rate. Economic Data – FRED®) (Federal Reserve Economic Data). Online. Available from <http://research.stlouisfed.org/fred2/> (accessed 15 May 2009).

Adding recessionary periods (Charts 1.1, 1.2, 1.3, 2.2, and 2.4)

Economists often want to indicate periods of recessions when plotting the time trend of an economic variable. We can add such information to a line chart after obtaining the dates for U.S. recessions.

The dates for the U.S. business cycle are set by the National Bureau of Economic Research (NBER) and can be found at <http://www.nber.org/cycles/cyclesmain.html>. NBER dates expansions and recessions according to the month in which each begins, and it indicates the corresponding quarter beside the month, making the construction of the recession time series straightforward for monthly and quarterly data series. (Creating an annual recession series is more challenging, since recessions may last only a part of a particular year. If including this information in a chart for annual data, be sure to include an explanatory note to the reader.)

The following steps outline how to add business cycle information to a time-series chart. Note that we have omitted the data source in the following charts to see more clearly the steps involved.

Add recessionary periods to a line chart

1. On the same worksheet as the variable data, create a new data series called Recession. Set the period value equal to **1** during times of recessions and **0** for the remaining periods:

	Recession
Jan-92	0
Feb-92	0
Mar-92	0
Apr-92	0
May-92	0
Jun-92	0
. . .	
Feb-01	0
Mar-01	0
Apr-01	1
May-01	1

2. Add the data series to the chart: From the **Design** tab under **Chart Tools**, choose **Select Data**. This will open a dialog box called **Select Data Source**. Under **Legend Entries (Series)**, choose **Add**. Include the label. Click **OK**.

   If the original *Y*-axis is not scaled from 0 to 1, as shown below, some additional formatting will be needed.

Step 2a: Adding recessionary periods to a line chart

Chart 1.1: U.S. unemployment rate, January 1992–May 2010

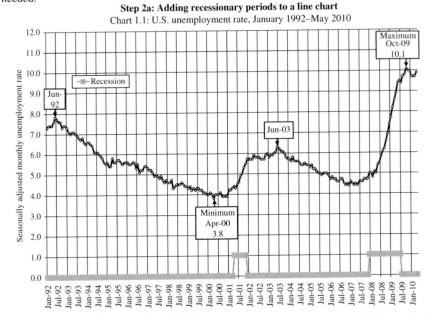

Right-click on any point of the new series, and select **Format Data Series**. Under **Series Options**, select **Secondary Axis** under **Plot Series On**. Your chart will now look like the following:

Step 2b: Adding recessionary periods to a line chart

Chart 1.1: U.S. unemployment rate, January 1992–May 2010

3. Rescale the secondary *Y*-axis so that its minimum value equals 0 and its maximum value equals 1.

   a. To **reset the minimum and maximum *Y*-axis values**, right-click on the secondary *Y*-axis, and select **Format Axis**.

   b. Under **Axis Options**, set the **Minimum** and **Maximum** values to 0 and 1, respectively.

Note: Be sure to change *both* the maximum and minimum values. If you do not, Excel will choose values so both axes are wider than the range of data values in the series.

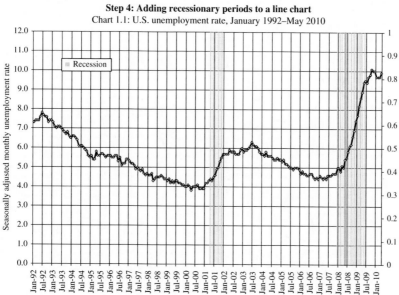

Step 3: Adding recessionary periods to a line chart

Chart 1.1: U.S. unemployment rate, January 1992–May 2010

4. Change the chart type. Activate a point on the new data series. Next, on the **Design** tab, select **Change Chart Type**. Select the **2-D Clustered Column** type, and click **OK**. The data series will now appear as a range of vertical bars on the chart.

Step 4: Adding recessionary periods to a line chart

Chart 1.1: U.S. unemployment rate, January 1992–May 2010

5. Next eliminate the space between the columns. Right-click on the Recession data series, and select **Format Data Series**. Under **Series Options**, reduce the **Gap Width** to **0**. Click **Close**.

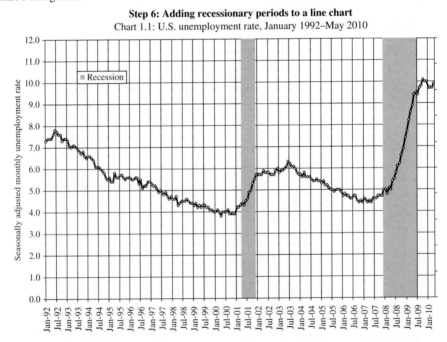

Step 5: Adding recessionary periods to a line chart
Chart 1.1: U.S. unemployment rate, January 1992–May 2010

6. To hide the secondary *Y*-axis values, activate the axis. Select a font color that is identical to the chart's background.

Step 6: Adding recessionary periods to a line chart
Chart 1.1: U.S. unemployment rate, January 1992–May 2010

Exercises

1. One section of the *Statistical Abstract* (see above) contains data on the place and behaviors of private business in the U.S. economy. Under "Establishments, Employees, Payroll," Table 743 in the 2010 edition compares the number of establishments across industries and across employment size classes. This latter categorization allows us to measure the number of employees and annual payroll for establishments according to the number of workers.

   Download the data from Table 743 into an Excel worksheet, and answer the following questions. For each, include both a descriptive paragraph(s) and an appropriate chart(s) to support your conclusions.

   (a) Politicians have argued that small businesses are a primary source of jobs in the U.S. Using data from the pre-recession year of 2007, determine in which industries the majority of employees work for small employers. Define "small" employers as those establishments with less than 100 total employees.

   (b) Conventional wisdom says that employees working for large establishments earn, on average, more than employees working for smaller entities. To evaluate this claim, use the annual payroll data to calculate the average annual earnings per employee by industry and employment size class, and then construct a chart illustrating the correlation between average earnings in establishments with less than 100 employees and establishments with less than 100 employees by industry in 2007.

   (c) As an economy changes over time, employment opportunities shift not only between industries but also between establishment size classes (measured by the number of employees). To further assess the claim regarding small business being the engine of potential U.S. job growth, examine how, for each industry, the share of employees working for small (less than 100 employees) businesses changed between 2000 and 2007. In light of your findings, comment on which industries would be most likely to benefit from policies to encourage the growth of small business and which industries would be least likely to benefit from such policies.

2. Economists argue that an individual's earnings are positively correlated with one's educational attainment. Download the data available in the table on "Educational Attainment by Selected Characteristics" in the most recent (2010) edition of *The Statistical Abstract of the United States* (U.S. Census Bureau 2011, Table 226), and construct a chart to illustrate:

   (a) the percentage of the population 25 years and older by highest level of education attained for that year;

   (b) the distribution of the population 25 years and older across the highest level of education attained by that age group for that year;

   (c) the percentage of the population 25 years and older who attained an advanced degree by sex, race, and Hispanic origin in that year;

   (d) the percentage of the population 25 years and older who were not a high-school graduate and those with a bachelor's degree between the first and final years available.

3. The U.S. Bureau of Labor Statistics publishes annual data related to union membership as part of the Current Population Survey. Available at http://www.bls.gov/cps/lfcharacteristics.htm#union, these data can help us paint a picture of trends and patterns in U.S. union membership.

In the October 2008 *Monthly Labor Review*, economist James A. Walker published "Union members in 2007: a visual essay" (available at http://www.bls.gov/opub/mlr/2008/10/art3full.pdf) in which he illustrated 11 statements about U.S. union membership.

Use Walker's article, the union membership data available at http://www.bls.gov/news.release/union2.toc.htm, and the historical data tabulated below to recreate the charts using Excel. (For statement 8, create a frequency distribution instead of a map.) For each, explain why Walker selected the particular chart type.

	Members of unions (percent of employed)*			Union membership rate		Median weekly earnings of earnings of full-time wage and salary workers by union affiliation (constant 2007 dollars)	
	Total, 16 years and over	Men, 16 years and over	Women, 16 years and over	Private sector	Public sector	Members of unions	Non-union
1992	15.8	18.7	12.7	11.5	36.7	792	598
1993	15.8	18.4	13.0	11.2	37.7	812	602
1994	15.5	17.9	12.9	10.9	38.7	819	598
1995	14.9	17.2	12.3	10.4	37.8	831	624
1996	14.5	16.9	12.0	10.2	37.7	809	608
1997	14.1	16.3	11.6	9.7	37.2	824	616
1998	13.9	16.2	11.4	9.5	37.5	837	634
1999	13.9	16.1	11.4	9.4	37.3	836	642
2000	13.5	15.2	11.5	9.0	37.5	838	653
2001	13.4	15.1	11.7	8.9	37.2	840	671
2002	13.2	14.7	11.6	8.5	37.5	853	677
2003	12.9	14.3	11.4	8.2	37.2	857	675
2004	12.5	13.8	11.1	7.9	36.4	857	672
2005	12.5	13.5	11.3	7.8	36.5	851	661
2006	12.0	13.0	10.9	7.4	36.2	857	660
2007	12.1	13.0	11.1	7.5	35.9	863	663

*Data refer to members of a labor union or an employee association similar to a union.

Source: Data obtained from U.S. Bureau of Labor Statistics (2010d), Current Population Survey.

Part I

Descriptive statistics of an economic variable

We begin our discussion of statistical methods used in economics by examining key descriptive concepts associated with an economic variable. These concepts are used to determine the central tendency and dispersion of values in a dataset with numerous observations. These statistics will also provide crucial information for the statistical inference methods we discuss later in the text.

Chapter 3 begins by examining various categories into which economic variables may be placed. Such categories are valuable for understanding which statistical methods can be applied to each type of variable. We also examine different methods for organizing many observations of a variable into manageable and informative classes.

Chapters 4 and 5 describe statistical measures of central tendency and dispersion, respectively, which are frequently used by economists. In Chapter 4, we also examine other positional measures, such as quintiles and percentiles, that economists use to make comparisons between groups within a dataset.

In Chapter 5, we bring together measures of central tendency and dispersion to create a new set of descriptive statistics, so we can compare "apples and oranges." These measures also allow us to comment on which variable exhibits the greatest variation or which is most equally distributed across the population.

The basic concepts presented in Part I are the foundations for statistical methods in economics. They provide economists with a statistical description of a particular economic variable, informing questions for further economic analysis. We will also note their importance when conducting tests of statistical inference in Part III.

3 Observations and frequency distributions

3.1 The design of observations: an introduction

Data for many economic variables can include a large number of observations, which must be systematically organized before we can identify and analyze their patterns and trends. Census data, for example, contains observations for all surveyed households in a particular geographical area, and, as we saw in Table 1.1, this translated into 117,181,000 households in the U.S. in 2008. Even if our analysis was confined to an individual census tract, we would be dealing with thousands of units (approximately 4,000 inhabitants in the 2000 U.S. Decennial Census, for example). Thus conducting an empirical analysis of large datasets requires us to find a manageable way to order large numbers of observations.

The most commonly used method for organizing large amounts of data is to construct a frequency distribution. As we shall develop further in Section 3.3, ordering and sorting the data into classes allows us to observe how the data is distributed across a range of values. While government agencies and non-governmental organizations often publish frequency distributions of economic variables, the increasing availability of public-use microdata series makes it important for economists to know how construct a frequency distribution from raw data.

Before we can organize large sets of data, the data must first be created – that is, we must have observations, or facts, for which we have identified who and what we want to measure. This process of observation must be carefully designed, because our statistical analyses can be no better than the data on which they are based; that is, we want to avoid problems of "garbage in, garbage out" when constructing economic data.

Formally, an **observation** refers to a specific data point for a particular element and a particular variable. In Table 1.1 in Chapter 1, this means that 117,181,000 observations of total money income for each U.S. household were used to create the frequency distribution.

An **element** is the unit we want to observe and describe. In Table 1.1 the element is U.S. households in 2008, although we might have selected instead U.S. families or persons. Other elements of interest to economists include: the firm, which might be the appropriate element for a study of market power; the plant or "establishment" for a study of industrial location; U.S. states for a study of economic growth; or the national economy for a study of unemployment. In all cases, we want the observed element to be the unit most relevant to the issue being investigated.

A **variable** is the characteristic of each element that we want to observe. In Table 1.1, the variable is U.S. household income before taxes in 2008, which we chose in Chapter 1 as a measure of economic welfare. Some scholars might take issue with our variable choice, arguing that factors such as employer-provided health benefits and realized capital gains should be included, or taxes excluded, in any definition of income used to assess economic well-being. Consequently, we must explain, as we did in Chapter 1, why we selected a particular variable to ensure effective communication with our audience.

Good design of observations is complex, requiring many considerations beyond the scope of this text, so we will let the experts determine and apply appropriate criteria. However, the clearly stated and generally accepted criteria used by such experts will help us determine which elements and variables are most appropriate for analysis of economic phenomena.

Let us mention one final point regarding the nature of observations. The fact that each element must be observed suggests the simple but significant characteristic of a variable – it takes on different values across different elements. This variability, or **heterogeneity**, among elements requires us to use statistical measures to fully describe the variable.

Suppose, for example, that household incomes in a nation were completely homogeneous, meaning that each household's income was exactly the same dollar amount, say $50,000. In such a case, we would need to observe only one family to determine the measured properties, such as the average, of household income. But such a situation is unlikely to obtain in the so-called real world, where we observe household incomes ranging from values below zero to values well above $100 million. Thus for us to fully understand household income, we must observe the incomes of all households[1] and base our descriptive measures on that set. This heterogeneity of the element called "household" thus leads us to use statistics to fully describe the economic variable, household income.

3.2 Attributes and measured variables

Given the heterogeneity of economic elements and the need for statistical methods to describe specific economic variables, it is useful to consider two basic types of variables – attributes and measured variables. An **attribute variable** (also called a **qualitative variable**) is a variable that leads to a classification of the element into specific categories, but not to a measurement of degree. For example, once we have clearly defined the variable "labor force," we can classify individuals as being either "in the labor force" or "not in the labor force." For those individuals "in the labor force," we can further classify them as being either "employed" or "unemployed," again using clearly established criteria for each category. In both cases, the attribute is said to be **binary** in nature, since the variable will be classified into one mutually exclusive category ("in the labor force" or "employed") or the other ("not in the labor force" or "unemployed").

It is also possible for an attribute variable to be **categorical**, which means that an observation is sorted into one category out of more than two mutually exclusive categories. For example, within employed members of the labor force, we can distinguish between employed whites, employed blacks, and employed Asians[2] if we want to examine questions about racial differences among employed workers.

In categorizing attribute (and measured) variables, economists often distinguish between the **scale** (or **level**) **of measurement**, which reflects different mathematical properties at each level. Economic variables that are attributes may be measured using either a nominal or an ordinal scale, while measured economic variables, defined below, may be measured in either interval or ratio scales. Each successive scale, beginning with nominal, then ordinal to interval and ending with ratio, builds on the mathematical properties of its predecessors. Another important characteristic is that the data is sorted into **mutually exclusive** categories such that each observation in the dataset is placed in at most one category regardless of the scale used.

The binary and categorical attribute variables described above are measured using a **nominal scale**. What this means is that we can assign to each category a numerical value that carries no economic meaning. In racial differences among the employed, for example, we could use a value of zero to denote those classified as "employed white" workers and a value of two to designate "employed Asian" workers. The fact that two is greater than zero has no meaning other than to distinguish the two different racial categories. Consequently, for variables measured on a nominal scale, the only statistical measures we can determine are the count (or number) of zeros (whites) and twos (Asians) observed among the employed elements as well as the proportion, or percentage, of each group among all workers employed.

Attribute variables may also signal rank, or relative value, across the dataset. In such cases, they are said to be measured using an **ordinal scale**. Surveys of people's opinions, for example, such as those used to assess consumer confidence, ask respondents to rate current and future economic conditions on a scale from "better" to "worse." The researcher can then assign values to each possible answer (say, a three for "better" and one for "worse") and place them into an economically meaningful order.

Other economic data, such as educational attainment measured by degrees earned, also indicate a rank order. If, for example, we assign a value of one for those with a high-school diploma, a two for a bachelor's degree, a three for a master's degree, and a four for a doctoral degree, we know that the category assigned a value of three (a master's degree) indicates a higher level of education attained than the category denoted by a value of two (a bachelor's degree).

Because variables measured on an ordinal scale indicate an economically meaningful order, we can do more than simply count the values assigned to each category. We can now also compare the values logically because we know, for example, that a doctoral degree represents a higher level of educational attainment than a master's degree. This means we can calculate positional measures for variables measured on an ordinal scale, such as the variable's median, since the data can be rank-ordered. We cannot, however, measure by *how much* the educational attainment associated with a doctoral degree "exceeds" the educational attainment associated with a master's degree. This inability to measure numerical differences between qualitative variables thus limits the statistical calculations we can perform on such variables, as indicated in Figure 3.1.

We now turn to the second basic variable type, measured or quantitative variables. A **measured variable** (also called a **quantitative variable**) enumerates the degree of difference between observations, such as the number of children per family or the federal funds rate over time. This variable type is the one more frequently used by economists, and, as listed in Figure 3.1, lends itself to a wide variety of descriptive statistical methods.

Quantitative variables can be measured using two scales, interval and ratio. The **interval scale** builds on the ordinal scale, but we can now also measure the numerical difference between two categories. In statistical terms, this means that zero is now a meaningful value but *only* in the sense that it provides an arbitrary point for comparison; that is, it does *not* signify the absence of the variable. Another feature of the interval scale is that the difference between one unit of the variable and the next unit is constant across the variable's range. For example, according to the Consumer Price Index (CPI), the difference between a monthly index value of 115 and the next month's index of 130 is 15 points, which is the same 15 point differential observed between two monthly indices with values of 165 and 180.

The **ratio scale** of measurement is generally considered superior to the interval scale because here zero *does* signify "none" or the total absence of the variable.[3] Ratio variables

Type of variable	Measurement scale type	Characteristics	Meaningful statistical calculations
Attribute/ qualitative variables	1. Nominal scale	Each category of the variable is assigned a numerical value for the sole purpose of demarcating it from the other categories. Ordering of values has no economic meaning	• Number of observations in each category • Percentage or proportion of observations in each category • Frequency distributions • Mode (most frequently observed) value
	2. Ordinal scale	Each category of the variable is assigned a numerical value that indicates a rank order among the categories. Numerical differences between categorical values have no economic meaning	• Number of observations in each category • Percentage or proportion of observations in each category • Frequency distributions • Mode value • Median value • Percentiles, quintiles, quartiles, and other positional measures
Measured/ quantitative variables	3. Interval scale	Each category of the variable is assigned a numerical value that indicates a rank order among the categories. The difference between categories can be measured and has economic meaning. Zero has no meaning; it merely serves as a reference point	• Number of observations in each category • Percentage or proportion of observations in each category • Frequency distributions • Histograms, frequency polygons, cumulative frequency polygons • Mode value • Median value • Percentiles, quartiles, quintiles, and other positional measures • Mean value • Standard deviation • Correlation coefficient • Regression coefficients • Z- and t-scores
	4. Ratio scale	Each category of the variable is assigned a numerical value that indicates a rank order among the categories. The difference between categories can be measured and has economic meaning. Zero is an economically-meaningful value and signifies absence of the variable	• Number of observations in each category • Percentage or proportion of observations in each category • Frequency distributions • Histograms, frequency polygons, cumulative frequency polygons • Mode value • Median value • Percentiles, quartiles, quintiles, and other positional measures

Figure 3.1 Scales of measurement for economic variables.

Type of variable	Measurement scale type	Characteristics	Meaningful statistical calculations
			• Mean value • Standard deviation • Correlation coefficient • Regression coefficients • Z- and *t*-scores • Coefficient of variation and index of dispersion • Gini coefficient and Lorenz curve

Figure 3.1 Continued.

have an additional desirable property – ratios calculated between two values are mathematically meaningful. For example, someone who is 60 years old has, in fact, lived twice as long as a 30-year-old. This second characteristic allows us to construct additional economically meaningful statistics such as the coefficient of variation and the Gini coefficient for variables measured on a ratio scale.

Economists also identify one other important distinction for measured variables, the difference between continuous and discrete variables. In theory, if not in practice, **continuous variables** are those that can be measured in smaller and smaller units until the unit of measurement is infinitely small. Time, for example, is a continuous variable, since it can be measured in years, months, days, hours, minutes, seconds, and so forth. Economic values measured by price may be expressed in terms of dollars or cents for convenience, but theoretically, value could be measured in infinitely smaller units. Many economic variables, including gross domestic product, wage rates, income, and commodity prices, are all classified as continuous variables.

Not all measured economic variables can, however, be measured in infinitely smaller units. Such variables are referred to as **discrete variables**. Examples of discrete measured variables include the number of people in the population and the number of housing units in an area. As we shall see in later chapters, the distinction between continuous and discrete measured variables will affect which statistical formulas can be applied for some statistical methods. But for now, let us explore how quantitative variables can each be better understood if we group the observations of each variable using the concept of a **frequency distribution**.

3.3 Organizing data: absolute frequency distributions

Frequency distributions organize the observations for a variable into a range of categories known as *classes*. When the frequency is stated as a simple count of the number of elements in each class, it is called an **absolute frequency distribution** because the **class frequencies** are reported in absolute numbers.

Table 3.1 describes the (dollar) value of owner-occupied housing units for the U.S. in 2000. Here the element observed is an owner-occupied housing unit defined as "a house, a mobile home, a group of rooms, or a single room that is occupied as separate quarters by the owner or co-owner, if the owner or co-owner lives in the unit even if it is mortgaged or

Table 3.1 Absolute frequency distribution of U.S. owner-occupied housing units, 2000

Value of housing unit	Number of units	Value of housing unit	Number of units	Value of housing unit	Number of units
Less than $10,000	235,097	$70,000 to $79,999	3,442,509	$200,000 to $249,999	4,018,468
$10,000 to $19,999	612,706	$80,000 to $89,999	3,986,707	$250,000 to $299,999	2,564,581
$20,000 to $29,999	1,013,346	$90,000 to $99,999	3,786,839	$300,000 to $399,999	2,442,848
$30,000 to $39,999	1,606,296	$100,000 to $124,999	6,852,290	$400,000 to $499,999	1,141,260
$40,000 to $49,999	1,990,372	$125,000 to $149,999	6,258,094	$500,000 to $749,999	973,014
$50,000 to $59,999	2,496,258	$150,000 to $174,999	4,711,681	$750,000 to $999,999	335,102
$60,000 to $69,999	3,066,658	$175,000 to $199,999	3,364,223	$1,000,000 or more	313,759
				Total units	55,212,108

Data source: U.S. Census Bureau (2000b), QT-H14, Value, Mortgage Status, and Selected Conditions: 2000.

not fully paid for" (U.S. Census Bureau 2007b). The variable observed is the dollar value of owner-occupied units, which is measured on a ratio scale.

According to this frequency distribution, we see that the largest number of owner-occupied housing units (6,852,290) were valued between $100,000 and $124,999 in 2000. We also observe that the smallest group of housing units (235,097) were valued at less than $10,000, while the second smallest group (313,759) were valued at $1,000,000 or more.

Assessing the variability of owner-occupied housing units from the frequency distribution presented in Table 3.1 is somewhat challenging. One reason is the large number of observations within each class. As we discussed in Section 2.3, many of us find it difficult to understand such large numbers, so we might choose to convert the "number of units" into the percentage of units in each class. Alternatively, we might translate the information in Table 3.1 into a graphic representation. We will examine both, beginning with the first option known as the relative frequency distribution.

3.4   Organizing data: relative frequency distributions

A **relative frequency distribution** uses class frequencies measured as the percentage of the total number of observations that fall into a class. Economists often prefer relative frequency distributions to absolute frequency distributions because they are easier to understand, a desirable characteristic for any statistic.

Returning to the value of owner-occupied housing units in 2000, let us now construct a relative frequency distribution by dividing the number of observations in each class by the total number of units and then multiplying the result by 100. (The latter step is consistent with the **convention for reporting percentages**, where we do not use the percent (%) symbol.)

Table 3.2 confirms our previous observation that the $100,000 to $124,999 class has the largest frequency, 12.4 percent of owner-occupied housing units, while the classes at each end of the distribution contain the lowest frequencies. We can also readily assess the frequency of housing units across multiple classes because of the additive property of percentages. For example, we can calculate that 3.3 percent of owner-occupied housing units were valued at less than $30,000 in 2000.

The tabular presentation of the relative frequency distribution of U.S. owner-occupied housing units is still a challenge to interpret. Trying to identify patterns across 21 classes which are not of equal width can easily overwhelm our analytical abilities. Let us now turn to a more effective way of communicating the overall distribution of a variable by presenting that information in a chart.

Table 3.2 Relative frequency distribution of U.S. owner-occupied housing units, 2000

Value of housing unit	Percentage of units	Value of housing unit	Percentage of units	Value of housing unit	Percentage of units
Less than $10,000	0.4	$70,000 to $79,999	6.2	$200,000 to $249,999	7.3
$10,000 to $19,999	1.1	$80,000 to $89,999	7.2	$250,000 to $299,999	4.6
$20,000 to $29,999	1.8	$90,000 to $99,999	6.9	$300,000 to $399,999	4.4
$30,000 to $39,999	2.9	$100,000 to $124,999	12.4	$400,000 to $499,999	2.1
$40,000 to $49,999	3.6	$125,000 to $149,999	11.3	$500,000 to $749,999	1.8
$50,000 to $59,999	4.5	$150,000 to $174,999	8.5	$750,000 to $999,999	0.6
$60,000 to $69,999	5.6	$175,000 to $199,999	6.1	$1,000,000 or more	0.6
				Total units	100.0

Source: Author's calculations from Table 3.1.

3.5 Visual presentations of frequency distribution: histograms

Using charts to depict economic variables can effectively provide us with important information about the variable. One kind of chart, the histogram, is particularly efficient for illustrating frequency distributions of quantitative variables.

A **histogram** is similar to a bar chart, but differs in several key characteristics. First, a histogram can only be constructed for continuous variables measured on either an interval or ratio scale, while bar charts can be constructed for any measurement scale. Each bar in a histogram measures the frequency of observations that fall into each class of the distribution, and, given the continuous nature of the variable, the bars in a histogram abut each other. (Technically each bar of a histogram should also correspond in size to the class interval, but computer software such as Excel cannot accommodate that requirement for unequal class intervals.)

Let us now translate the relative frequency distribution of the value of owner-occupied housing units from Table 3.2 into the histogram displayed in Chart 3.1 (p. 66). Here we clearly observe which class ($100,000 to $124,999) contained the most owner-occupied housing units in 2000. We can also readily see, for example, that two classes, $20,000 to $29,999 and $500,000 to $749,999, each accounted for 1.8 percent of the 2000 housing units. Such information might be more challenging to recognize from Table 3.2.

We can also quickly determine the ranges into which housing values are clustered from Chart 3.1. For example, it is easy to calculate that almost one-quarter (23.7 percent) of all owner-occupied housing units were clustered in two of the middle classes ($100,000 to $149,999), while 66.2 percent (or two-thirds) of units had values in the more broadly defined middle range of $40,000 to $174,999.

Creating a frequency distribution from previously organized data is a straightforward process. Of the three methods just described, histograms are generally the most easily understood, while the absolute frequency distribution is perhaps the most difficult to interpret. But what if our data has not been sorted into classes? That is, how do we construct a frequency distribution from raw data?

3.6 Classes in a frequency distribution

Economic data in raw form is increasingly available to those conducting economic analyses, but large numbers of observations are unwieldy, and little can be learned by simply scanning columns of values. Thus it is important that we learn how to organize large sets of data into

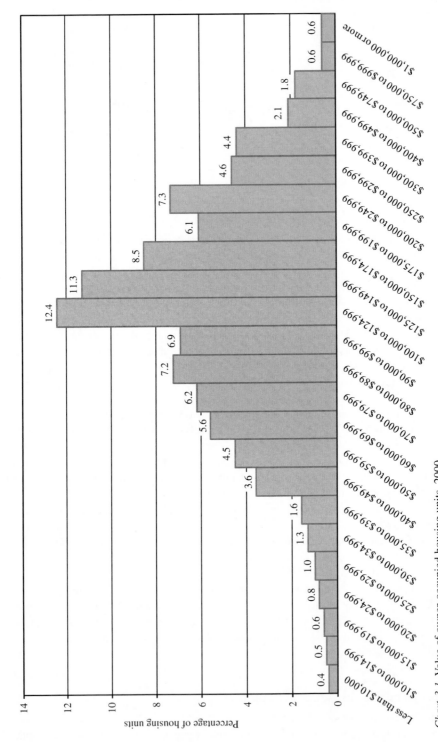

Chart 3.1 Value of owner-occupied housing units, 2000.

Data source: Table 3.2.

meaningful frequency distributions. Before turning to the specific steps involved, we must first introduce several concepts associated with frequency distributions.

Frequency distributions organize a dataset of observations into a series of categories, or **classes**. One property of a frequency distribution is that each observation is sorted into one, and only one, class. That is, a frequency distribution's classes must be **mutually exclusive**, meaning no observation can fall into more than one class. A second property is that a frequency distribution's classes are **all inclusive** (or **exhaustive**), meaning we have accounted for all values in the dataset. The frequency distributions in Tables 3.1 and 3.2, for example, exhibit both properties: the housing value for each of the 55,212,108 elements is contained in one and only one class, each class contains housing values not found in any other class, and the group of all classes encompasses all housing values observed.

Let us now describe the features of a frequency distribution's class. The **class interval** is defined as the range of all values contained within the class. In Table 3.1, for example, the class interval for the class labeled "$20,000 to $29,999" is $10,000, which indicates that we want to count all observations with values from $20,000 *through* $29,999. To remember that the class interval is not a simple difference between $20,000 and $29,999, we can subtract the lower value in the given class ($20,000) from the lowest value of the next highest class ($30,000) to determine the interval.

When we determine the class interval, we simultaneously establish the class limits. **Class limits** are defined as the lowest and highest values of the variable included in a given class. For the "$20,000 to $29,999" class, for example, the lower class limit is $20,000, while the upper class limit is $29,999.

You might now wonder into which class a housing unit valued at $29,999.50 might be sorted. For the data presented in Table 3.1, the U.S. Census Bureau measured the value of a housing unit in whole dollars, but it is almost certain that many of the observed housing values were not reported to the nearest dollar. Consequently, Census statisticians had to determine how they would round such observations before they could sort them into classes.

Rounding numbers is common practice in statistical studies, and in this text we will use the following **conventions for rounding**. For numbers with decimal places less than 0.5, we will round down to the nearest whole number; for decimal places equal to 0.5 or higher, we will round up to the nearest whole number. Thus a housing value of $29,999.50 would be rounded up to $30,000, while a value of $29,999.49 would be rounded down to $29,999.

Another statistic we can calculate is the **class midpoint**, the variable's central value for each class. The class midpoint is found by adding the class limits together and then dividing that sum by 2. For example, the class midpoint for the "$20,000 to $29,999" class is $24,999.50. If we were reporting housing values in whole numbers, we would then round this class midpoint to $25,000.

We now raise two issues associated with frequency distributions. The first is whether the class intervals in a frequency distribution should be of equal size across all classes. Because economists construct frequency distributions to observe how a variable is distributed across all observed elements, imposing this constraint on the distribution may obscure the patterns we wish to observe.

Suppose, for example, we want to use equal class intervals in Table 3.2 (or Table 3.1). If we choose an interval of, say, $10,000, the resulting frequency distribution would have over 100 classes, an unmanageable number for analysis. On the other hand, suppose we set the class interval equal to $50,000. In this case, the class interval of "$50,000 to $99,999" would sort 30.4 percent of the total observations into a single class, eliminating important details about housing values for almost one-third of the elements. Consequently, economists accept that it may be necessary to use unequal class intervals to avoid such problems.

A second concern relates to the first and last classes in Table 3.2. Both classes are said to be **open-end classes**, which means we do not know one of the class limits. The last class of "$1,000,000 or more," for example, has no upper class limit, and given the very high price of some homes in the United States, we cannot even begin to guess at the value of the highest-value observations in this class. The first class, "less than $10,000," is also an open-end class. Although we might assume that this interval's lower limit must be zero, some home-owners actually reported negative housing values in the 2000 Decennial Census. Thus we cannot assume the lower limit is $0.

Open-end classes create problems for statistical analysis, particularly when we want to calculate a single measure for the variable's central tendency or dispersion. However, many datasets in economics contain extreme data values known as *outliers* which are *far* removed from the main body of data. When outliers appear, we may choose to use open-end classes in the frequency distribution, just as we might use unequal class intervals, if doing so avoids the problem of having empty (or nearly empty) classes. Because empty classes tell us little of significance about the overall distribution of the variable, excluding them from a frequency distribution is more important than forcing the distribution's classes to exhibit equal widths or closed ends.

3.7 Constructing a frequency distribution

Let us now turn to the steps for systematically organizing observations into a frequency distribution.

Step 1: Determine the number of classes

We want to use enough classes to reveal the overall shape of the distribution. One commonly used guideline for determining the number of classes is applying equation (3.1) to the dataset:

$$y = \frac{\log(n)}{\log(2)} \tag{3.1}$$

where y is the number of classes and n is the number of observations in the sample.

It is important to note that equation (3.1) is a guideline, not a rule to follow slavishly. Indeed, as we gain experience and practice, we may find that fully describing the data's distribution requires us to use fewer or more classes than the number calculated from equation (3.1). A good rule of thumb to follow is to begin with the number of classes determined by equation (3.1) and then adjust the number of classes as necessary. This means we want to have enough classes so we can identify where the variable's values cluster without any one class accounting for a large percentage of the observations. We also do not want so many classes that some classes are empty or nearly empty.

Step 2: Determine the class interval

After we select the number of classes, we next determine the class interval using equation (3.2):

$$\text{class interval} = \frac{(\text{largest observation value} - \text{smallest observation value})}{\text{number of classes}} \tag{3.2}$$

Equation (3.2) may generate a non-integer value or a value whose units are difficult to readily grasp, like $136. As a result, we typically round the class interval to a convenient and easily understood whole number to facilitate interpretation of the interval.

The procedure used in step 2 ensures that our class intervals will be of equal width. In the presence of outliers in the dataset, we may wish to adjust the intervals to obtain a more effective frequency distribution.

Step 3: Set the individual class limits

Because each observation must be sorted into one and only one class, we must establish the class limits so none overlap. In addition, class limits must be consecutive in value; that is, we want the lower limit of a class interval to be the next unit after that for the upper limit of the previous class.

Step 4: Count the observations in each class

This step is most easily performed by first sorting the observations in rank order and then tallying the number of observations in each class. While we could perform this step manually, we will use Microsoft Excel to execute this step. (See "Using Excel to create a frequency distribution" at the end of this chapter for detailed instructions.)

Step 5: Present the frequency distribution in a table or a histogram

The final step is to present the results in an easily understood table or histogram.

Let us now illustrate the steps for constructions a frequency distribution using a sample of monthly housing costs[4] from the 46,716 U.S. households participating in the 2004 American Housing Survey Metropolitan Sample.[5]

Step 1: Determine the number of classes using equation (3.1):

$$y = \frac{\log(n)}{\log(2)} = \frac{\log(46,716)}{\log(2)} = 15.5116$$

Because the number of classes must be a whole number, we will initially set the number of classes equal to 16. After generating results for the 16 classes, we might try out different numbers of classes, such as 15 or 20, to see if they more fully capture the details in the data.

Step 2: Determine the class interval. To do so, we must first identify the smallest and largest values in the dataset. In this sample, the highest monthly household cost observed was $9667 and the lowest was $0. We can now apply equation (3.2):

$$\text{class interval} = \frac{(\$9,667 - \$0)}{16} = \$604.19$$

The result is not an integer nor is it a value that is as easily grasped as a dollar amount ending in zero. Given the goal of effective communication, we will therefore select the more readily understood interval of $600.

There are no fixed rules selecting the "best" class interval value in step 2. We could, for example, have selected $650 or even $500. As when determining the number of classes for a frequency distribution, we may want to experiment with different values for the class interval before determining the interval most effective for our analysis. In this process, keep

Table 3.3 Relative frequency distribution of 2004 monthly housing costs from AHS Metropolitan
Sample

Monthly housing costs	Percentage of households	Monthly housing costs	Percentage of households	Monthly housing costs	Percentage of households
$0 to $599	38.049	$3,600 to $4,199	0.383	$7,200 to $7,799	0.002
$600 to $1,199	36.936	$4,200 to $4,799	0.285	$7,800 to $8,399	0.006
$1,200 to $1,799	15.451	$4,800 to $5,399	0.150	$8,400 to $8,999	0.000
$1,800 to $2,399	5.760	$5,400 to $5,999	0.068	$9,000 to $9,599	0.000
$2,400 to $2,999	1.909	$6,000 to $6,599	0.021	$9,600 to $9,667	0.002
$3,000 to $3,599	0.955	$6,600 to $7,199	0.021		
				Total	46,716

Data source: U.S. Census Bureau (2005), American Housing Survey, 2004 Metropolitan Sample.

in mind that we want each class to include more than just a few observations but not so many observations that most observations are clustered in only a few classes.

Step 3: Set the individual class limits. We begin with the first class, whose lower limit is the minimum value in the dataset and whose upper limit is one unit less than the class interval. In this sample, the first class is defined to as "$0 to $599". Each subsequent class begins with a readily understood value; the second class, for example, begins at $600, while the third begins at $1200.

Step 4: Count the number of observations in each class. As for any large sample, we use Excel to perform this task.

Step 5: Present the frequency distribution in a table or a histogram. For this sample, the results are presented in Table 3.3 and Chart 3.2 (p. 71). Note that we would typically use only one or two decimal places for the frequencies, but here we use three to distinguish between the empty and the nearly empty classes.

We observe that the vast majority of metropolitan households in the sample (just under 75 percent) had monthly housing costs between $0 and $1,200. We also see that only 25 households (0.03 percent) had housing costs of $6,000 or more per month, and that two of those seven classes have zero observations in them.

The presence of outliers accounts for the poor results. Less than 2 percent (885 out of 46,716) of the households surveyed had monthly housing costs of $3,000 or more, while less than 10 percent (9.6 percent, or 4,468 households) had monthly costs of $1,800 or higher. This suggests we might want to modify our results to obtain a more informative frequency distribution.

One approach is to eliminate the outliers and reapply steps 1 through 5 to the remaining data. Once completed, we must restate the last class as an open-end class so it includes all observations from the original dataset. We must defer this option until Chapter 5, where we explain how the outliers are identified.

A second approach is to see if there are published frequency distributions available for the particular variable. We know that the data for our example were obtained from the American Housing Survey (AHS) conducted by the U.S. Census Bureau. Searching the Census Bureau's homepage (<www/census.gov>) for the terms "American Housing Survey", "monthly housing costs", and 2004 yields a link to the 2004 Current Housing Reports (Publication H170) for 12 metropolitan areas. A further search reveals that Table 3–13 of the report presents the classes used to create a frequency distribution of monthly housing costs.

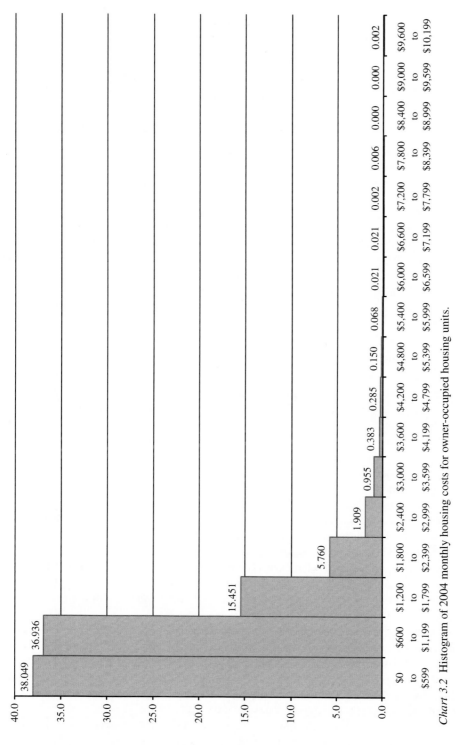

Chart 3.2 Histogram of 2004 monthly housing costs for owner-occupied housing units.

Data source: Table 3.3.

Table 3.4 Relative frequency distribution of monthly housing costs from 2004 AHS Metropolitan Sample with class intervals defined by the U.S. Census Bureau

Monthly housing costs	Percentage of households	Monthly housing costs	Percentage of households	Monthly housing costs	Percentage of households
Less than $100	1.1	$350 to $399	4.1	$700 to $799	7.8
$100 to $199	4.5	$400 to $449	4.1	$800 to $999	12.1
$200 to $249	3.4	$450 to $499	4.2	$1,000 to $1,249	10.5
$250 to $299	4.0	$500 to $599	8.5	$1,250 to $1,499	7.4
$300 to $349	4.1	$600 to $699	8.4	$1,500 or more	15.8
				Total	100.0

Data source: Author's calculations based on U.S. Census Bureau (2005), American Housing Survey, 2004 Metropolitan Sample.

We will now use these classes to sort the observations from our sample into the AHS classes. The results are presented in Table 3.4 and Chart 3.3 (p. 73).

While the different class intervals make it difficult to compare Tables 3.3 and 3.4 (or the histograms in Charts 3.2 and 3.3), we clearly see that the observations are more equally distributed across the classes in Chart 3.3 and that we have no empty or nearly empty classes. Thus the frequency distribution presented in Table 3.4 (p. 74) is more effective than that in Table 3.3 in communicating how monthly housing costs were distributed across U.S. households in 2004.

3.8 Frequency polygons

Histograms are one chart type for visualizing frequency distributions, but they may not be appropriate for some analysis. Suppose, for example, we want to determine how monthly housing costs changed between 2004 and 2005. If we assume that real housing costs did not change significantly between 2004 and 2005, how might we compare the 2004 sample results with the relative frequency distribution of monthly housing costs published in the 2005 American Housing Survey?

The first requirement is that the two frequency distributions use the same class intervals. Since we do not have the raw data from the 2005 survey, we will use the monthly housing costs classes published by the Census Bureau, which are those used in Table 3.4. Next we must decide if it is more effective to present the two frequency distributions in tabular form or in a chart. Since the visual depiction of information in a chart is easier for most of us to analyze than a table of numbers, we will select the chart form for the comparison.

We must now determine what chart type is most effective. While it is mechanically possible to plot two frequency distributions on the same chart, what results is not a histogram but a bar chart. In other words, a histogram illustrates the frequencies of a *single* continuous variable over a range of values.

We can, however, create a line chart of both frequency distributions. This construction is known as a **frequency polygon**, and it plots each distribution as a continuous line showing the frequencies associated with each class. A frequency polygon is most popularly used for comparing the frequency distributions of a variable for more than one group of elements or across different time periods. In Chart 3.4, we present the frequency polygons for the monthly housing costs variable in 2004 and 2005.

According to Chart 3.4, monthly housing costs did differ between the two years. We see, for example, that more households in 2005 incurred costs less than $400 per month

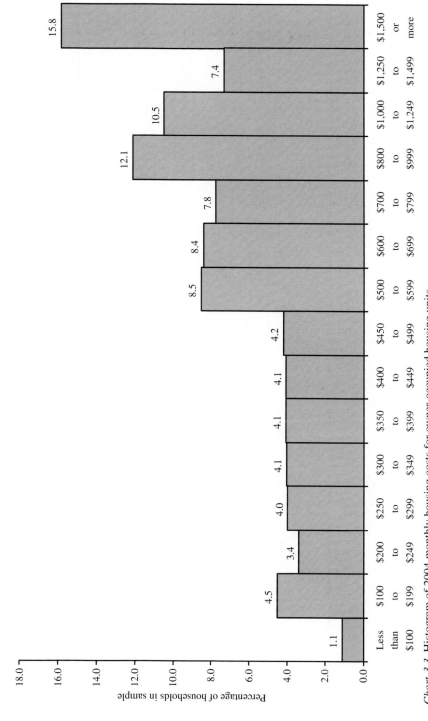

Chart 3.3 Histogram of 2004 monthly housing costs for owner-occupied housing units.

Data source: Table 3.4.

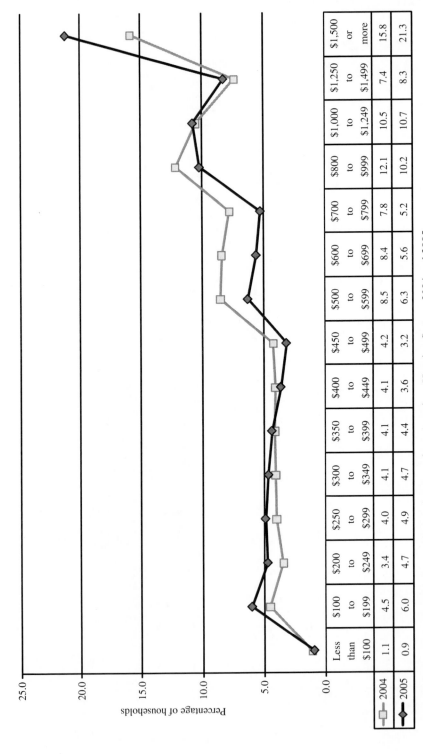

Chart 3.4 Frequency polygon of monthly housing costs from American Housing Survey, 2004 and 2005.

	Less than $100	$100 to $199	$200 to $249	$250 to $299	$300 to $349	$350 to $399	$400 to $449	$450 to $499	$500 to $599	$600 to $699	$700 to $799	$800 to $999	$1,000 to $1,249	$1,250 to $1,499	$1,500 or more
2004	1.1	4.5	3.4	4.0	4.1	4.1	4.1	4.2	8.5	8.4	7.8	12.1	10.5	7.4	15.8
2005	0.9	6.0	4.7	4.9	4.7	4.4	3.6	3.2	6.3	5.6	5.2	10.2	10.7	8.3	21.3

Data sources: Table 3.5, U.S. Census Bureau (2005), American Housing Survey (2005), American Housing Survey, 2004 Metropolitan Sample Table 3-13, U.S. Census Bureau (2006), American Housing Survey for the United States: 2005.

Note that the 2004 AHS was national in scope, so the geographical areas are not strictly comparable. The 2005 AHS was for select metropolitan areas.

(25.6 percent of households compared to 21.2 percent in 2004), while a larger percentage of households in 2004 (45.1 percent to 34.1 percent) incurred costs between $400 and $1,000. We also observe that more households in 2005 (40.3 percent) had housing costs greater than $1,000 per month, compared to 33.7 percent of households in 2004. These results suggest that monthly housing costs may be more evenly distributed in 2004 than in 2005.

We must add one caveat here. The 2004 sample represented 12 metropolitan areas whereas the 2005 results were for the entire nation. Thus the elements are not strictly comparable between the two years. It is important that we both note this point on Chart 3.4 and make it explicit in our discussion of the results.

3.9 Cumulative frequency distributions and ogives

We now turn to one final type of frequency distribution, the cumulative frequency distribution and its graphical representation, the cumulative frequency polygon, or ogive. We have already seen the analytical value of this type of distribution in Section 3.8, where we compared, for example, the frequencies of monthly housing costs in the "less than $400" range for 2004 and 2005.

Formally, a **cumulative frequency distribution** adds the frequency of each class to the sum of all frequencies from the preceding classes. A cumulative frequency distribution may be constructed from either an absolute or relative distribution for either an attribute (measured in ordinal scale) or a measured variable, and it facilitates comparisons of "how many" fall above or below specific values along the continuum of class intervals. For example, a cumulative relative frequency distribution would tell us that 61.5 percent of metropolitan households had monthly housing costs less than $1,000 in 2004.

Let us now consider the percentage of females and males 25 years and older in the United States who had attained various levels of education as of June 2009. According to the relative frequency distribution in Table 3.5, we see that 19.0 percent of women had attained "less than a high-school diploma" compared to 18.3 percent of men, while 19.5 percent of adult females and 25.4 percent of adult males had attained a "bachelor's degree or higher."

We might now ask what percentage of adult women and adult men had acquired education *through* a certain level, say, the high-school diploma. We can quickly read the answer from the cumulative frequency distribution in Table 3.5: 56.3 percent of females and 51.4 percent of males 25 years and older had at most a high-school degree in June 2009. We can also determine, by subtracting the "Some college, no degree" cumulative frequency of 73.3 percent from 100, that 26.7 percent of adult women had an associate degree or higher, while

Table 3.5 Relative and relative cumulative frequency distributions of persons over 25 years of age by educational attainment and sex in June 2009

Highest level of education completed	Females		Males	
	Percentage	Cumulative frequency	Percentage	Cumulative frequency
Less than a high-school diploma	19.0	19.0	18.3	18.3
High-school graduates, no college	37.3	56.3	33.1	51.4
Some college, no degree	17.0	73.3	16.9	68.3
Associate degree	7.2	80.5	6.3	74.6
Bachelor's degree or higher	19.5	100.0	25.4	100.0

Data source: U.S. Census Bureau (2009b), Current Population Survey, June 2009.

Table 3.6 U.S. males 25 years and over by last year of school completed and race, June 2009

Last year of education completed	Relative frequencies		Cumulative relative frequencies	
	American Indian, Alaskan native only	Asian only	American Indian, Alaskan native only	Asian only
Less than 1st grade	1.5	0.9	1.5	0.9
1st through 8th grade	12.8	5.1	14.3	6.0
9th through 11th grade	15.0	3.5	29.3	9.5
12th grade, no diploma	2.6	1.8	31.9	11.3
High-school graduate-diploma or equivalent (GED)	29.3	17.7	61.2	29.0
Some college, but no degree	16.0	11.0	77.2	40.0
Associate degree	10.6	6.6	87.8	46.5
Bachelor's degree (e.g., BA, AB, BS)	6.8	30.2	94.6	76.8
Master's degree (e.g., MA, MS, MEng, MEd, MSW)	3.6	14.6	98.2	91.4
Professional school degree (e.g., MD, DDS, DVM) or doctorate degree (e.g., PhD, EdD)	1.8	8.6	100.0	100.0

Data source: U.S. Census Bureau (2009b), Current Population Survey, June 2009.

31.7 percent of adult males fell into that category. These results indicate that adult females had, on average, lower levels of educational attainment than adult men in June 2009.

We can visually represent a cumulative frequency distribution in a chart known as a **cumulative frequency polygon**, or **ogive**. The ogive looks very much like the frequency polygon describe above, except we now plot the *cumulative* frequencies on the Y-axis.

Let us now turn to another set of data on educational attainment to demonstrate the use of ogives. One striking feature in the U.S. economy is differences in educational attainment across racial groups. Table 3.6 presents the relative and relative cumulative frequency distributions by last year of school completed for adult American Indian and Alaskan Native (AIAN) Only males and adult Asian Only males.[6] We use this information to construct ogives for the two racial groups, which are depicted in Chart 3.5 (p. 77).

What does Chart 3.5 tell us about the educational differences between adult AIAN men and adult Asian men in June 2009? We quickly note that the two groups experience very different patterns in the last year of school completed. We observe, for example, that more than twice as many AIAN males (14.3 percent) had completed at most the 8th grade, compared to just 6.0 percent of Asian males. We also observe that the disparity between the educational attainment of American Indian and Alaskan Native males and Asian males persists as the years of education completed increase. For example, the ogives indicate that almost two-thirds (61.2 percent) of AIAN males had not completed schooling beyond high school, while that was true for only 29.0 percent of Asian males. In fact, because the ogive for adult AIAN males lies above that for adult Asian males, we can see that a smaller percentage of AIAN males have completed each level of schooling as compared with Asian males.

Another observation is the striking differences in college degrees obtained by males in the two racial groups. We can determine, for example, that only 22.8 percent of AIAN males had attained some type of college degree in June 2009, while 60.0 percent of Asian males had a college degree of some kind. The disparity is even greater at the bachelor's degree level.

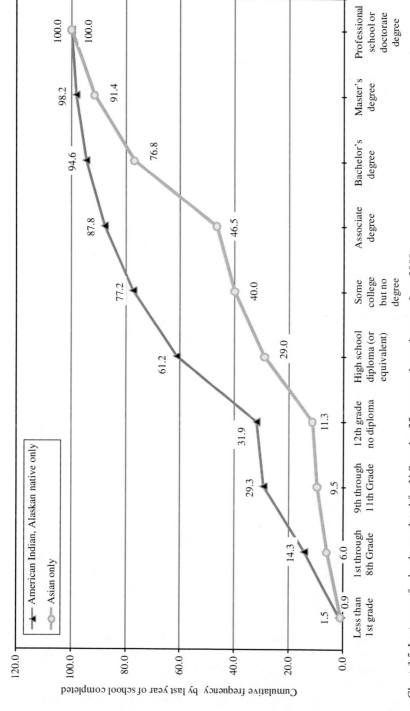

Chart 3.5 Last year of school completed for U.S. males 25 years and over by race, June 2009.

Data source: U.S. Census Bureau (2009b), Current Population Survey, June 2009.

Table 3.7 Relative and cumulative relative frequency distributions for U.S. owner-occupied housing units, 2000

	Relative frequency	Cumulative relative frequency
Less than $19,999	1.5	1.5
$20,000 to $39,999	4.7	6.2
$40,000 to $59,999	8.1	14.3
$60,000 to $79,999	11.8	26.1
$80,000 to $99,999	14.1	40.2
$100,000 to $124,999	12.4	52.6
$125,000 to $149,999	11.3	63.9
$150,000 to $174,999	8.5	72.4
$175,000 to $199,999	6.1	78.5
$200,000 to $249,999	7.3	85.8
$250,000 to $299,999	4.6	90.4
$300,000 to $399,999	4.4	94.8
$400,000 to $499,999	2.1	96.9
$500,000 to $749,999	1.8	98.7
$750,000 to $999,999	0.6	99.3
$1,000,000 or more	0.6	99.9

Data source: Author's calculations from Table 3.3.

Here we see that 54.5 percent of Asian males had at least a bachelor's degree compared to only 12.2 percent of AIAN males.

Such empirical findings point to one potential source of the economic disparity between racial groups in the United States. Indeed, economists frequently cite educational attainment as one of the key factors in economic success and upward mobility, so they are likely to find the differences noted above significant. Policy-makers, too, would find our results interesting, particularly for identifying who might most benefit from programs to improve educational attainment.

The two previous examples of cumulative frequency distributions both use attribute variables measured on an ordinal scale, but we can also construct cumulative frequency distributions and ogives for measured variables. Let us return to the data for values of owner-occupied housing from the 2000 U.S. Census (Table 3.7) to construct a cumulative frequency distribution and accompanying ogive. (Note that, for illustration purposes, we have redefined the class intervals from Table 3.2 to reduce the number of classes.)

Chart 3.6 (p. 79) illustrates that over half (52.6 percent) of U.S. owner-occupied housing units had values less than $125,000 in 2000. It also shows that over 90 percent (90.4 percent) of the units were valued below $300,000, or that approximately 10 percent of Americans had homes valued at $300,000 or higher. Finally, we observe that just over one-quarter (26.1 percent) of U.S. owner-occupied housing units in 2000 were valued under $80,000, while just over three-quarters (78.5 percent) were valued at less than $200,000. From this, we can calculate that the middle 50 percent of the housing units occupied by their owners were valued between roughly $80,000 and $200,000 in 2000. While we could continue our empirical analysis of Chart 3.6 (as well as the others in this chapter), we will defer further discussions until we have discussed the measures of central tendency and dispersion associated with the various datasets presented here.

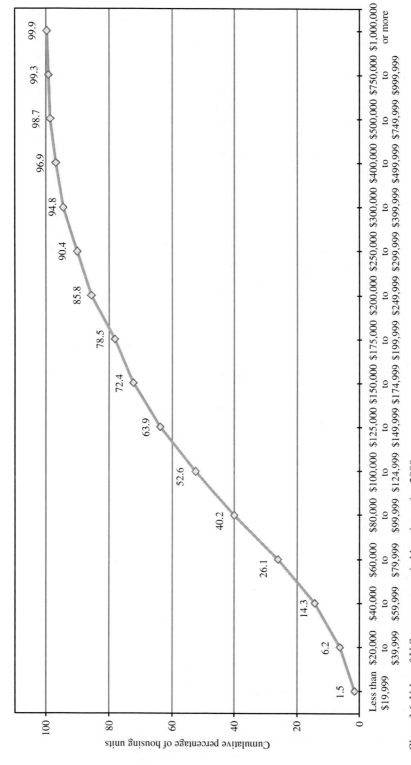

Chart 3.6 Value of U.S. owner-occupied housing units, 2000.

Data source: Table 3.7.

Summary

All statistical data encountered in empirical economics requires researchers to identify the specific elements and variables appropriate to the economic analysis at hand. For the analyst who designs data collection procedures, these elements and variables must be carefully defined. Likewise, those of us who use data collected by others must understand the definitions of all elements and variables we use so we can assess the appropriateness of the data for our analyses and so we can alert our audience to any data limitations.

Both qualitative and quantitative variables are widely used in empirical economics. The values of such variables usually differ between elements, producing the general characteristic of data known as heterogeneity. This variability of elements with respect to any given variable requires that we use statistical methods to describe the mathematical features of a group of elements.

One common method for describing this variability is the frequency distribution, which can be expressed in absolute or relative form. In both cases, the frequency distribution counts the elements that fall into a range of classes defined by the researcher. In particular, the class intervals for measured variables define a range of values for each class, while class intervals for an attribute correspond to the elements exhibiting particular characteristics.

Frequency distributions can also be depicted using several different charts. The histogram utilizes vertical bars to denote the number, or percentage, of elements within each class, while the frequency polygon utilizes lines connecting points that represent the number of elements in each class. Because of its construction, the frequency polygon lends itself to visual comparisons of multiple groups, while histograms are appropriate for depicting the distribution of a single variable. In cases where we want to know the total number or percentage that fall above or below a particular value, we can create cumulative frequency distributions and ogives.

Selecting the "best" frequency distribution, or its graphical representation in a chart, ultimately depends on the analysis we are conducting. While we may have to try several different possibilities before we determine the most effective way, the methods presented in this chapter will guide us in creating "order out of chaos" (Guthrie 1966, p. 1).

Concepts introduced

Absolute frequency distribution
All-inclusive (classes)
Attribute variable
Binary
Categorical
Class
Class frequencies
Class interval
Class limits
Class midpoint
Continuous variable
Convention for reporting percentages
Conventions for rounding
Cumulative frequency distribution
Cumulative frequency polygon
Discrete variable
Element

Exhaustive (classes)
Frequency distribution
Frequency polygon
Heterogeneity
Histogram
Interval scale
Level of measurement
Measured variable
Mutually exclusive (classes)
Nominal scale
Observations
Ogive
Open-end class
Ordinal scale
Qualitative variable
Quantitative variable
Ratio scale
Relative frequency distribution
Scale
Variable

Box 3.1 Where can I find. . . ? DataFerrett

DataFerrett is a data extraction and analysis tool from the U.S. Census Bureau (FERRETT stands for Federated Electronic Research, Review, Extraction, and Tabulation Tool). According to its website (<http://www.thedataweb.org/what_ferrett.html>):

DataFerrett helps you locate and retrieve the data you need across the internet to your desktop or system, regardless of where the data resides. DataFerrett:

- lets you receive data in the form in which you need it (whether it be extracted to an ASCII, SAS, SPSS, Excel/Access file); or
- lets you move seamlessly between query, analysis, and visualization of data in one package; or
- lets data providers share their data easier, and manage their own online data.

In DataFerrett, you can also:

- perform basic statistical manipulations of the variables selected across data sets, and
- search for topics in DataFerrett (see <http://dataferrett.census.gov/topics.html> for the list of searchable terms).

These features make DataFerrett a valuable source for analysis of U.S. economic data.

Accessing DataFerrett

To use DataFerrett, you will need to install the Beta DataFerrett on your computer. Directions are available at <http://dataferrett.census.gov/applet.html>. Note that you will need run the SecurityPolicy the first time you use Ferrett.

Datasets available through DataFerrett

The following descriptions of each dataset were obtained from the DataFerrett website (<http://www.thedataweb.org/datasets.html>) on 16 July 2010. The links are to the Survey/Sponsor's website, with further documentation for that dataset. These sites may also provide access to additional data. Do note that not all of the datasets at DataFerrett are from the most recent survey year.

American Community Survey (ACS)

<http://www.census.gov/acs/www/>

   The American Community Survey provides annual estimates of demographic, housing, social, and economic characteristics for all U.S. states, cities, counties, metropolitan areas, and population groups of 65,000 people or more.

American Housing Survey (AHS)

<http://www.census.gov/hhes/www/housing/ahs/ahs.html>

The American Housing Survey (AHS) provides biannual data on characteristics of the U.S. housing sector, including group and single-family housing, vacant housing units, household characteristics, income, housing and neighborhood quality, housing costs and maintenance, and recent movers. Data for each of 46 selected metropolitan areas are collected about every four years, with an average of 12 metropolitan areas included each year. The national sample covers on average 55,000 homes, while each metropolitan areas sample covers 4,800 or more homes.

Consumer Expenditure Survey (CEX)

<http://www.bls.gov/cex/home.htm>

   The Consumer Expenditure Survey (CEX) program consists of two surveys – the quarterly Interview survey and the Diary survey – that provide information on the buying habits of American consumers, including data on their expenditures, income, and consumer unit (families and single consumers) characteristics.

County Business Patterns (CBP)

<http://www.census.gov/econ/cbp/>

   County Business Patterns provides annual sub-national economic data by industry for most of the country's economic activity. The series excludes data on self-employed individuals, employees of private households, railroad employees, agricultural production employees, and most government employees.

Current Population Survey (CPS)

<http://www.census.gov/cps/>

   The Current Population Survey (CPS) is a monthly survey of about 50,000 households conducted by the Bureau of the Census for the Bureau of Labor Statistics. This survey includes variables on employment, unemployment, earnings, educational attainment, income, poverty, health insurance coverage, job experience and tenure, school enrollment, voting and registration, and veterans.

Decennial Census of Population and Housing (Census 2000)

<http://www.census.gov/main/www/cen2000.html>

   DataFerrett offers the data for the following Census products: Census 2000 Summary Files 1 and 3; Census 2000 Public Use Microdata Sample (1 percent); and 1990 Census Public Use Microdata Sample (1 percent Housing and Person Records).

National Ambulatory Medical Care Survey (NAMCS)

<http://www.cdc.gov/nchs/ahcd.htm>

   The National Ambulatory Medical Care Survey (NAMCS) provides national data about the provision and use of ambulatory medical care services in the United States. The survey was conducted annually from 1973 to 1981, in 1985, and annually since 1989.

National center for health statistics. Mortality – Underlying Cause of Death (MORT)

<http://www.cdc.gov/nchs/deaths.htm>

This annually updated longitudinal dataset is a key source of U.S. demographic, geographic, and cause-of-death information for small geographic areas and a long time period.

National Health Interview Survey (NHIS)

<http://www.cdc.gov/nchs/nhis.htm>

This annual survey interviews respondents about the following core categories: general health and personal care needs, availability and quality of health-care services, vaccination history, healthy living habits, and chronic diseases. Additional questions are added annually to obtain information on topics not covered by the core questions.

National Hospital Ambulatory Medical Care Survey (NHAMCS)

<http://www.cdc.gov/nchs/ahcd.htm>

This annual survey collects data on the utilization and provision of ambulatory care services in hospital emergency and outpatient departments.

National Survey of Fishing, Hunting and Wildlife-Associated Recreation (FHWAR)

<http://www.census.gov/prod/www/abs/fishing.html>

This survey collects data on individuals involved in fishing, hunting, and other wildlife-associated recreation, such as wildlife observation, photography, and feeding. Data include the state in which these activities occurred; number of trips taken; duration of trips; and expenditures for food, lodging, transportation, and equipment.

Model-based Small Area Health Insurance Estimates (SAHIE) for counties and states

<http://www.census.gov//did/www/sahie/>

The Census Bureau program produces estimates of health insurance coverage for states and all counties across demographic groups in the U.S.

Model-based Small Area Income and Poverty Estimates (SAIPE) for school districts, counties, and states

<http://www.census.gov/did/www/saipe/>

This program provides current estimates of selected income and poverty statistics for U.S. school districts, counties, and states.

Survey of Income and Program Participation (SIPP)

<http://www.census.gov/sipp/>

This annual national survey of approximately 14,000 to 36,700 interviewed households collects information on the source and amount of income, labor force information, program participation and eligibility data, and general demographic characteristics. Because this survey has been conducted since 1984, it is a particularly rich source of longitudinal data for the United States.

Survey of Program Dynamics (SPD)

<http://www.census.gov/spd/>

The SPD survey was designed to collect cross-sectional and longitudinal national data for the purpose of assessing the effectiveness of U.S. federal welfare programs between 1997 and 2002.

Box 3.2 Using Excel to create a frequency distribution

Pre-determined classes

When the data are already distributed into classes, use an Excel column chart to create a histogram. To eliminate the gap between the columns:

- Right-click on any column, and select **Format Data Series**.
- Under **Series Options**, reduce the **Gap Width** to 0.

Researcher-generated classes

If you are working with unsorted data, follow the instructions in Section 3.7 to determine the number of classes and class intervals. When you reach step 4, let Excel sort the data into the classes.

You will first need to tell Excel where to place each observation. In Excel's lingo, you must define a **Bin Range** in which each **bin number** corresponds to a class interval. An observation, or **data point**, is sorted into a particular bin if

> lower limit of class interval < **data point** ≤ upper limit of class interval

Returning to Table 3.3, we would construct the following column of values for the classes indicated:

Value of housing unit	Excel bin range values
Less than $10,000	9,999
$10,000 to $14,999	14,999
$15,000 to $19,999	19,999
$20,000 to $24,999	24,999
. . .	. . .
$750,000 to $999,999	999,999
$1,000,000 or more	†

Notes on bin ranges

- Do not include a value for the final class. Excel creates a bin labeled "More" that will contain the number of values greater than the upper limit of the penultimate bin.
- Provide a label in the row above the first value. The label, "Excel Bin Range values" was used above.
- If you do not specify the bin range, Excel creates a set of evenly distributed bins between the minimum and maximum values of the input data. This is analogous to applying equation (3.2) and will rarely generate economically meaningful bins.

Once we have defined the Bin Range, we can use Excel's **Histogram Tool** to sort the data and to construct a rudimentary histogram.

1. On the **Data** tab, click on the **Data Analysis** group.

> *Accessing Excel's Histogram Tool the first time*
> If the **Data Analysis** option does not appear on the **Data** ribbon, click on the Microsoft Office button and then, at the bottom of the drop-down menu, select **Excel Options**.
> On the left-hand menu, select **Add-Ins**.
> At the bottom of the Add-In screen, select **Manage Excel Add-Ins**, and click **Go**.
> Check the boxes for both the **Analysis ToolPak** and the **Analysis ToolPak-VBA**. Click **OK**. The **Data Analysis** option will now appear at the end of the **Data** tab.

2. Next select **Histogram**.

- Highlight the unsorted data as the **Input Range**. (You do not need to sort the data series; Excel will do that as part of its calculations.)
  It is highly recommended that you include the name of the data series so it is clear what variable is being analyzed. Put the name in the first row of the column, and be sure to check the **Labels** box.
- Highlight the **Bin Range** including the label. Omitting the label for either the Input or Bin Range will return the error message, Input Range/Bin Range "cannot have non-numeric data."
- Under **Output** options, select where you want the frequency distribution (and histogram) to appear.
- Check the **Chart Output** box if you want Excel to generate the histogram. Note that you will need to reformat the result to conform to the guidelines established for a histogram and those for effective communication.

Additional formatting of frequency distribution charts

Displaying data labels for multiple data series can make reading the chart difficult when the data points are close in value. That was the case in Chart 3.4, so a Data Table was used instead of data labels.

Add a Data Table

- Activate the chart.
- Select **Layout** and then **Data Table**.
- From the drop-down menu, select the desired option. (The chart below places the title above the chart.)
  - If you select the **Show Data Table** option, you will still need to include a legend on the chart.
  - If you select the **Show Data Table with Legend Keys** option, the legend will be unnecessary.

Chart 3.5 contained ten classes where each class interval indicated the last year of school completed. Some labels were quite lengthy, causing them to dominate the Excel-generated chart. Fortunately, we can manually force hard returns in an Excel cell so the labels appear on more than one line.

Add a hard return in an Excel cell

- Activate the cell where you want to insert a hard return.
- Go to the point in the cell where you want to insert a hard return.
- Simultaneously press the **Alt** and **Enter** keys. Excel will place the information following the point of insertion on the next line.

Exercises

1. Tracking the income in a nation's economy is important for assessing the well-being of that nation's residents. Each fall, the U.S. Census Bureau releases its annual P60 report on income, poverty, and health insurance: *Income, Poverty, and Health Insurance Coverage in the United States*. In it, both cross-section and time-series money income statistics are presented for the purpose of describing changes across U.S. households and people.

This exercise uses data from the report's Table A-1; the 2009 version may be found at <http://www.census.gov/prod/2010pubs/p60-238.pdf>. (The data may also be found electronically in Table H-17. *Households by Total Money Income, Race, and Hispanic Origin of Householder* at the following link: <http://www.census.gov/hhes/www/income/data/historical/household/index.html>.)

(a) Identify the element(s) and the variable(s) that are the focus of the table. For each variable(s), explain if it is a measured/quantitative or attribute/qualitative variable; if its measurement scale is nominal, ordinal, interval, or ratio; and if it is a discrete or continuous variable.

(b) For the All Races category, construct an absolute frequency distribution for the percentage distribution of money income across households for the most recent year.

(c) Construct graphical representations of the absolute and relative frequency distributions of money income for all U.S. households. After identifying the chart types you used, discuss what they tell us about the distribution of money income in U.S. households.

(d) We often hear that racial disparities in income persist in the United States. Select three race-ethnic groups from Table A-1 (or H-17), and compare the income distributions for the most recent year. For two of these groups, discuss how the distribution of their money income has changed between 2002 and the present.

(e) Economists have identified 1973 as a peak year in U.S. productivity. Construct an appropriate chart for comparing the distribution of money income for U.S. households in that year with the most recent year. Explain what you observe.

2. Rising gasoline prices in recent years have led Americans to reconsider how much they drive for pleasure and for work. In anticipation of results from the 2010 Decennial Census, suppose we want to know how long it took residents of three metropolitan statistical areas (MSAs) (Los Angeles, CA; District of Columbia–Baltimore; and New York City, NY) to travel to work when the previous Census was conducted in 2000.

(a) Using the data from P31. Travel Time to Work for Workers 16 Years and Over, construct a frequency distribution of the travel time to work that is appropriate for comparing the three MSAs in 2000. Comment on how work travel time varied across the three geographical areas in 2000.

(b) Construct the most effective chart(s) of the frequency distribution in part (a) for capturing the differences in travel time for the three MSAs. Discuss whether the table or chart(s) more effectively showed those differences.

(c) Construct a cumulative frequency distribution of the travel time to work for the three MSAs, and comment on what you observe.

(d) Compare the travel time to work for the District of Columbia MSA by state and the District using the most effective method from Chapter 3.

(e) Provide empirical evidence to identify which state's residents in the New York MSA need to be the most, and the least, concerned about higher gas prices based on the data from 2000.

3. Go to the American Housing Survey at DataFerrett (see above), select a year other than 2005, and retrieve data on monthly housing costs for owner-occupied housing.

(a) Follow the procedure outlined in the text to construct a relative frequency distribution, histogram, cumulative frequency distribution, and ogive for the sample.

P31. Travel Time to Work for Workers 16 Years and Over [15]
2000 decennial census of population and housing, summary file 3 (summary file 3)

State	Did not work at home:												Worked at home
	Less than 5 minutes	5 to 9 minutes	10 to 14 minutes	15 to 19 minutes	20 to 24 minutes	25 to 29 minutes	30 to 34 minutes	35 to 39 minutes	40 to 44 minutes	45 to 59 minutes	60 to 89 minutes	90 or more minutes	
CA – Los Angeles, Riverside, Orange County	116,601	516,794	829,713	978,495	932,267	366,881	1,023,493	179,916	268,919	585,546	485,339	242,204	241,451
DC MSA	57,461	212,084	346,953	423,879	455,290	219,059	533,950	132,887	185,213	434,834	329,290	123,139	124,129
DC – District of Columbia	3,782	12,404	23,202	34,257	40,556	17,102	50,138	9,271	11,995	25,167	15,935	7,145	9,930
DC – Maryland	37,113	139,094	231,278	280,673	303,672	145,917	348,135	86,427	116,707	276,233	212,389	83,422	77,710
DC – Virginia	18,633	68,021	108,665	135,064	144,148	69,843	178,977	44,900	66,685	154,480	112,517	35,658	45,082
DC – West Virginia	1,715	4,969	7,010	8,142	7,470	3,299	6,838	1,560	1,821	4,121	4,384	4,059	1,337
NY MSA	183,047	661,059	984,197	1,084,048	1,083,781	432,118	1,258,393	249,519	421,311	1,022,181	1,111,889	550,525	277,150
NY – Connecticut	22,063	89,130	139,880	140,158	120,961	45,735	91,640	20,605	25,358	54,404	48,532	29,089	30,381
NY – New Jersey	73,735	264,978	371,574	395,604	383,981	161,067	394,198	87,248	126,435	289,106	287,407	151,193	86,945
NY – New York	86,652	305,491	471,152	546,678	577,321	224,654	770,647	141,040	268,654	676,574	773,723	366,758	159,165
NY – Pennsylvania	597	1,460	1,591	1,608	1,518	662	1,908	626	864	2,097	2,227	3,485	659

Data source: U.S. Census Bureau (2000a), Census 2000. Summary File 1 (SF 1) 100-Percent Data.

(b) Using the cost classes in Table 3.5, construct a relative frequency distribution, histogram, cumulative frequency distribution, and ogive for the sample.

(c) Compare the two sets of results, and comment on the effectiveness of each set of classes.

4. The distribution of the U.S. labor force by age group and sex has changed dramatically since 1960. Use the data below to create the most appropriate charts for answering the following questions.

U.S. Civilian Labor Force – Percent Distribution by Sex and Age, 1960 to 2005
[69,628 represents 69,628,000. Civilian non-institutional population 16 years old and over.
Annual averages of monthly figures. Based on Current Population Survey; see text,
Section 1, Population, and Appendix III].

Sex and year	Civilian labor force (1,000)	Percent distribution						
		16 to 19 years	20 to 24 years	25 to 34 years	35 to 44 years	45 to 54 years	55 to 64 years	65 years and over
Male								
1960	46,388	6.0	8.9	22.1	23.6	20.6	13.8	4.9
1970	51,228	7.8	11.2	22.1	20.4	20.3	13.9	4.2
1980	61,453	8.1	14.0	27.6	19.3	16.1	11.8	3.1
1990*	69,011	5.9	11.4	28.8	25.3	16.1	9.6	2.9
2000*	76,280	5.6	9.9	23.4	26.3	21.3	10.2	3.3
Female								
1960	23,240	8.8	11.1	17.8	22.8	22.7	12.8	3.9
1970	31,543	10.3	15.5	18.1	18.9	20.7	13.2	3.3
1980	45,487	9.6	16.1	26.9	19.0	15.4	10.4	2.6
1990*	56,829	6.5	12.0	28.3	25.8	16.1	8.7	2.6
2000*	66,303	6.0	10.2	22.5	26.4	22.3	9.9	2.7

*Data not strictly comparable with data for earlier years. See text, this section, and February 1994, March 1996, February 1997–99, and February 2003–08 issues of Employment and Earnings Online.

Data source: Table 577, U.S. Census Bureau (2007c), Statistical Abstract of the United States: 2007.

(a) How did the distribution of the U.S. labor force in 1960 compare for males and females? In 1980? In 2000?

(b) How did the distribution of women in the U.S. labor force change between 1960 and 2000? Between 1970 and 1990?

(c) How did the distribution of men in the U.S. labor force change between 1960 and 2000? Between 1970 and 1990?

(d) How many women under the age of 25 participated in the U.S. labor force in 1960? Under the age of 35?

(e) Suppose you wanted to compare the results in (d) to those for 2000. Construct the most appropriate frequency distribution for the comparison. What might be the reasons for what you observe?

(f) How did the percentage of men under the age of 25 and over 55 change between 1960 and 1990? What might be the reasons for what you observe?

4 Measures of central tendency

4.1 Desirable properties for descriptive statistics

We have seen how frequency distributions organize many observations of an economic variable into a reduced and ordered form while still presenting the original data in its entirety. It is often useful, however, to describe economic phenomena using a single measure. One popular example is presenting data in per-capita, or per-person, terms. For example, if we know that the per-capita gross national income (GNI) in India in 2008 was US $1,040 and the per-capita GNI in the United States was US $47,930, we immediately recognize that the average level of economic well-being in the U.S. was much larger (approximately 46 times) than that in India.

Reducing many observations to a single, unique statistic is just one desirable property of a descriptive statistic. We would like our descriptive statistics to exhibit several other properties (Figure 4.1), so the statistics are easy to comprehend and take full advantage of the information available in the dataset.

The first desirable property of a descriptive statistic is that it be easily understood. This characteristic not only facilitates the literal interpretation of the result, but also improves our understanding of the economic variable in the context of our analysis. In other words, if we do not have to digress into a complicated explanation of what the descriptive statistic measures, we can focus our attention on the economic meaning of the statistic.

We have already noted the second desirable property – a descriptive statistic should take on a single value, which we would also like to be unique. In addition, we would like the statistic not to be distorted by extreme values, or outliers, in the dataset. A fourth property is that the statistic be "algebraically tractable." This means that we want the statistic to be easily manipulated by ordinary arithmetic operations such as addition, multiplication, and logarithms.

The next two properties are related. Property 5 states that the descriptive statistic should consider every observation of the variable, while property 6 acknowledges the importance of

---

Property 1: The statistic is easily understood
Property 2: The statistic is a single (unique) value
Property 3: The statistic's value is not affected by extreme observations
Property 4: The statistic is algebraically tractable
Property 5: The statistic utilizes all values in the dataset
Property 6: The statistic utilizes the frequencies of all values in the dataset

---

Figure 4.1 Desirable properties for descriptive statistics.

considering the frequency of every observed variable. Statistics that do not exhibit these properties increase the loss of information that accompanies any summary of a set of observations.

Let us now turn to the three most frequently used measures of central tendency – the mean, median, and mode. We will begin with definitions of the three measures and then use an example to illustrate how each is calculated. We will also evaluate each in terms of the six desirable properties just identified.

4.2   Three measures of central tendency: mean, median, and mode

A **measure of central tendency** is used to represent a total set of observations in a single numerical value that signifies the location around which the variable's observations tend to cluster, the so-called **average** value. The most commonly used measures of central tendency in economics are, in order, the mean (technically, the arithmetic mean), the median, and the mode. Although the three measures are similar in their general purpose, they do differ by definition. Specifically, given the values of a variable for a set of observed elements, we can calculate each measure as follows.

The **mean** or **arithmetic mean** sums all values of the variable and then divides by the number of observations. Technically, the mean is determined by the magnitude of the observed values and their frequencies. The mean can only be calculated for quantitative variables.

The **median** identifies the value of the variable for the middle element when the elements have been arranged in increasing (or decreasing) rank order. Technically, the median is determined entirely by the rank order of observed values. The median can be determined for qualitative variables measured on a nominal scale and for quantitative variables measured on both interval and ratio scales.

The **mode** identifies the value of the variable that has the highest frequency among the elements observed. Technically, the mode is determined entirely by the frequency with which the observed values occur. The mode is also the only measure of central tendency that can be determined for all scales of measurement.

Table 4.1 presents the frequency distribution for simulated observations of the duration of unemployment for ten unemployed persons. The sample is used here to illustrate, in much simplified form, the central tendency of the duration of unemployment for the almost eight million unemployed persons in the United States in November 2004.

Let us begin with the mode, or the most frequently observed value of the variable, the duration of unemployment. We quickly observe that four of the ten persons were unemployed for five weeks, while only one person was unemployed for each of the other durations shown. Because more persons were unemployed for five weeks than for any other number of weeks observed, five weeks is the modal duration of unemployment.

We next turn to calculating the median, or the middle value in the dataset. Here the median is the weeks of unemployment (the variable) associated with the middle person (element) where the observations have been ordered from the smallest (5 weeks) to largest (68 weeks) values.

Determining the median means that we must first locate the position of the middle value in the dataset. To establish that location, or **rank**, for ungrouped data, we can use equation (4.1):

$$\text{median rank } (R_M) = \frac{n+1}{2} \text{th item in the rank-ordered data} \tag{4.1}$$

Table 4.1 Duration of unemployment for ten hypothetical unemployed persons

Person	Number of weeks unemployed
1	5
2	5
3	5
4	5
5	9
6	11
7	15
8	27
9	50
10	68
Total	200

Data source: Simulated data for observations. The arithmetic mean and median approximate the BLS-calculated arithmetic mean and median for November 2004 (given in Table A-12, U.S. Bureau of Economic Analysis (2010e), Data Retrieval: Labor Force Statistics (CPS)).

where n equals the number of observations. In our example, we obtain a median rank equal to the 5.5th person, who, of course, does not appear among the actual observations of unemployed workers.

This result will obtain any time we have an even number of observations in the dataset. (If n is an odd number, equation (4.1) does yield a clearly defined (integer) location among the rank-ordered elements.) Thus to determine the median value, we must interpolate a value between the middle pair of observations identified by equation (4.1).

After determining the median rank (R_M) using equation (4.1), we can parse that number into two separate parts as shown in equation (4.2):

$$R_M = IR + FR \tag{4.2}$$

Here IR is the **integer portion** of the median rank that lies to the left of the decimal point, and FR is the **fractional portion** of R_M to the right of the decimal point. In our example, IR equals 5 and FR equals 0.5, telling us that the median duration of unemployment will lie halfway between the weeks of unemployment incurred by the 5th and 6th persons.

Now that we have identified the location of the median element, we can calculate the value of the median using equation (4.3):

$$\text{median } (M) = X_{IR} + [FR \times (X_{IR+1} - X_{IR})] \tag{4.3}$$

We first observe that the integer (5th) person was unemployed for 9 weeks, such that $X_{IR} = 9$. Thus we know that the median duration of unemployment will be greater than 9 weeks.

We next determine the difference in the weeks of unemployment experienced by the 5th and 6th persons. Since the 6th person was unemployed for 11 weeks, the difference in the duration of unemployment between the two persons is 2 weeks. In equation (3.3), we denote this component as $(X_{IR+1} - X_{IR})$ where X_{IR+1} is the value of the variable associated with the element immediately following the "integer" element.

We now multiply the difference by the fractional portion (*FR*) of the median rank. In our example, the number of weeks is halfway (or 0.5) between two weeks, so $[FR \times (X_{IR+1} - X_{IR})]$ equals 1 week.

The last step is to add this result to the corresponding integer value's weeks of unemployment (X_{IR}). For this example, we thus obtain a median duration of unemployment equal to

$$M = 9 + [0.5 \times (11 - 9)] = 10 \text{ weeks}$$

What does the median tell us about the "middle" unemployed worker? One way to interpret the median is it indicates that half of the ten unemployed people observed were unemployed for a period of ten weeks or less; an alternative interpretation is that the other half were unemployed for more than ten weeks.[1] As we see, this measure of central tendency gives us a different result from that indicated by the five weeks of unemployment measured by the mode.

Let us now calculate the third measure of central tendency, the arithmetic mean. As its definition indicates, the arithmetic mean sums the entire set of observations and then divides that sum by the number of elements observed. In our example, the arithmetic mean equals the total weeks of unemployment incurred by all elements (200 weeks) divided by the number of persons (10). In other words, for the ten people in the sample, the average number of weeks of unemployment was 20.

We can now generalize the calculation of the **arithmetic mean for ungrouped data** (denoted as $\overline{X}$, called "X-bar") using equation (4.4):

$$\text{arithmetic mean } (\overline{X}) = \frac{\sum X}{n} \tag{4.4}$$

where *n* is again the number of observations in the dataset.

The calculations of the three measures of central tendency give us three quite different results. The mean is double the value of the median, which is twice the value of the mode. Thus we must ask the following question. Which measure most accurately describes the duration of unemployment for the ten elements in our sample: 20 weeks, 10 weeks, or 5 weeks?

First, let us note that all three measures are correct according to their definitions. That is, we have accurately determined the mode, median, and mean for our sample of ten unemployed persons. However, the different values obtained illustrate important differences between the three concepts of central tendency, which, in turn, will influence which measure we decide to use.

Selecting the most appropriate measure of central tendency ultimately depends on the context of the questions we want to answer. Suppose, for example, we want to know the most likely number of weeks a person had been unemployed in May 2004. Here the mode gives us that information, so our answer would be five weeks. What if, instead, we wanted to know the number of weeks of unemployment that distinguished the number of people who were newly unemployed versus those who had been unemployed longer than average? Here, the median value of ten weeks would indicate an easily understood result. Finally, if we wanted to know the per-person duration of unemployment in May 2004, the arithmetic mean of 20 weeks would be the appropriate statistic to use.

Desirable statistical properties		*Holds for these measures of central tendency*
Property 1:	The statistic is easily understood	Mean, median, mode
Property 2:	The statistic is a single (unique) value	Mean, median
Property 3:	The statistic's value is not affected by extreme observations	Median, mode
Property 4:	The statistic is algebraically tractable	Mean
Property 5:	The statistic utilizes all values in the dataset	Mean
Property 6:	The statistic utilizes the frequencies of all values in the dataset	Mean

Figure 4.2 Three measures of central tendency by desirable statistical properties.

Let us now examine the three measures in terms of the desirable properties for descriptive statistics that we identified in Figure 4.1. As shown in Figure 4.2, we first note that none of the three measures exhibits all six properties. We also see that the mode has the fewest desirable properties. A major reason is that the mode may not return a unique value; that is, a set of observations can have multiple modes. This statistic is also not algebraically tractable because it is determined solely by the frequency of its appearance in a dataset. And even though we use the data's frequency distribution to locate the mode, the statistic itself does not use the other frequencies or any of the actual observations in its calculation. However, because it is the only measure of central tendency that can calculated for both levels of qualitative variables, the mode does provide important economic information about attribute variables. It is also an easily understood statistic.

The choice between the two remaining measures of central tendency is less clear-cut, and we will, in fact, not be able to make a definitive selection until we have discussed the shapes of frequency distributions in Chapter 5. However, we can still discuss the desirable statistical properties exhibited by the median and the mean.

The median, like the mode, is an easily understood statistic. However, like the mode, the median is not algebraically tractable nor does it consider the variable's frequencies in its calculation. In addition, neither the median nor the mode utilize all values in the dataset, but, because of this property, neither are affected by extreme values. The median does have one clear advantage over the mode – it will generate a unique value.

Of the three measures, the arithmetic mean exhibits the most (four) desirable statistical properties. It is easily understood, it is a single unique value, and it is algebraically tractable. The mean also utilizes all observations in the dataset and, implicitly, their frequencies, but the presence of outliers can distort the result. In the unemployment data in Table 4.1, for example, we see that the 9th and 10th persons have been unemployed for almost one (50 weeks) and one-and-a-half (68 weeks) years, respectively. These two observations obviously contribute to the mean being double the size of the sample's median.

Let us now return to an actual (rather than simulated) sample of economic data, the 2004 Metropolitan Sample from the American Housing Survey, to measure the central tendency of another economic variable, monthly housing costs.

Table 4.2 Measures of central tendency for monthly housing costs, 2004 Metropolitan Sample, American Housing Survey

Measure of central tendency	Calculated value
Mode	400.00
Median	744.00
Arithmetic mean	922.85

Data source: Author's calculations, using data from U.S. Census Bureau (2005), American Housing Survey, 2004 Metropolitan Sample.

As shown in Table 4.2, the sample's mode was 400.00. Since the data is measured in U.S. dollars, we have determined that the most frequently observed value of monthly housing costs incurred by 46,716 households in the 2004 was $400.

We next rank-order the observations and locate the 23,358.5 household. Interpolating, we determine that the median of the sample was 744.00. This statistic indicates that half of the 46,716 metropolitan households surveyed in 2004 had monthly housing costs that were $744 or less, while the other half had housing costs greater than $744 per month.

Our last computation is the sample's arithmetic mean, which is calculated to be 922.85. This indicates that the per-household monthly housing cost for the 46,716 metropolitan households sampled in 2004 was $922.85.

As with the unemployment data, we obtain three quite different "average" values for the 2004 monthly housing costs. If we were required to select only one of the three measures as the "best" measure of central tendency, most economists would choose the median value for this sample. As the frequency distribution in Chapter 3 (Table 3.4) suggested, and as we shall confirm in Chapter 5, the small number of households incurring high monthly housing costs signals the likely presence of outliers in the sample.

4.3 Measures of central tendency for frequency distributions

The statistics in Section 4.2 were generated for ungrouped data. However, published economic data is often presented in frequency distributions that may not include measures of central tendency. In those cases, we employ a different set of procedures to estimate the mean, median, and mode.

Table 4.3 illustrates a frequency distribution for the variable, "size of U.S. families," in 2008. The variable is clearly defined for the first five classes such that we know, for example, the variable equals two persons in the first class and four persons in the third class. However, the variable's value for the last class is not apparent because that class is open-ended. This feature of the distribution will affect only our calculation of the arithmetic mean, and we shall explore two methods for dealing with the open-end class below.

To determine the measures of central tendency from a frequency distribution, we must also account for the frequency associated with each class. Here we can use either the absolute frequency (the number of families) in a particular family size, or the relative frequency, which is the percentage of total families within a family size. As we observe in Table 4.3, 35,075,000 families, or 45.0 percent of all families, for example, were in "two persons" families in 2008. For reasons that will become apparent, we will use relative frequencies, which we will denote as f_i, when calculating the three measures of central tendency.

Let us now determine the measures of central tendency for the size of U.S. families in 2008. The modal value for the variable is obvious from the relative frequencies, or percentage

Table 4.3 Size of U.S. families, 2008

Size of family (X_i)	Number of families (in thousands)	Percentage of families (f_i)	Cumulative percentage of families	$f_i \times X_i$	$\sum(f_i \times X_i)$
2 persons	35,075	45.0	45.0	90.1	
3 persons	17,308	22.2	67.3	66.7	
4 persons	14,960	19.2	86.5	76.8	
5 persons	6,820	8.8	95.2	43.8	
6 persons	2,442	3.1	98.4	18.8	
7 or more persons	1,269	1.6	100.0	–	
7 persons	–	–	–	11.4	307.6
8 persons	–	–	–	13.0	309.2
9 persons	–	–	–	14.7	310.9
10 persons	–	–	–	16.3	312.5
Total	77,874	100.0			

Data source: Table F1, U.S. Census Bureau (2008b), Current Population Survey, March 2008.

of families, column (f_i). In 2008, 45 percent of U.S. families were classified as a family with two persons. Since this is more than double the percentage of families in the next largest class (three persons), the most frequently observed U.S. family size in 2008 was two persons.

The median family size can be determined using simple arithmetic. We first recognize that the median rank corresponds to the 50th percentile of the frequency distribution. Next we consult the cumulative percentage of families column in Table 4.3, where we see that the first class of the two-person family size accounts for 45 percent of all families. We now need just 5 percent additional families to reach the median family size. Since the second class contains 22.2 percent of all families, we see that the median family size will be three persons. This result tells us that in 2008, half of all U.S. families contained three persons or less, and half had more than three persons. (We know from Table 4.3 that technically, 67.3 percent of families contained three persons or less, but this is not what the median measures.)

We now turn to the arithmetic mean "size of family" in the U.S. in 2008. To obtain that statistic, we will modify equation (4.4) to account for the varying frequencies across classes. The result is equation (4.5), the formula for the **arithmetic mean for grouped data**:

$$\overline{X} = \frac{\sum f_i \times X_i}{\sum f_i} \tag{4.5}$$

The numerator of equation (4.5) multiplies each family size (X_i) by that class's frequency (f_i). As shown in the last column of Table 4.3, this calculation is straightforward for the first five classes, but not for the open-end final class (X_6). We are now faced with one of the statistical challenges associated with open-end classes that we noted in Chapter 3.

There are no definitive rules for calculating the arithmetic mean from a frequency distribution with open-end classes. The approach we will use here is to perform a **sensitivity analysis**, for which we calculate the arithmetic mean for a range of (realistic) values of the variable. Once we determine how the mean is affected by the different values, we can select a value for the variable in the open-ended class.

Table 4.3 presents the product of ($f_6 \times X_6$) for four different family sizes of seven to ten persons. If we set $X_6 = 7$, we obtain an arithmetic mean of 3.08 persons. If we increase the variable's value to 8, the arithmetic mean increases to 3.09 persons, while a value of

Table 4.4 Percentage distribution of total money income across U.S. households, 2008

Total number of households (thousands)		117,181	
Household income before taxes	Class midpoint (X_{Mi})	Percentage of all household units (f_i)	$f_i \times X_{Mi}$
Under $5,000	$2,500	3.0	$75
$5,000 to $9,999	$7,500	4.1	$308
$10,000 to $14,999	$12,500	5.8	$725
$15,000 to $24,999	$20,000	11.8	$2,360
$25,000 to $34,999	$30,000	10.9	$3,270
$35,000 to $49,999	$42,500	14.0	$5,950
$50,000 to $74,999	$62,500	17.9	$11,188
$75,000 to $99,999	$87,500	11.9	$10,413
$100,000 and over	$175,000	20.5	$35,875
Total		100.0	
		Calculated mean	$70,163
	Median household income from Census Bureau		$50,303
	Mean household income from Census Bureau		$68,424

Data source: Table 1.1.

10 generates a mean of 3.12 persons. Rounding each result to one decimal place, we obtain the same arithmetic mean, 3.1 persons. In fact, if we continue the sensitivity analysis using values greater than ten, we will obtain a mean equal to 3.1 until we reach a 12-person family, a very large, and infrequently observed,[2] size among U.S. families in 2008. Because the mean is not appreciably affected by the different values, we can be relatively confident that the mean size of U.S. families in 2008 was 3.1 persons.

It is standard practice to report arithmetic mean values to at least one decimal place. Since the arithmetic mean will almost always result in decimals beyond the natural units of measurement, it may inspire some to make derisive quips about statistics. (What does a 0.1 person look like?) In response, we can remind them that economists employ measures of central tendency not to identify the literal number of people in the average family, but rather to obtain a statistical description of the representative family in the population. Thus, a 3.1-person average family size is a widely accepted and acceptable statistical result.

Based on our calculations of the three measures of central tendency, the "average" size of U.S. families in 2008 was between 2 and 3.1 persons, depending on the measure used. The most frequently observed family size was two persons; the middle family size was three persons; and the average size for all 77,874,000 U.S. families was 3.1 people.

Let us now consider how to determine the mean for frequency distributions with more complex class intervals[3] using the 2008 distribution of U.S. household incomes. As shown in Table 4.4, the first step is to determine the class midpoint (X_{Mi}), as defined in Section 3.6, for each interval. This calculation is straightforward for the middle classes, but the open-ended nature of the first and last class intervals again present a challenge.

For the first class interval, we will assume a lower limit of $0 even though we know some households reported negative incomes in 2008. This yields a class midpoint of $2,500.

Identifying a reasonable upper end value for the final class is more difficult, particularly when we think of very wealthy entertainers, athletes, and corporate executives. In this case, however, we do find additional information about the distribution of income from the U.S. Census Bureau. According to Table HINC-06: Income Distribution to $250,000 or More for Households, 2008 (U.S. Census Bureau 2009c), we discover that only 2.1 percent of U.S. households had income of $250,000 or more in 2008. Even though we understand that the presence of those 2.1 percent households will certainly raise the mean, we will select $250,000 as the upper end value of the last class, which then yields a midpoint of $175,000. (Economists refer to this as *top-coding* the variable.)

We now have the information required to apply equation (4.4) to the 2008 distribution of household income. Here we determine that the arithmetic mean income for U.S. households in 2008 was $70,163.

We observe that our calculated mean is higher than that published by the Census Bureau. Since Census statisticians have access to the entire sample of data, their published arithmetic mean, calculated directly from the ungrouped data, is more accurate than the one we computed from the frequency distribution. As a result, always use the published measures of central tendency if they are available. We apply the above procedure only when we have access to just a frequency distribution for the variable, and we need to measure its central tendency for the economic analysis at hand.

4.4 Weighted arithmetic means

The method for calculating the arithmetic mean for a variable presented in a frequency distribution introduced a key principle in reducing a set of observations to a single descriptive statistic – accounting for the frequency with which each unique value appears in the data. Thus we now introduce a new arithmetic mean formula, the **weighted arithmetic mean** ($\overline{X}_W$), in equation (4.6):

$$\text{weighted arithmetic mean } (\overline{X}_W) = \frac{\sum W_i X_i}{\sum W_i} \tag{4.6}$$

Here W_i corresponds to the frequency associated with the unique value of the variable denoted by X_i. The frequencies in equation (4.6) are referred to as the **weights** by which each unique value is balanced according to its relative importance (or appearance) in the data.

One familiar application of a weighted arithmetic mean is a student's grade-point average (GPA). This statistic is generated by weighting each credit hour by the grade earned in the course. In the U.S., for example, an A is often the highest earnable grade, so it receives the greatest weight in calculations of a GPA. Once each possible grade has been assigned a weight that reflects its relative value, we multiply each grade earned (X_i) by the corresponding credit hour weight (W_i). After summing the weighted grades, we then divide that result by the sum of the weights, which here is the number of credits taken, to obtain the GPA.

Economists often use the weighted average formula to calculate the arithmetic mean of a variable. Suppose, for example, we wanted to calculate how much, on average, U.S. consumers spent on necessities in a given year. After deciding that necessities comprise food, clothing, and shelter,[4] we turn to the Consumer Expenditure Survey (CEX) at the U.S. Bureau of Labor Statistics. There we discover that the three expenditure categories are available only for subgroups of all consumer units, such as the number of persons in each unit. We present this data for 2007 in the first six columns of Table 4.5.

We see, not surprisingly, that, as the number of persons in a consumer unit increases, so do the dollars spent on necessities. We further observe that, when we move from a one-person unit to two persons, expenditures on food and apparel almost double, but expenditures in

Table 4.5 Calculation of mean consumer expenditures on necessities by size of consumer unit, 2007

Size of consumer unit	Number of consumer units (in thousands)	Average annual expenditures				Weighted expenditures on necessities
		Food	Apparel and services	Shelter	Total on necessities	
1 person	35,740	$3,328	$971	$7,212	$11,511	$411,403,140
2 persons	38,260	$6,209	$1,848	$9,923	$17,980	$687,914,800
3 persons	18,175	$7,251	$2,330	$11,116	$20,697	$376,167,975
4 persons	16,496	$8,671	$2,859	$13,123	$24,653	$406,675,888
5 or more persons	11,499	$9,220	$2,719	$12,914	$24,853	$285,784,647
All consumer units	120,170				$99,694	$2,167,946,450

Data source: Table 4, Size of consumer unit: average annual expenditures and characteristics, U.S. Bureau of Labor Statistics (2008c), Consumer Expenditure Survey, 2007.

those categories increase more incrementally for units larger than two persons. We also note that expenditures on apparel and shelter decline for consumer units of five or more persons, suggesting the presence of economies of scale for some items in those categories.

Now let us consider the average expenditures on necessities for all consumer units in 2007. If we were to calculate the average using the arithmetic mean for ungrouped data, we would apply equation (4.4) to obtain

$$\text{unweighted arithmetic mean } (\overline{X}) = \frac{\sum X_i}{n} = \frac{\$99,694}{5} = \$19,939$$

This result says that the average annual expenditure on necessities by U.S. consumers in 2007 was $19,939 per person. But this calculation ignores the fact that over 35 million of the 120,170,000 consumer units consisted of only one person whose average annual expenditures were less than two-thirds the expenditures for two-person units. Thus, to accurately represent the diverse average expenditures across consumer units, our arithmetic mean statistic must weigh the different total expenditures on necessities by the number of consumers in each size of consumer unit.

Let us now apply equation (4.6), the weighted arithmetic mean formula, to measure average consumer expenditures on necessities:

$$\overline{X}_W = \frac{\sum W_i X_i}{\sum W_i}$$

$$= \frac{(\$11,511 \times 35,740) + (\$17,980 \times 38,260) + \cdots + (\$24,853 \times 11,499)}{120,170}$$

$$= \frac{\$2,167,946,450}{120,170} = \$18,040.66$$

According to our results, the average annual expenditures on necessities by U.S. consumers in 2007 was $18,040.66, a value almost $2,000 less than the unweighted average. Given the high percentage (61.5 percent) of consumer units with only one or two persons who spend less on necessities, this result is expected. It is also a more accurate summary of expenditures because it balances the relative importance of each consumer unit's size in its contribution to total expenditures on necessities.

In the above example, we used the absolute frequencies for each class as the weights, but we could instead use the relative frequency. In such cases, the denominator of equation (4.6), ($\sum W_i$), will equal 100.

Let us now illustrate the differences between the unweighted and weighted arithmetic averages using relative frequencies for a subcategory of shelter expenditures, household operations, which largely comprises the services needed to maintain a household.[5] Suppose we suspect that consumers in a region's largest cities spend more on household operations than the average amount spent by all consumers in the region. Such services are more likely to be available to consumers in metropolitan areas, and metropolitan consumer units typically have higher average incomes to pay for them.

Turning to the Consumer Expenditure Survey (CEX) for the Midwest region of the United States,[6] we find the average amount spent on household operations for the entire region in 2006–2007 was $891. The CEX also collected household operation expenditure data for the region's four largest cities: Chicago, Cleveland, Detroit, and Minneapolis–St. Paul. To compare metropolitan expenditures with regional expenditures, we must first calculate the average expenditures for the four cities,[7] which we will designate as the metropolitan average.

Table 4.6 shows how average annual expenditures on household operations in 2006–2007 vary across the four cities. We also note substantial differences in the number of consumer units. Thus, to obtain an accurate measure of the four-city average annual expenditure on household operations, we must balance the expenditures per consumer unit for each city by the percentage of consumers units for that city.

Let us first ignore the difference in the number of consumer units between the four cities, and calculate the unweighted average expenditures on household operations. Here the $1,328 result weighs the (below-average) household operations expenditures for the relatively large number of consumer units in Detroit and the (above-average) expenditures of the relatively small number of consumer units in Minneapolis–St. Paul equally. This has the effect of pulling the four-city arithmetic mean upward.

Table 4.6 Expenditures on household operations in four Midwestern cities, 2006–2007

	Expenditures on household operations (annual averages) (X_i)	Number of consumer units (in thousands)	Percentage of consumer units (W_i)	Weighted expenditures on household operations ($X_i W_i$)
Chicago	$1,063	3,224	39.8	$42,326
Detroit	$602	2,251	27.8	$16,736
Minneapolis– St. Paul	$2,871	1,467	18.1	$52,016
Cleveland	$775	1,155	14.3	$11,055
Total		8,097	100.0	$122,133
Unweighted city average	$1,328			
Weighted city average				$1,221
Midwest	$891			

Data source: Table 22, Selected Midwestern metropolitan statistical areas: average annual expenditures and characteristics, U.S. Bureau of Labor Statistics (2008b), Consumer Expenditure Survey, 2006–2007.

Now let us calculate the weighted arithmetic mean for the four metropolitan areas. Our first task is to compute the frequency of the consumer units in each city relative to the total number of consumer units across all four cities. As we see in Table 4.6, Detroit accounts for over one-fourth (27.8 percent) of all consumer units, while Minneapolis–St. Paul represents less than one-fifth (18.1 percent).

The next step is to calculate the weighted expenditures for each city. Multiplying the annual average expenditures by the corresponding city's percentage of total consumer units yields, for example, weighted expenditures on household operations of $42,326 in Chicago compared to weighted expenditures of $11,055 in Cleveland. Summing the weighted expenditures across the four cities and dividing that result by the sum of the relative frequencies (100) yields a weighted arithmetic mean of $1,221 spent on household operations by consumers in the four largest Midwestern cities in 2006–2007. As we expected, it is lower than the unweighted average of $1,328.

We now have a statistic that we can compare with the regional average of $891. According to our results, expenditures on household operations in Midwestern metropolitan areas exceed those for the entire region by approximately 40 percent. We must, however, defer discussion regarding the statistical significance of this difference until Part III.

We conclude our discussion of the arithmetic means by reiterating a previous point. The procedures used to calculate both the unweighted and weighted average expenditures on household operations for the four Midwestern metropolitan cities were technically correct. But while we used two technically correct procedures, one method is more *meaningful* than the other for analytical purposes. As the examples here demonstrate, the weighted arithmetic mean of consumer expenditures provides a more meaningful measure of average consumer expenditures than does the unweighted mean when unique observations have different frequencies in the dataset.

4.5   Geometric means

Arithmetic means are appropriate for calculating the average of economic variables that are measured in units such as dollars, number of persons, or weeks of unemployment. For these variables, we can sum the variable's values and then divide by the number of observations or the sum of weights. However, some economic activity is described by variables defined as rates of change, indices, or ratios. Because such variables have different mathematical properties, we must use a different formula, the geometric mean, to calculate the average value.

Equation (4.7) is one formula for calculating the **geometric mean** for ungrouped data:

$$GM = \sqrt[n]{X_1 \times X_2 \times \cdots \times X_n} \qquad (4.7)$$

Here we multiply each of the variable's n observations together and then take the nth root of the product. Note that we can calculate the geometric mean only when the product of all the X_i is positive. If the product is negative or equal to zero, we cannot take the nth root. In these cases, economists may employ a different variable construction, such as an index number, which we discuss in Chapter 6.

An alternative procedure for calculating the geometric mean is to compute the logarithm of each observation and then take the arithmetic mean to obtain the logarithm of the geometric mean:

$$\ln(GM) = \frac{\sum \ln(X_i)}{n}$$

We can then solve for the geometric mean (*GM*) by taking the anti-log, or exponential, of ln(*GM*) using equation (4.8):

$$GM = e^{\ln(GM)} \tag{4.8}$$

To understand why the geometric mean must be used for certain variables, let us turn to an example of its application. Recall that economic growth is a major macroeconomic goal of many nations, which is typically assessed by determining if real gross domestic product increases from one year to the next. Now suppose we want to determine under which of the two recent U.S. presidents, Bill Clinton or George W. Bush, that goal of greater economic growth, as measured by rising real GDP, was achieved.

Table 4.7 presents the annual rates of growth in GDP for each president's tenure in office, along with the arithmetic and geometric means calculated for each eight-year term. According to both statistical measures of central tendency, we observe that the U.S. economy grew more quickly during the Clinton Administration. (Again, we still need to test if this is a statistically significant result.)

Let us now explain why the geometric mean is the more appropriate formula for calculating the average annual change in real GDP. Suppose we want to determine the average rate of change between 2006 and 2008. We see in Table 4.7 that real GDP grew by 2.1 percent between 2006 and 2007, but grew more slowly, 0.4 percent, between 2007 and 2008. If we were to take the arithmetic average of the two growth rates, we would obtain a value of 0.0129, or 1.29 percent.

Let us now apply this arithmetic average rate of change to real GDP in 2006 to calculate the value of real GDP in 2008. Assuming GDP grew at a constant 1.29 percent annual rate between 2006 and 2008, we find that

$$GDP = \$12,976.2 \times (1 + 0.0129) \times (1 + 0.0129) = \$13,228.9 \text{ billion}$$

Table 4.7 Annual percentage change of real gross domestic product (billions of chained (2005) dollars)

	Clinton			Bush		
	Year	*Real GDP*	*Annual percentage change*	*Year*	*Real GDP*	*Annual percentage change*
	1992	8,287.1		2000	11,226.0	
	1993	8,523.4	2.9	2001	11,347.2	1.1
	1994	8,870.7	4.1	2002	11,553.0	1.8
	1995	9,093.7	2.5	2003	11,840.7	2.5
	1996	9,433.9	3.7	2004	12,263.8	3.6
	1997	9,854.3	4.5	2005	12,638.4	3.1
	1998	10,283.5	4.4	2006	12,976.2	2.7
	1999	10,779.8	4.8	2007	13,254.1	2.1
	2000	11,226.0	4.1	2008	13,312.2	0.4
Arithmetic mean			3.9			2.2
Geometric mean			3.8			1.8

Data source: Table 1.1.1, Percent change from preceding period in real gross domestic product, U.S. Bureau of Economic Analysis (2010c), National Economic Accounts.

This value is more than $83 billion higher than $13,312.2 billion, the observed value of real GDP in 2008. Thus by applying the arithmetic mean formula, we overestimate the actual change in real GDP.

Now let us calculate the geometric mean rate of change in real GDP for the same period. Here we obtain an average annual growth rate of 0.97 percent. Again assuming that GDP grew at this constant rate, we obtain

$$\text{GDP} = \$12,976.2 \times (1 + 0.0097) \times (1 + 0.0097) = \$13,313.1 \text{ billion}$$

This value is only slightly higher (due to rounding) than the actual value of GDP. Clearly, then, the geometric mean is the correct procedure for calculating the average rate of growth.

The technical reason is how rates (as well as indices and ratios) are calculated. The rate of change for the current period depends not only on this period's value, but also on that in the previous period. Thus each successive periodic rate of change has a different starting point. To calculate the 2.1 percent rate of change in GDP between 2006 and 2007, for example, we begin with a value of GDP of $12,976.2 billion, while we use a value for GDP of $13,254.1 billion to calculate the rate of change between 2007 and 2008. As we will discuss further in Chapter 7, the geometric mean accounts for the process of *compounding* that occurs when a variable changes logarithmically over time. The compounding process also explains why the geometric mean will always be less than or equal to the arithmetic mean.

The various interest rates that prevail in an economy represent another group of variables that engage economists and policy-makers. In recent years, mortgage rates have received much attention, particularly those for home loans in which interest rates change over the life of the mortgage. These adjustable rate mortgages (ARMs) typically offer a low interest rate at the beginning of the loan period, leading to a monthly payment that is lower than one associated with a fixed-rate mortgage. However, at some future period, the ARM's interest rate increases, thereby raising, often significantly, the borrower's monthly payment. As we now know, many U.S. home-owners were enticed during the mid-2000s to take out adjustable rate mortgages without fully understanding how their monthly payments would increase over time. Once the higher rates kicked in, owning a home became unaffordable and, for many, led to foreclosure.

Table 4.8 presents hypothetical data for a ten-year adjustable rate mortgage[8] where the first year's mortgage rate is set at 5 percent. The lender also sets a 2 percent annual adjustment cap on the mortgage with a lifetime cap of 15 percent. As we observe, the borrower will pay a rate of 5 percent for the first year, 7 percent in the second year, 9 percent during the third year, and so forth, until year 6, when the interest rate reaches the 15 percent lifetime cap. Given this variable rate schedule, what is the average annual mortgage rate for the loan?

If we erroneously use the arithmetic mean to arrive at the average, we obtain an annual value of 12.0 percent. However, because interest payments for a mortgage are calculated through the process of multiplication (the principal times the interest rate), the correct formula for the average is the geometric mean. Accordingly, the annual (geometric) average interest rate for our hypothetical mortgage is 11.31 percent, which, as we expect, is lower than the arithmetic mean.

We notice another feature about the mortgage rate schedule in Table 4.8. Half (or 50 percent) of the rates have the same value, 15 percent. This suggests that a *weighted* geometric mean could be used to calculate the average annual mortgage rate.

As with the unweighted geometric mean, the **weighted geometric mean** is based on multiplicative processes, but now we must also account for the different frequencies with which

Table 4.8 Annual interest rate schedule for hypothetical ARM

	Mortgage interest rate
Year 1	5.0
Year 2	7.0
Year 3	9.0
Year 4	11.0
Year 5	13.0
Year 6	15.0
Year 7	15.0
Year 8	15.0
Year 9	15.0
Year 10	15.0
Arithmetic mean	12.00
Geometric mean	11.31

each unique value appears. As with the weighted arithmetic mean, we will use weights. However, unlike the weighted arithmetic mean, the weights (W_i) are now *powers* to which each observed value (X_i) is raised. In other words, the weighted geometric mean can be expressed by equations (4.9) or (4.10):

$$GM_W = \sqrt[\sum w_i]{X_1^{w_1} \times X_2^{w_2} \times \cdots \times X_n^{w_n}} \tag{4.9}$$

so that

$$\ln(GM_W) = \frac{\sum W_i \ln X_i}{\sum W_i}$$

which is equivalent to

$$GM_W = e^{\ln GM_W} \tag{4.10}$$

Let us apply equation (4.9) to the ten annual rates in Table 4.8. In this example, the number of years serve as the weights, so the average annual mortgage rate equals

$$GM_W = \sqrt[10]{0.05^1 \times 0.07^1 \times 0.09^1 \times 0.11^1 \times 0.13^1 \times 0.15^5} = 0.1131$$

Once we multiply by 100 to put the result in percentage terms, we find that the average interest rate on the adjustable rate mortgage was 11.31 percent, the same result as obtained above.

The geometric mean has many applications in economics since it is the correct mean for averaging ratios or percentages and for determining average rates of change. Its applications include calculating average rates of return on investments, average daily changes in the Dow Jones Industrial Average and other financial statistics, and average quarterly growth rates of economic variables. Like the arithmetic mean, the geometric mean is a unique value that utilizes all observations and frequencies, and it is algebraically tractable. It has the added feature of diminishing (although not eliminating) the effects of extreme values in its calculation. It is, unfortunately, not an easily understood statistic, but we should not let that discourage us from using it when it is appropriate to do so. Once we more thoroughly understand the concept of compounding, we can explain the reasons for the geometric mean's usage to our audience.

There are other types of means that may have economic applications. The harmonic mean, for example, is used to average weighted indices such as the Consumer Price Index or the GDP deflator. Because these means *per se* are used infrequently by economists, we will not discuss them here.

4.6   Positional measures for ungrouped data

The mean is the most frequently used measure of central tendency because of its statistical superiority over the median (and mode), with one exception – when outliers are present among the observations. In those cases, economists use the median as the more appropriate measure of central tendency.

The median is also known as a **positional measure** in that it locates the center position among the rank-ordered observations of a variable, dividing the variable's observations into two equally sized groups. Economists find other positional measures useful in economic analysis as well, so we now turn to procedures for determining the most frequently used positional measures.

Let us continue with the example of 2008 U.S. household income from Table 4.4. The published median value of $50,303 was calculated by splitting the 117,181,000 households, rank-ordered according to household income, into two groups of equal size, or 58,590,500 households each. Those households whose income was more than $50,303 put them into the upper half of the income distribution, while the other group, consisting of households with incomes of $50,303 or less, fell into the bottom half of the income distribution.

Now suppose we want to compare the household income received by the wealthiest 20th percent of households with that earned by the poorest 20th percent of households. We begin by dividing the households, still ranked according to household income, into five equally populated groups. Thus for the 117,181,000 U.S. households in 2008, we would parcel them into groups of 23,436,200 households each.

Next we determine the location of the household that demarcates the first fifth (or 20th percent) of all households, which here is the 23,436,200th household, and record its income. We then find the 93,744,800th household, which demarcates the 80th percentile, and record its income. We can then construct a ratio of the two households' income to tell us how many times greater the income of the wealthier household is compared to that of the poorer one.

This income ratio, commonly labeled P80/P20 (where the P indicates percentile), is one of several ratios economists use to assess income inequality. In 2008, for example, the P80/P20 ratio calculated by the Census Bureau was 4.84 (DeNavas-Walt *et al.* 2009, p. 40). This indicates that the household at the 80th percentile (the threshold for the wealthiest 20 percent of U.S. households) had a household income almost five times that of the household at the 20th percentile (the point that demarcates the poorest 20 percent of U.S. households). Other 2008 household income ratios, such as the P90/P10 (11.37), the P95/P50 (3.58), and the P20/P50 (0.41), give us additional information about the distribution of household income. The P20/P50 ratio, for example, indicates that the income for the 20th percentile household was 41 percent of the income received by the median, or 50th percentile, household.

Before examining the procedures for determining various positional measures, let us first define those measures most frequently used by economists. To construct the 80th/20th ratio, for example, we divide the rank-ordered observations into five equal-sized groups known as **quintiles**, and then use the first and last quintiles to calculate the ratio. The 90th/10th ratio

Positional measure	Rank formulas	Equation
Median (R_M)	$R_M = \dfrac{(N+1)}{2}$ item	(4.1)
ith quartile (R_{qi})	$R_{qi} = \dfrac{i \times (N+1)}{4}$ item	(4.11)
ith quintile (R_{Qi})	$R_{Qi} = \dfrac{i \times (N+1)}{5}$ item	(4.12)
ith decile (R_{di})	$R_{di} = \dfrac{i \times (N+1)}{10}$ item	(4.13)
ith percentile (R_{Pi})	$R_{Pi} = \dfrac{i \times (N+1)}{100}$ item	(4.14)

Figure 4.3 Positional locators of variables for ungrouped data.

can be found by dividing the total observations into ten equal-sized groups called **deciles**, while the 95th/50th ratio uses **percentiles** to divide the number of total observations into 100 equal-sized groups. **Quartiles**, where the total observations are divided into four equal-sized groups, are also used by economists.

Determining any positional measure is a two-step process. The first is to establish which element corresponds to the desired position. Once we know where the rank-ordered observation is located, we can then determine the value of the variable for the target element.

Figure 4.3 presents formulas for identifying the positional measure's rank (R). As we observe, the only difference across the five equations is the denominator's value, which corresponds to the number of equally sized groups into which we divide the observations.

We note that the i in equations (4.11) through (4.14) will be the numerical value of the particular positional measure we seek to locate. Suppose, for example, we want to locate the 7th decile (d_7) in a 49-observation data series of housing prices. Here we would use equation (4.13) with $i = 7$ and $N = 49$. Accordingly, the 7th decile would correspond to the 35th observation in the rank-ordered series, which is the 35th smallest housing price. Similarly, the 8th decile (d_8) corresponds to the 40th smallest housing price, while the 9th decile (d_9) corresponds to the 45th observation.

Note that once we have calculated decile ranks one through nine, we have identified all positions need for dividing the rank-ordered observations into ten equal groups; that is, we only need to determine nine decile limits to divide the data into ten equal groups. We can now generalize this result for all positional measures – to divide a dataset into g equal-sized groups, we only need to determine $g - 1$ ranks to locate the **positional limits**. Thus we need only to determine three quartile limits to divide the observations into four equal-sized groups, four quintile limits to obtain five equal-sized groups, and 99 limits to divide the data into 100 equal-sized groups.

Let us now return to the ten "duration of unemployment" observations from Table 4.1 to locate a new positional measure, the quintile. We determine the element associated with the fourth quintile, for example, by applying equation (4.12):

$$\text{4th quintile rank } (R_{Q4}) = \frac{4(10+1)}{5} = \frac{44}{5} = 8.8$$

This result tells us the 8.8th observation will divide the rank-ordered observations such that four-fifths of the observations fall below that element's duration of unemployment and one-fifth lie above it.

At this point, it is useful to recognize that the median, quartiles, quintiles and deciles can all be expressed as percentiles. Thus we will continue our discussion using percentiles, first by applying equation (4.14) to define the **positional locator**, or rank, associated with the 80th percentile:

$$\text{80th percentile rank } (R_{P80}) = \frac{80(10+1)}{100} = \frac{880}{100} = 8.8$$

As we would expect, we obtain the same location in the dataset as we did for the 4th quintile.

We realize, of course, there is no 8.8th person among the elements, so we cannot observe the corresponding weeks of unemployment from Table 4.1. Consequently, we will use the same process of interpolation as we used for determining the median for an even number of observations. We begin by modifying equation (4.3) so it now applies to percentiles:

$$i\text{th percentile (P}i\text{) for ungrouped data} = X_{IR_{Pi}} + [FR \times (X_{IR_{Pi}+1} - X_{IR_{Pi}})] \qquad (4.15)$$

Let us return to the observations in Table 4.1 to identify the values in equation (4.15).

The first component is the number of weeks of unemployment ($X_{IR_{Pi}}$)associated with the integer element, here the 8th person, which is 27 weeks. We next determine the difference in the weeks of unemployment between the 8th person and the 9th person ($X_{IR_{Pi}+1}$). Since the 9th person was unemployed for 50 weeks, that difference equals 23 weeks. We then multiply the 23 weeks by 0.8, the fractional portion of the rank (*FR*), to obtain 18.4 weeks. This then yields a duration of unemployment for the 80th percentile (or fourth quintile) of 45.4 weeks of unemployment.[9] In less technical language, this result indicates that 80 percent of people in the sample were unemployed for 45.4 weeks or less, while 20 percent were unemployed more than 45.4 weeks.

Determining a variable's value at various positions among rank-ordered observations can be time-consuming when dealing with large datasets, but most software packages, including Excel, can calculate the positional measures identified here. Excel's procedure for determining the positional rank differs slightly from that described here,[10] but its results are comparable, particularly when we have a large number of observations. Consequently, we will use Excel's PERCENTILE function to facilitate the calculation of positional measures.

4.7 Positional measures for a frequency distribution

Determining various positional measures is more complex when our data is organized into a frequency distribution. While we discussed an informal procedure for locating the median value in grouped data in Section 4.3, we can now formalize the process for any positional measure. Again using the 2008 distribution of U.S. household income, let us find the incomes associated with the first four quintiles.

The first step is to locate the demarcation position for each quintile. Because we no longer have individual observations, but rather class intervals, we will use the cumulative relative frequency distribution to identify the class into which each quintile will fall (Table 4.9).

Recognizing that the first quintile corresponds to the 20th percentile, we see that the 20th percentile observation will fall into the $15,000 to $24,999 class interval; the second quintile,

Table 4.9 Percentage distribution of income across U.S. households, 2008

Total number of households	117,181,000		
Household income before taxes	Relative frequency in class	Cumulative frequency distribution	Quintile rank class interval
Under $5,000	3.0	3.0	–
$5,000 to $9,999	4.1	7.1	–
$10,000 to $14,999	5.8	12.9	–
$15,000 to $24,999	11.8	24.7	1
$25,000 to $34,999	10.9	35.6	–
$35,000 to $49,999	14.0	49.6	2
$50,000 to $74,999	17.9	67.5	3
$75,000 to $99,999	11.9	79.4	–
$100,000 and over	20.5	99.9	4
Total	100.0	100.0	
Median household income from Census Bureau	$50,303		
Mean household income from Census Bureau	$70,163		

Source: Table 1.1.

or 40th percentile, will appear in the $35,000 to $49,999 class; the third (60th) in the $50,000 to $74,999; and the fourth (80th) in the final class, $100,000 and over.

We next calculate the value of the ith percentile using equation (4.16):

$$i\text{th percentile } (Pi) \text{ for grouped data} = X_{PiL} + \left(\frac{i - [\sum f_{Pi-1}]}{f_{Pi}}\right) I_{Pi} \qquad (4.16)$$

where X_{PiL} is the lower endpoint of the percentile class, $i = i$th percentile, $[\sum f_{Pi-1}]$ is the cumulate frequency up to, but not including, the percentile class, f_{Pi} is the frequency of the percentile class, and I_{Pi} is the interval of the percentile class.

We will begin with the first quintile, or 20th percentile, by identifying the various components of equation (4.16):

- $X_{PiL} =$ lower endpoint of percentile class $= \$15,000$
- $(i \times \sum f)/100 = i$ percentile $= 20$
- $(\sum f)_{Pi-1} =$ cumulate frequency up to, but not including, percentile class $= 12.9$
- $f_{Pi} =$ frequency of percentile class $= 11.8$
- $I_{Pi} =$ interval of percentile class $= \$10,000$.

We now have

$$20\text{th percentile} = \$15,000 + \left(\left(\frac{20.0 - 12.9}{11.8}\right) \times \$10,000\right) = \$21,017$$

This indicates that the poorest 20 percent of U.S. households in 2008 had incomes of $21,017 or less. We can also say that the top 80 percent of U.S. households in 2008 had incomes greater than $21,017.

Table 4.10 summarizes the calculations of the four quintile limits:

Table 4.10 Quintile calculations for the percentage distribution of U.S. household units by income classes, 2008

	Quintile 1	Quintile 2	Quintile 3	Quintile 4*
X_{PiL} = lower endpoint of the percentile class	$15,000	$35,000	$50,000	$100,000
$i = i$th percentile	20	40	60	80
$[\sum f_{Pi-1}]$ = cumulative frequency up to, but not including, percentile class	12.9	35.6	49.6	79.4
f_{Pi} = frequency of the percentile class	11.8	14.0	17.9	20.5
I_{Pi} = interval of the percentile class	$10,000	$15,000	$25,000	$250,000
ith percentile	$21,017	$39,714	$64,525	$107,317

*As in Table 4.5, we assumed an upper limit for the last class interval of $250,000.

Let us make several observations about our results. First, we see that the wealthiest 20 percent of U.S. households had incomes greater than $107,317. This value is close to that calculated by the U.S. Census Bureau, which determined the 80th percentile limit to be $100,240 in 2008 (DeNavas-Walt *et al.* 2009, p. 40). The calculation of the 20th percentile limit is also close to that obtained by the Census Bureau, $21,017 here compared to their $20,712. Remember that if we need only these two percentiles for our analysis, we should use those determined by the Census Bureau. If, however, we also need the 40th and 60th percentiles, as we might to construct the Gini coefficient described in Chapter 5, we must calculate those limits from the above frequency distribution.

We also note that a positional measure may, or may not, correspond to income received by an actual household. Since the purpose of determining positional measures is to divide the observations into equally sized groups, this outcome does not create issues for the calculations.

Let us now consider what desirable statistical properties the positional measures exhibit. Like the median, the positional measures discussed in Section 4.7 will generate a unique value. They are also typically not as affected by extreme values as is the mean. Positional measures are, after a bit of practice, also easily understood measures. But they do suffer the same flaws as the median for ungrouped data, and some can be difficult, if not impossible, to calculate when we have open-end classes. Despite these limitations, economists do find these positional measures useful in applying other quantitative methods, as we shall see shortly in Chapter 5.

Summary

Chapter 4 has described the measures of central tendency most frequently used in economics. All serve the purpose of summarizing a total set of observed values into a single value. While this is not the only characteristic of a series of observations that interest economists, we do typically want to identify the location around which the variable's observations tend to cluster.

Measures of central tendency fall into three general categories – means, medians, and modes. Economists generally prefer to use means because of their desirable statistical properties, but they also recognize that extreme values can distort the statistic, rendering it less effective. In such cases, the median is typically chosen as the better measure of central tendency. While the mode gives us limited information, it too is useful for describing the statistical characteristics of qualitative variables in particular.

We introduced two statistical means in addition to the arithmetic mean. Economists use the weighted average mean when observations in a data series differ in terms of their relative importance. The geometric mean, while less easily understood than the arithmetic mean, is the proper measure of central tendency to use when an economic variable is measured in percents, growth rates, or ratios.

In addition to the median, economists find other positional measures useful for certain kinds of economic analysis. Here we examined how to calculate them for ungrouped data and frequency distributions. Such measures allow us to compare groups of elements across different portions of a population.

Concepts introduced

Arithmetic mean
Arithmetic mean for grouped data
Arithmetic mean for ungrouped data
Average
Deciles
Fractional portion (of a positional rank)
Geometric mean
Integer portion (of a positional rank)
Mean
Measure of central tendency
Median
Mode
Percentiles
Positional limits
Positional locators
Positional measures
Quartiles
Quintiles
Rank
Sensitivity analysis
Weight
Weighted arithmetic mean
Weighted geometric mean

Box 4.1 Using Excel to calculate measures of central tendency

Excel facilitates the calculations for many of the measures of central tendency discussed in Chapter 4. For most measures, Excel has built-in functions that require only the data's range to generate results. For the remaining measures, Excel streamlines our calculations.

Excel's built-in functions for measures of central tendency

MODE calculates the highlighted data's mode. Note that Excel will generate a mode only if a unique value exists. If all values in the series are unique *or* if the dataset has multiple modes, Excel will return "#N/A".

MEDIAN calculates the highlighted data's median value.

AVERAGE calculates the arithmetic mean of the data series.

GEOMEAN calculates the geometric mean. If the data series has either a zero or negative value, Excel will return "#NUM!" to indicate an unacceptable argument in the function.

PERCENTILE calculates the specified percentile, which means this function has two arguments: the **array**, or values in the range, and k, which is the percentile of interest. Note that k must have a value between 0 and 1, inclusive.

QUARTILE calculates the specified quartile. The function's two arguments are the **array**, or values in the range, and **quart**. If quart equals 0, QUARTILE returns the minimum value among the observations; quart $= 1$ returns the first quartile, or 25th percentile; quart $= 2$ returns the median, or 50th percentile; quart $= 3$ returns the third quartile, or 75th percentile; and quart $= 4$ returns the maximum value in the dataset.

Using Excel for more efficient calculation of the weighted arithmetic mean

Excel can simplify calculating the weighted arithmetic mean of a variable by using the appropriate formulas for the steps involved. We illustrate this using household operations expenditures for the four Midwestern cities in Table 4.6.

- In cells A4 through A7, enter the four metropolitan areas.
- In column B, enter the values for each city's annual average expenditures on household operations, and in column C, enter the number of consumer units (CUs) for each city.
- In column D, calculate the product of average expenditures and the number of CUs for each city. This yields the total expenditures on household operations for the cities.
- Sum the total expenditures across the four cities. This is the numerator value for equation (4.6).
- Sum the total number of CUs across the four cities. This is the denominator of equation (4.6).
- Divide the total (weighted) expenditures by the total number of CUs. This is the weighted arithmetic average expenditure on household operations for the four Midwestern cities.

	A	B	C	D
1		TABLE 4.6		
2		Expenditures on household operations in four Midwestern cities, 2007		
3		Expenditures on household operations (annual averages)	Number of consumer units (in thousands)	Weighted expenditures on household operations
4	Chicago	$1,063	3,224	=B4*C4
5	Detroit	$602	2,251	=B4*C5
6	Minneapolis-St. Paul	$2,871	1,467	=B4*C6
7	Cleveland	$775	1,155	=B4*C7
8	Total		=SUM(C4:C7)	=SUM(D4:D7)
9	Midwest	$891	27,334	
10	Unweighted average for cities	=AVERAGE(B4:B7)		
11	Weighted average			=D8/C8

Exercises

1. Each of the following economic variables may be used to track an economic recession. For each, explain which one or more of the three major measures of central tendency would be appropriate for understanding the variable's central tendency:
   (a) Prices for new homes in the New York metropolitan area.
   (b) Incomes of workers in management occupations.
   (c) Number of people who identified themselves as American Indian or Alaskan Native in the most recent American Community Survey.
   (d) Marital status of persons classified as "not in the labor force."
   (e) Ages of adults in Florida.
2. Explain why quartiles and quintiles are referred to as positional measures but not measures of central tendency.
3. Use the following data to discuss how the "average" size of family in the United States varied across race and ethnic groups in 2008. Be sure to calculate all appropriate measures of central tendency and to comment on which measure is most appropriate for the comparisons.

Size of family	Number (in thousands) of families					
	All races	White only	Black only	Asian only	White non-Hispanic	Hispanic
2 persons	35,075	29,666	3,705	1,056	27,035	2,849
3 persons	17,308	13,596	2,425	870	11,409	2,409
4 persons	14,960	12,072	1,743	793	9,700	2,515
5 persons	6,820	5,495	807	327	3,988	1,596
6 persons	2,442	1,864	345	151	1,252	640
7 or more persons	1,269	872	231	101	517	385

Source: Table F1, U.S. Census Bureau (2008b), Current Population Survey, March 2008 (Housing and Household Economic Statistics Division, Fertility and Family Statistics Branch), data obtained from <http://www.census.gov/population/www/socdemo/hh-fam/cps2008.html>.

4. Perform a sensitivity analysis for the midpoint value used in of the final class in Table 4.4 using the following values: $150,000, $250,000, and $500,000. Be sure to discuss how sensitive the mean value of U.S. household income is to the choice of the midpoint value, and explain which value (including the $175,000 used in the text) you would use to calculate the mean.
5. Transportation in the American West is one key to understanding the region's development since World War II. As the population has grown substantially, different metropolitan areas have made different choices regarding the types of transportation available to their residents. The following table presents consumer expenditure data on transportation, by category, for 2007. Use it to answer the following questions.

Item	All consumer units in West	Los Angeles, CA	San Francisco, CA	San Diego, CA	Seattle, WA	Phoenix, AZ
Number of consumer units (in thousands)	26,800	5,049	2,956	1,009	1,954	1,556
Transportation	$10,116	$10,141	$10,792	$7,258	$10,047	$12,424

Item	All consumer units in West	Los Angeles, CA	San Francisco, CA	San Diego, CA	Seattle, WA	Phoenix, AZ
Vehicle purchases (net outlay)	$3,976	$3,338	$3,564	$2,090	$3,917	$6,383
Gasoline and motor oil	$2,386	$2,712	$2,489	$2,504	$2,288	$2,594
Other vehicle expenses	$2,992	$3,364	$3,502	$1,962	$2,747	$2,846
Public transportation	$762	$727	$1,238	$702	$1,094	$601

Source: Table 24, U.S. Bureau of Labor Statistics (2008c), Consumer Expenditure Survey, 2007; data obtained from <http://www.bls.gov/cex/tables.htm>.

(a) Compare how annual average expenditures on gasoline and motor oil in the three California cities compared with those expenditures in Seattle, Phoenix, and the West region.

(b) Compare how annual average expenditures on vehicle purchases in Phoenix compared with those in the other major West cities.

(c) Two of the major cities, San Francisco and Seattle, have historically invested in public transportation to a greater extent than the remaining three cities due in part to their unique geographical locations. Compare the annual average expenditures on public transportation in San Francisco and Seattle with those in the other three cities.

(d) Other vehicle expenses, which include vehicle finance charges, maintenance and repairs, vehicle insurance, and vehicle rental, leases, licenses, and other charges, vary widely across the five major western cities. Compare the five-city annual average expenditures on other vehicle expenses with the entire West region for 2007.

6. The World Bank collects data for many socio-economic variables for countries ranked according to gross national income (GNI) per capita. Use an appropriate measure to determine the world "average" among the four income groups for the variables in the following table.

Variable	High income	Upper middle income	Lower middle income	Low income
GNI per capita, Atlas method (current US$)	37,571.64	7,106.97	1,905.05	574.39
Population growth (annual percent)	0.71	0.74	1.03	2.15
Population, total	1,056,272,785.44	823,682,932.49	3,434,515,788.48	1,295,785,123.80
Time required to start a business (days)	23.36	55.97	40.64	55.09

Variable	High income	Upper middle income	Lower middle income	Low income
GDP growth (annual percent)	2.54	5.83	10.24	6.41
Internet users (per 100 people)	65.69801331	26.64500844	12.44283527	5.211385769

High-income economies are those in which 2007 GNI per capita was $11,456 or more. Upper-middle-income economies are those in which 2007 GNI per capita was between $3,706 and $11,455. Lower-middle-income economies are those in which 2007 GNI per capita was between $936 and $3,705. Low-income economies are those in which 2007 GNI per capita was $935 or less. See <http://go.worldbank.org/IEH2RL06U0> for information on the World Bank's Atlas method. Data obtained through the World Bank's Quick Query; permanent URL: <http://go.worldbank.org/1SF48T40L0>.

7. College students in the U.S. are well aware that college costs have been increasing at a greater rate than the overall price level for a number of years, and they know that acquiring a college education from a private institution generally costs more than that at a public institution. Use the data below and the appropriate statistical measure to answer the following question: Have the costs for public and private college educations been increasing at the same average annual rate or is one set of costs growing more quickly than the other?

Percentage change in U.S. college tuition at public and private four-year institutions, 2000–01 through 2006–07

	Percentage change from previous year	
	Public four-year institutions of higher education, in-state tuition and required fees	Private four-year institutions of higher education, tuition and required fees
2000–01	4.5	6.0
2001–02	6.7	4.8
2002–03	8.3	3.8
2003–04	13.4	5.6
2004–05	9.6	4.7
2005–06	6.4	3.7
2006–07	6.2	6.2

Data source: Table 320, U.S. Department of Education, National Center for Education Statistics (2008), Average undergraduate tuition and fees and room and board rates charged for full-time students in degree-granting institutions, by type and control of institution: 1964–65 through 2006–07, available at <http://nces.ed.gov/programs/ digest/d07>.

8. In Table 4.7, we compared the average annual percentage growth in real GDP during the Clinton and Bush administrations for the purpose of evaluating under which administration greater economic growth was achieved. But what about the other two macroeconomic goals discussed in Chapter 1 – full employment and price stability?

(a) Go to FRED®, and collect monthly unemployment rate (URATE) data for the two administrations. Use January 1993 through December 2000 for the Clinton administration, and January 2001 through December 2008 for the Bush administration. After determining the average periodic unemployment rate for each administration, discuss which administration had the lower average unemployment rate.

(b) Go to FRED®, and collect monthly data for the Consumer Price Index for all Urban Consumers (CPIAUCSL) from January 1992 through December 2008. Calculate the average annual rate of inflation for the two administrations. (To obtain the average annual rate of inflation, first calculate the annual CPI index. Next, determine the percentage change in the annual CPI, which yields the average annual rate of inflation. Finally, determine the average annual percentage change in the CPI for each seven-year period.) Using this variable as the measure of price stability, discuss under which administration prices were more stable.

9. Use the 2008 household income data in Table 4.9 to calculate the following percentiles: 10th, 25th, 75th, and 90th. Interpret the results, and comment on any concerns you have about the results.

10. Return to the American Housing Survey monthly housing costs for owner-occupied housing data that you extracted from DataFerrett for Exercise 3 in Chapter 3.

(a) Calculate the mean, median, and mode for the sample. Interpret the results, and then explain which of the three measures likely provides the best measure of central tendency for the data series.

(b) Calculate the quintiles for the sample, and interpret the results.

5 Measures of dispersion

5.1 The concept of dispersion

Measures of central tendency capture one important characteristic of economic data, but they are generally insensitive to how observations of the variable are distributed. Since many issues in economics involve not only aggregates and averages, but also questions of ranges and variations, we now turn to statistical measures of dispersion.

In Chapter 1, we introduced the distribution of household income as an important factor when considering economic welfare. This makes empirical descriptions of how income is distributed key for assessing a nation's economic health. In 2008, for example, U.S. mean household income was $68,424, while the median was $50,303, both of which may seem reasonable amounts for meeting basic needs and improving the quality of one's life. However, both the mean and the median measures obscure the fact that 21.5 percent of all household income was concentrated in the hands of the wealthiest 5 percent of U.S. households (U.S. Census Bureau 2010c). As this example demonstrates, any measure of central tendency describes only part of the total picture, leading us to seek additional empirical measures. Thus we now examine statistics that describe the **dispersion**, or distance from which each observation falls from a variable's central values.

Let us now consider the hypothetical data for annual incomes in two occupations shown in Table 5.1. If we were to calculate only the measures of central tendency discussed in Chapter 4, we would conclude that incomes in occupations A and B were identical, since the arithmetic mean, median, and mode for both occupations equal $50,000. However, further examination of the distributions reveals a clear difference in how income is distributed across the two occupations in that the incomes in occupation B are spread over a wider range of values than the incomes observed in occupation A.

How, then, do we measure the degree of dispersion for the incomes earned in each occupation? We will examine four specific measures of the absolute dispersion in a variable: the range, the interquartile range, the average deviation, and the standard deviation. In addition, we will evaluate each measure based on the desirable statistical properties for a measure of dispersion. Before we turn to the specific measures, however, we first need to discuss a very important distinction in statistical analysis, the difference between a population and a sample.

5.2 Populations and samples

Empirical analysis in economics always originates with a question. Some questions are fairly straightforward. For example, did the economy experience inflation or increased unemployment over the last month? Other questions are more complex. How, for example, would

Table 5.1 Annual incomes in two hypothetical occupations

	Occupation A	Occupation B
	$45,000	$10,000
	$50,000	$15,000
	$50,000	$50,000
	$50,000	$50,000
	$55,000	$125,000
Total	$250,000	$250,000
Mean	$50,000	$50,000
Median	$50,000	$50,000
Mode	$50,000	$50,000

eliminating agricultural price supports affect the economic well-being of farmers, or what are the effects of increased economic growth on a country's income distribution? Without exception, such questions arise from interest in the workings of the economy or concerns about effective economic policy. While statistical methods are not involved in the original formulation of a question, they are the means by which empirical analysis can help answer it.

Once a particular economic question has been posed, an important statistical question arises immediately. What is the universe of elements, or **population**, to which the question applies? Suppose, for example, we wanted to know the price paid for a gallon of milk in 2009. Before we could obtain that information, we would first need to specify if we wanted the average price of milk paid by all urban consumers in the United States (information that is collected by the U.S. Bureau of Labor Statistics), or if we were interested in the local prices paid, say, by students of the College of Saint Benedict and St. John's University in St. Joseph, Minnesota (data we would likely have to collect ourselves).

To obtain the information on milk prices regardless of geographical location, we might, if the necessary resources are made available, conduct a **census** of all members of the specified population, asking each member how much they paid for a gallon of milk in 2009. Such a census might be feasible if we have defined our population as the approximately 3,800 college students in St. Joseph, but it would be much more costly to conduct a similar census for the entire U.S. urban population, which numbered 256,758,711 persons in 2009 (Economic Research Service, U.S. Department of Agriculture 2010b). Consequently, we would be more likely to conduct a **survey**, whereby we draw a **sample** from the population, and then ask that representative group what prices they paid for a gallon of milk.

Drawing a representative sample from a population is beyond the scope of this text. (Indeed, economists as a rule generally leave the design and execution of surveys to other professionals.) As a result, we will use economic data gathered by others to calculate the appropriate descriptive statistics for our analyses. But even if we do not collect the data ourselves, we can still pursue the objective of our empirical study, which is to describe the population (generally not observed) by way of the measures based on the sample (observed).

Two further observations about populations and samples are warranted. First, we must recognize that identifying the appropriate population is a crucial first step in any empirical analysis. Fortunately, the most relevant population is usually suggested by the context of our analysis.

A second crucial point relates to our statistical results. We must not forget that our results will apply, in a strict sense, only to the particular population specified and to no other. That is, the average price of milk calculated for college students in St. Joseph describes only the average price paid by that specific population (at a given point in time), while the average price obtained for all urban consumers is the average price paid by those individuals only. In other words, we must not generalize the statistical results from one population to all populations.

5.3   Range

We now turn to our discussion of the specific measures of dispersion most frequently used by economists beginning with the **range**. The range is the most obvious way to indicate dispersion because it identifies the lowest (**minimum**) and highest (**maximum**) values among the observations of the variable. In Table 5.1, for example, we see that the incomes for occupation A range from a minimum of $45,000 to a maximum of $55,000, while incomes in occupation B range from $10,000 to $125,000. Since occupation B's range of incomes ($115,000) is greater than the range in occupation A ($10,000), we conclude that the incomes for occupation B are more widely dispersed than those in occupation A.

The range statistic may be reported in one of two ways. We could simply report the minimum and the maximum values of the variable and let the reader determine the difference, or we could use equation (5.1) to calculate the range as a single value:

$$\text{range} = (\text{maximum} - \text{minimum}) \tag{5.1}$$

While either measure is technically correct, we will use the convention of reporting *both* the single range value and the minimum and maximum observations in our analyses to provide more information about the variable's dispersion.

As a concept of dispersion, the range is simple to calculate and easy to understand. Its descriptive power, however, is limited by the fact that it is based on only two values of the variable, and it can be affected by extreme values. Consequently, although the range is often reported, it is rarely used as the *only* measure of dispersion in empirical analysis.

5.4   Interquartile range

The range can be misleading if the minimum or maximum is an extreme value. As a result, economists may prefer to measure the dispersion of a variable by the **interquartile range** (*IQR*). As shown in equation (5.2), the *IQR* is the difference between the first and third quartile values. Thus it captures the dispersion of the middle 50 percent of the variable's observations:

$$\text{interquartile range } (IQR) = Q_3 - Q_1 \tag{5.2}$$

As with the range, we will report each quartile value and the result from equation (5.2) when we discuss the interquartile range.

Let us now calculate the *IQR* for occupations A and B in Table 5.1. Applying the procedure in Section 4.6, we determine that the first quartile value for occupation A is $47,500 and the third quartile, $52,500, yielding an *IQR* for occupation A of $5,000. Thus the range of income for the middle 50 percent of earners in occupation A is $5,000. For occupation B, we determine Q_1 to be $12,500 and Q_3, $87,500, thus generating an *IQR* of $75,000. Here the middle 50 percent of earners receive incomes over a range of $75,000.

The range of income for the middle group of workers in occupation A is clearly less than the income range for a comparable group of workers in occupation B. Thus, based on the interquartile range statistic, we conclude that the annual incomes in occupation B are more widely dispersed than those in occupation A.

The interquartile range is statistically superior to the range because it is not affected by extreme values in a dataset, but it too is limited since it utilizes only two values from the population or sample. The *IQR* does, however, serve two valuable purposes when describing the empirical characteristics of a variable. As we shall see in later sections, the *IQR* can be used to identify extreme values, or outliers, in a data series. In addition, if such extreme values do obtain, or if the variable is not symmetrically distributed, the *IQR*, along with the median, may be chosen as the better measures of dispersion and central tendency, respectively, for an economic variable.

5.5   Average deviation

We now turn to a measure of dispersion that is rarely used by economists, but which demonstrates the statistical inferiority of the range and interquartile range. As we have noted, both the range and *IQR* are determined using only two values of the variable, but, as we previously established in Chapter 4, a descriptive statistic that includes all observed values in its calculation is superior to one that does not.

We also want a point of reference when calculating measures of dispersion. That is, we need to know how far each observation is from the measure of central tendency. In the case of the range, we have only the minimum or the maximum value to use as a reference point, and the first or third quartiles for the *IQR*. Using positional statistics as the reference point not only ignores most values in the dataset, it may also misrepresent the variable's central tendency.

One obviously superior reference point is the arithmetic mean. Let us now define a new measure of dispersion that incorporates the mean, a statistic known as the **average deviation** (*AD*):

$$\text{average deviation } (AD) = \frac{\sum |X_i - \bar{X}|}{n} \tag{5.3}$$

The average deviation measures the average absolute difference of each observation from the arithmetic mean. (If we omitted the absolute value signs from equation (5.3), the result would always equal zero, which, of course, is a useless measure of dispersion.)

Let us now return to the income distributions for occupations A and B, and calculate the average deviations. According to the results in Table 5.2, in occupation A, the absolute amount by which a worker's income differs, on average, from the mean income is $2000, while in occupation B that absolute amount is $30,000. Because the average deviation for occupation B is 15 times that for occupation A, incomes in occupation B are said to be more widely dispersed about the mean than those in occupation A.

The average deviation is statistically superior to the range and interquartile range because it utilizes all values in the dataset and because it uses a superior (non-positional) measure of central tendency as its reference point. It is, however, affected by extreme values, and taking the absolute value of the deviations renders the statistic algebraically intractable.

The average deviation has never really caught on in economics, so we will not include it among our measures of dispersion. But our discussion of the statistic did establish another

Table 5.2 Average deviations for two hypothetical occupations

Earner	Occupation A		Occupation B	
	Earnings	$|X_i - \bar{X}|$	Earnings	$|X_i - \bar{X}|$
1	$45,000	$5,000	$10,000	$40,000
2	$50,000	0	$15,000	$35,000
3	$50,000	0	$50,000	0
4	$50,000	0	$50,000	0
5	$55,000	$5,000	$125,000	$75,000
Sum	$250,000	$10,000	$250,000	$150,000
Mean	$50,000		$50,000	
	AD	$2,000		$30,000

desirable property for a descriptive statistic, using a reference point when calculating a variable's dispersion, and it introduced the arithmetic mean as that point of reference. Let us now turn to another measure of dispersion which uses the arithmetic mean as its reference point, the standard deviation.

5.6   The concept of the standard deviation

The standard deviation is the most frequently used measure of dispersion in economics because of its statistical superiority and its use in statistical inference. Like the average deviation, the standard deviation includes all observations in its calculation, and it uses a superior statistical measure (the mean) as its reference point. It has the added advantages of indirectly retaining the direction of each observation's deviation from the mean and being algebraically tractable.

The standard deviation is, however, more difficult to interpret than the previously discussed measures of dispersion. While it measures how far observations are from their mean, the economic meaning of this distance is not easily ascertained from a single standard deviation. For example, as shown in Table 5.3, the standard deviation of incomes in occupation B is calculated to be $45,962. Without having another standard deviation for comparison, we cannot fully assess what this tells us about the variability of occupational incomes about a mean of $50,000. In fact, the only value of a single standard deviation that has any economic meaning is zero (0), which indicates no variability about the variable's mean.

Table 5.3 Calculations of dispersion for annual incomes in two hypothetical occupations

$\bar{X} = \$50,000$	Occupation A			Occupation B		
	X_i ($)	$X_i - \bar{X}$ ($)	$(X_i - \bar{X})^2$ (^{2})	X_i ($)	$X_i - \bar{X}$ ($)	$(X_i - \bar{X})^2$ (^{2})
1	45,000	−5,000	25,000,000	10,000	−40,000	1,600,000,000
2	50,000	0	0	15,000	−35,000	1,225,000,000
3	50,000	0	0	50,000	0	0
4	50,000	0	0	50,000	0	0
5	55,000	5,000	25,000,000	125,000	75,000	5,625,000,000
Sum	250,000	0	50,000,000	250,000	0	8,450,000,000

	Population		Sample	
	Symbolic notation	Algebraic definition	Symbolic notation	Algebraic definition
Variation	$V(X)$	$V(X) = \sum (X_i - \mu)^2$	$V(x)$	$V(X) = \sum (X_i - \bar{X})^2$ (5.4)
Variance	σ^2	$\sigma^2 = \dfrac{\sum (X_i - \mu)^2}{N}$	s^2	$s^2 = \dfrac{\sum (X_i - \bar{X})^2}{n-1}$ (5.5)
Standard deviation	σ	$\sigma = \sqrt{\sigma^2}$ $= \sqrt{\dfrac{\sum (X_i - \mu)^2}{N}}$	s	$s = \sqrt{s^2}$ $= \sqrt{\dfrac{\sum (X_i - \bar{X})^2}{n-1}}$ (5.6)

Figure 5.1 Algebraic definitions and symbolic notation* for variations, variances, and standard deviations.

*We follow the convention of denoting the population mean by the Greek symbol mu (μ) and the sample mean by "X-bar" ($\bar{X}$).

Let us now turn to several concepts that will help explain how the standard deviation measures dispersion. As we proceed, we want to remember that these concepts have been arbitrarily constructed by statisticians to describe how observations are scattered about the mean. Thus the values obtained from these concepts are unlikely to be found among the actual observations of the variable.

The first concept is the **variation**, which is defined as the sum of the *squared* deviations of a variable's values about the mean. While this concept utilizes the arithmetic mean as the reference point, it gives greater weight to observations far from the arithmetic mean and can thus be affected by extreme values.

The **variance** of the variable is obtained by dividing the variable's variation by the number of observations. This concept measures the average variation (rather than deviation) of the variable about its mean. If we take the square root of the variance, we obtain the **standard deviation** of the variable. Thus the standard deviation is defined in terms of the variance, and the variance is defined in terms of the variation.

Figure 5.1 presents the algebraic definitions and symbolic notation for these three conceptual measures. As we see, we use different formulas for a population versus a sample. (We will discuss the reason for this distinction in Chapter 9.) Because economists rarely have data for an entire population (even the U.S. Decennial Censuses do not capture everyone), we will utilize the sample formulas – equations (5.4) through (5.6) – for our calculations.

We might now wonder why the standard deviation is the most frequently used measure among the three presented in Figure 5.1. Quite simply, it is because the units of measurement for the standard deviation are identical to the units of measurement for the variable, which makes that statistic more easily understood. Take, for example, the variation, variance, and standard deviation of the incomes for occupation A. Since income is measured in dollars, the units of measurement for the variation and the variance are "dollars squared," which has no economic meaning. However, once we calculate the standard deviation by taking the square root of the variance, the measurement units will again be in dollars.

Let us conclude Section 5.6 by (re-)evaluating the three measures of dispersion most frequently used by economists with regard to their desirable properties. As indicated in Figure 5.2, the range may be considered the least desirable measure of dispersion since it exhibits only two of the seven properties – it is easily understood, and it has a unique value.

Desirable statistical properties	Holds for
Property 1: The statistic is easily understood	Range, interquartile range
Property 2: The statistic is a single (unique) value	Range, interquartile range, standard deviation
Property 3: The statistic's value is not affected by extreme observations	Interquartile range
Property 4: The statistic is algebraically tractable	Standard deviation
Property 5: The statistic utilizes all values in the dataset	Standard deviation
Property 6: The statistic utilizes the frequencies of all values in the dataset	Standard deviation
Property 7: The statistic uses a non-positional point of reference	Standard deviation

Figure 5.2 Three measures of dispersion by desirable statistical properties.

The interquartile range is a better measure than the range in that it is not as affected by extreme observations, but it, too, is not algebraically tractable, it uses only two observations in its calculation, and it does not use a non-positional point of reference.

The standard deviation is clearly a superior measure of dispersion when compared to the range and the interquartile range, but it is not perfect. As noted, the standard deviation is affected by extreme values in the dataset, and it is less readily understood than the other two measures. These limitations are, however, outweighed by its desirable statistical properties, and, as we shall see later, it can be used in combination with the mean to construct easily interpretable measures of *relative* dispersion.

5.7   Calculating the standard deviation for ungrouped data

Let us now return to the hypothetical occupational income data, and determine the standard deviation for each occupation using the calculations in Table 5.3. According to equation (5.5), the variance for occupation A equals

$$s_A^2 = \frac{50,000,000\$^2}{4} = 12,500,000\$^2$$

while the variance for occupation B equals

$$s_B^2 = \frac{8,450,000,000\$^2}{4} = \$2,112,500,000\$^2$$

Applying equation (5.6) to these results yields the following standard deviations for the two occupations:

$$s_{\text{occupation A}} = \$3,535.53$$

$$s_{\text{occupation B}} = \$45,961.94$$

We can deduce from these results that the incomes in occupation B are more widely dispersed than those in occupation A. However, we can reach this conclusion *only* because the means for the two occupations are equal. If they were not, we could not comment on which variable was more widely dispersed about the mean.

Even though the two occupations have the same mean, their standard deviations have little economic meaning. Thus the descriptive value of the standard deviation derives solely from comparing the size differential between the statistics from different samples. However, when used in conjunction with other descriptive statistics, the standard deviation can tell us a great deal about a variable's overall distribution. We will explore those applications after we discuss how to calculate various measures of dispersion for frequency distributions.

5.8   Calculating measures of dispersion for frequency distributions

As for measures of central tendency, we can calculate measures of dispersion for a frequency distribution, at least under certain conditions.

We begin with the range. If the frequency distribution clearly defines the minimum and maximum values in the distribution, we can compute the range using equation (5.1). If the frequency distribution contains one or more open-end classes, we cannot determine the range.

Determining the interquartile range for a frequency distribution is usually feasible. Once we have determined the first and third quartiles using the procedures described in Section 4.7, we can apply equation (5.2) to obtain the *IQR*.

Determining the standard deviation of a frequency distribution requires us to incorporate the frequencies with which each unique observation occurs. We then apply equation (5.7) to obtain the sample standard deviation:

$$\text{standard deviation for grouped data } (s) = \sqrt{s^2} = \sqrt{\frac{\sum f_i (X_{Mi} - \bar{X})^2}{n - 1}} \tag{5.7}$$

Table 5.4 Calculating the standard deviation from a frequency distribution: U.S. household income in 2008

Household income before taxes (X)	Proportion of household units (in thousands) (f_i)	Class midpoint (X_{Mi})	$X_{Mi} - \bar{X}$	$f_i(X_{Mi} - \bar{X})^2$
Under $5,000	3,515	$2,500	$(67,664)	16,092,897,551,359
$5,000 to $9,999	4,804	$7,500	$(62,664)	18,863,935,171,729
$10,000 to $14,999	6,796	$12,500	$(57,664)	22,597,238,462,371
$15,000 to $24,999	13,827	$20,000	$(50,164)	34,793,941,076,821
$25,000 to $34,999	12,773	$30,000	$(40,164)	20,604,212,291,029
$35,000 to $49,999	16,405	$42,500	$(27,664)	12,554,241,755,061
$50,000 to $74,999	20,975	$62,500	$(7,664)	1,231,845,646,444
$75,000 to $99,999	13,945	$87,500	$17,337	4,191,228,768,726
$100,000 and over	24,022	$175,000	$104,837	264,020,915,180,518
			Sum	394,950,455,904,058
Mean household income	$70,163		n − 1	117,180.999
Total households (in thousands)	117,181		Standard deviation	$58,055

Source: Author's calculations based on Table 1.1.

where f_i is the proportion (not percentage) of elements in class i, X_{Mi} is the midpoint for the ith class, and n is the total number of elements.

Let us return to the frequency distribution of U.S. household income for 2008 to calculate its standard distribution. Here we will continue with the assumptions previously made about the two open-end classes: that the lower limit of the first class equals $0, and that the upper limit of the last class equals $250,000. These assumptions permit us to calculate a midpoint for the two classes, which in turn yields an arithmetic mean of $70,163.

As shown in Table 5.4, we determine that the standard deviation for U.S. household income in 2008 was $58,056. While this statistic has no economic meaning in and of itself, we will use it to construct other measures that allow us to compare the relative dispersion of economic variables.

5.9 Locating extreme values

Throughout our discussion of various descriptive statistics, we have mentioned that the presence of extreme values can distort our results. The arithmetic mean and standard deviation, for example, may both be significantly affected by even a few observations that fall at either end of the variable's distribution, thereby reducing their reliability as measures of central tendency and dispersion, respectively.

Extreme values in a frequency distribution are called **outliers**. In general, outliers are values that differ "greatly" from the mean. In more technical terms, outliers are values that lie outside the so-called "normal" variation of the data. Determining which observations will be classified as outliers does, however, depend on the method chosen to detect them. Here we will use a procedure that uses the positional measures of the first and third quartiles and the interquartile range (none of which are not distorted by extreme values) to locate such values among the observations.

Let us return to the hypothetical earnings data in Table 5.1, where three of the observations for occupation B are quite different in value compared to the mean. To determine if these differences lie outside the norm, we apply the following formal test to identify the presence of two types of outliers, mild and extreme.

An observation is considered to be a **mild outlier** if its value lies one-and-a-half times the interquartile range above or below the third and first quartiles, respectively. In equation form, this means that an observation (X_i) is categorized as a mild outlier if

$$X_i < Q_1 - (1.5 \times IQR) \quad \text{or} \quad X_i > Q_3 + (1.5 \times IQR) \tag{5.8}$$

Applying equation (5.8) to the occupational earnings data, we obtain values for demarcating the location of mild outliers. As shown in Table 5.5, a mild outlier is present if any income in occupation A is less than $37,500 or greater than $60,000. For occupation B, income must be less than a negative $100,000 or greater than $200,000 to be classified as a mild outlier. Since all incomes in occupations A and B fall between the two respective points of demarcation, there are no mild outliers in either occupational group.

An observation is classified as an **extreme outlier** if its value lies three times the interquartile range above or below the third and first quartiles, respectively. In equation form, this means that an observation (X_i) is categorized as an extreme outlier if

$$X_i < Q_1 - (3.0 \times IQR) \quad \text{or} \quad X_i > Q_3 + (3.0 \times IQR) \tag{5.9}$$

Table 5.5 Determining the presence of mild outliers among the earnings for two hypothetical occupations

	Q_1	Q_3	IQR	Lower-end outlier if	Upper-end outlier if
Occupation A	$47,500	$52,500	$5,000	$X_i < \$47,500 - (1.5 \times \$5,000)$ $= \$37,500$	$X_i > \$52,500 + (1.5 \times \$5,000)$ $= \$60,000$
Occupation B	$12,500	$87,500	$75,000	$X_i < \$12,500 - (1.5 \times \$75,000)$ $= \$(100,000)$	$X_i > \$87,500 + (1.5 \times \$75,000)$ $= \$200,000$

Table 5.6 Determining the presence of extreme outliers among the earnings for two hypothetical occupations

	Q_1	Q_3	IQR	Lower-end outlier if	Upper-end outlier if
Occupation A	$47,500	$52,500	$5,000	$X_i < \$47,500 - (3 \times \$5,000)$ $= \$32,500$	$X_i > \$52,500 + (3 \times \$5,000)$ $= \$67,500$
Occupation B	$12,500	$87,500	$75,000	$X_i < \$12,500 - (3 \times \$75,000)$ $= \$(212,500)$	$X_i > \$87,500 + (3 \times \$75,000)$ $= \$312,500$

Applying equation (5.9) yields the demarcation values for extreme outliers shown in Table 5.6. Here we see that any income in occupation A that is less than $32,500 or greater than $67,500 would be classified as an extreme outlier. Since all incomes in occupation A fall between those two values, we do not have any extreme outliers among those incomes.

Different boundaries for extreme outliers obtain for occupation B due to its larger *IQR*. Here any income in occupation B that is less than a negative $212,500 or greater than $312,500 would be classified as an extreme outlier. Since all incomes in occupation B fall between the two demarcation points, we conclude that there are also no extreme outliers for this group of earners.

Economists often find outliers present among the observations of many economic variables. Let us now return to the 2004 AHS monthly housing costs sample that we cited in Chapters 3 and 4 to determine if outliers are present in that dataset. Applying equations (5.8) and (5.9) to the 46,716 observations, we obtain the demarcation values presented in Table 5.7. As we observe, monthly housing costs would need to be negative in value to be counted as either a mild or an extreme lower-end outlier. Since the minimum monthly housing costs reported in the 2004 AHS sample was zero, we conclude that the 2004 AHS sample contains no lower-end outliers of either type.

Table 5.7 Determining the presence of outliers in the 2004 AHS Metropolitan Survey of monthly housing costs

	Q_1	Q_3	IQR	Lower-end outlier if	Upper-end outlier if
Mild outlier	$445	$1,200	$755	$X_i < \$445 - (1.5 \times \$755)$ $= \$(688)$	$X_i > \$1,200 - (1.5 \times \$755)$ $= \$2,333$
Extreme outlier				$X_i < \$445 - (3.0 \times \$755)$ $= \$(1,820)$	$X_i > \$1,200 + (3.0 \times \$755)$ $= \$3,465$

We obtain a different result at the upper end, where the demarcation values for the upper-end mild and extreme outliers are $2,333 and $3,465 respectively. Since the maximum monthly housing costs in the 2004 sample was $9,667, we determine that the 2004 AHS sample of monthly housing costs *does* contain outliers at the upper end of the distribution. This result leads us to ask how the presence of upper-end outliers distorts our descriptive statistics.

We begin by counting the number of mild and extreme upper-end outliers in the sample. We find that 1,960 observations have values greater than $2,333 such that 4.2 percent of the observations can be classified as mild outliers. We also determine that 498 elements, or 1.1 percent of the observations, had monthly housing costs greater than $3,465.

What is our next step? After we have identified the presence of outliers, we first need to verify that no measurement or data-recording errors are responsible for finding outliers in the dataset. Since we have identified a large number of mild and extreme upper-end outliers in the 2004 AHS sample (rather than only a few), we can be fairly confident that the sample's outliers are not the result of human error.

We next consider if the presence of outliers can be explained by the economic conditions associated with the elements in the sample. According to the Census Bureau, the metropolitan areas in the 2004 sample include Sacramento and Seattle. Given the higher housing costs in the West,[1] we might expect some households (think California and Bill Gates) to have monthly housing costs much greater than the mean. Recognizing this possibility may lead us to retain the outliers in the dataset and acknowledge their presence in our discussion of the variable's descriptive statistics. Alternatively, we may choose to omit the outliers from our calculations if we determine that their presence distorts the descriptive statistics of interest for the analysis at hand. We will consider this further in Section 5.11. But, first, there is one additional feature of a frequency distribution that we must consider, the distribution's overall shape.

5.10   The shape of frequency distributions

A frequency distribution can be described using four parametric statistics, two of which we have already discussed. The first is the distribution's central tendency, which we can measure using the mean, median, and mode (assuming the last is a single unique value). The second statistic is the dispersion of the observations about the distribution's central point, which we will measure using the standard deviation. The two remaining parametric statistics relate to the distribution's symmetry and kurtosis. The symmetry of a distribution indicates whether the distribution's observations lie equally on either side of its central location, while the kurtosis (or peakedness) indicates whether the distribution is relatively flat or pointed. These four statistics will define the shape of the frequency distribution.

For our purposes, we will focus here only on the distribution's symmetry because it helps us identify which measures of central tendency and dispersion are more appropriate for describing a set of observations. A frequency distribution is said to be **symmetric** if the mean, median, and mode are equal, which ensures that half the observations lie to the left of the mean and the other half lie to the right. If this does not obtain, the distribution is said to be **asymmetric** or **skewed**.

A skewed distribution may be **positively skewed**, in which case outliers at the upper end of the distribution pull the arithmetic mean above the median and mode. A skewed distribution may also be **negatively skewed**, in which case outliers at the lower end of the distribution pull the arithmetic mean below the median and mode. Figure 5.3 illustrates these three possible shapes for a frequency distribution *vis-à-vis* their skewness.

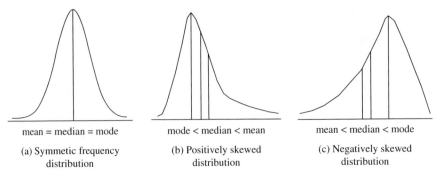

mean = median = mode   mode < median < mean   mean < median < mode

(a) Symmetic frequency   (b) Positively skewed   (c) Negatively skewed
distribution   distribution   distribution

Figure 5.3 Shapes of frequency distributions.

A frequency distribution's skewness can be measured using many different formulas (with statisticians continuing to debate which is the statistically "best" measure). Since we want to assess a frequency distribution's skewness to determine which measures of central tendency and dispersion are more appropriate, we will use the easy-to-compute and intuitively understandable **Pearson coefficient of skewness** (which we note is *not* the same statistic generated by Excel):

$$\text{Pearson coefficient of skewness} = S_k = \frac{3 \times (\bar{X} - \text{median})}{s} \qquad (5.10)$$

If the frequency distribution is symmetric, its mean and median are equal, yielding a Pearson coefficient of skewness of zero. If the distribution is positively skewed, the mean will be greater than the median, making the value of S_k a positive number. If the distribution is negatively skewed, the mean will be less than the median, generating a negative value for S_k. Note, too, that the Pearson skewness coefficient is a unit-free number, and it has no economic meaning in and of itself.

Let us now revisit the 2004 American Housing Survey statistical results. As previously determined, the mean and median (and mode) do not equal one another, suggesting that the distribution of monthly housing costs is skewed. We can formally test this apparent result by calculating the Pearson skewness coefficient using equation (5.10):

$$S_k = \frac{3 \times (\$922.85 - \$744.00)}{\$708.18} = 0.76$$

The non-zero value obtained indicates that the distribution of monthly housing costs is indeed skewed, and the positive value tells us that the distribution is positively skewed. This result is consistent with the values calculated for the three measures of central tendency as well as our finding that the distribution contains upper-end outliers. We are now ready to address the following question. Which descriptive statistics are most appropriate for describing the monthly housing costs in the 2004 AHS Metropolitan Sample?

5.11 Choosing the appropriate descriptive statistics

When selecting the most statistically appropriate and economically meaningful descriptive statistics for a given variable, we must first calculate the six measures of central tendency and dispersion – the appropriate mean, the median, the mode (if available), the range, the interquartile range, and the standard deviation. In addition, we must determine if any outliers are present among the observations and if the distribution is symmetric. Only then can we determine which statistics best capture the variable's central tendency and dispersion.

Let us begin with the case of a frequency distribution that contains no outliers and is symmetric. Here the appropriate mean, the median, and the (unique) mode all provide useful statistical information about the distribution's central tendency. In addition, all three measures of dispersion – the range, interquartile range, and standard deviation – offer statistical information about the underlying distribution. Thus the full complement of descriptive statistics is appropriate for describing the observations, and all, save the standard deviation, will have clear economic meaning.

Now suppose the frequency distribution is symmetric but contains outliers. Here we might remove the outliers and then calculate what is called a **trimmed mean**. For this statistic, we remove the same number (or percentage) of observations from *both* ends of the distribution and then apply the formula for the arithmetic mean to those observations remaining.

If the trimmed mean and the arithmetic mean are close in value, this indicates that the arithmetic mean is not greatly distorted by the presence of outliers. If, however, the trimmed mean and the arithmetic mean differ significantly in value, we must consider the trade-off involved with removing the outliers from our calculations. One problem with this approach is what constitutes "close in value."

Let us return to the monthly housing costs from the 2004 American Housing Survey sample to discuss these considerations further. (For the moment, we will assume these costs are symmetrically distributed.) As we previously determined, the 498 extreme upper-end outliers account for 1.1 percent of the observations, which suggests that the outliers are not a negligible portion of the observations.

Let us now exclude the 498 extreme upper-end outliers *and* the first 498 observations to calculate the trimmed mean. Applying the arithmetic mean formula from equation (4.4) to the truncated dataset, we obtain average monthly housing costs per household of $905.45, a value that is $17.40 less than the arithmetic mean for the entire sample.

Let us now ask what we gain and what we lose by truncating the sample. We first note that the trimmed arithmetic mean is 98.1 percent of the untrimmed mean, so eliminating almost 1,000 observations changes this average statistic by a small amount. We further note that excluding those observations from the sample may eliminate important information about the distribution of monthly housing costs. That is, the omitted observations may be entirely consistent with the economic realities facing some households in "expensive" metropolitan areas. For example, three of the five largest values of monthly housing costs were for households in the Denver, Colorado, area. We also find that upper-end outliers comprise approximately 2 percent of the observations for Denver, Sacramento, Seattle, and Atlanta.[1]

Given these results, we may decide that the extreme outliers are consistent with the locations in the sample and that omitting them has little effect on the mean. Thus we could report the (untrimmed) arithmetic mean and median statistics determined from the complete sample, and note that, while over 1 percent of the observations in the sample have extreme upper-end values based on our outlier detection method, such values may be economically relevant for the particular sample.

Let us next consider the third case, where the underlying frequency distribution of the variable not only contains outliers but is also asymmetric. In such cases, economists generally consider the median to be the "better" measure of central tendency (even though they will typically also report the mean and include a note about the effect of outliers among the observations). As for the distribution's dispersion, the interquartile range is the best measure to cite since it is less affected by outliers. Again, the range and standard deviation may also be reported with a note about their limited usefulness.

As we have determined, the 2004 American Housing Survey sample is consistent with the third case. How, then, would we report its descriptive statistics for the monthly housing costs? Beginning with the measures of central tendency, we would state that the average value of monthly housing costs was $922.85 but that this statistic is affected by the presence of 498 upper-end outliers. Consequently, the more accurate measure of central tendency is the median, which equals $744 and indicates that half of the households surveyed had monthly housing costs of $744 or less while the other half had monthly housing costs greater than $744. In addition, the most frequently observed value of monthly housing costs was $400.

According to the measures of dispersion, the interquartile range is a better indicator than the range because of outliers in the data. While the range tells us that the maximum costs observed was $9,667 and the minimum was $0, the interquartile range indicates that the middle 50 percent of households incurred costs ranging from $445 to $1,200, or $755. We also mention that the standard deviation for the sample was $708.18.

As the previous paragraphs suggest, choosing which descriptive statistics to report is not simply a mechanical process of calculation. That is, we must use our understanding of each statistical measure of central tendency and dispersion, as well as information about the presence of outliers and the shape of the underlying frequency distribution, to guide which measures we report. More importantly, we must consider the economic context of the elements and variables being examined because, if we do not, our analysis may provide inaccurate or misleading conclusions about the economic realities underlying the empirical evidence.

5.12   Assessing relative dispersion: coefficient of variation

Our discussion of descriptive statistics thus far has focused on the characteristics of an individual variable. We now turn to several descriptive measures which will allow us to compare variables, even if those variables are measured using different units or have measures of central tendency and dispersion far apart in value. In the next two sections, we will examine two statistical measures, the coefficient of variation and the index of dispersion, which we can use to assess the *relative* dispersion across variables.

The **coefficient of variation** (*CV*) measures relative dispersion by calculating a **pure number** (unit-less number) that is defined as the ratio of a variable's standard deviation divided by its arithmetic mean:

$$CV = \frac{s}{\bar{X}} \times 100 \tag{5.11}$$

Here we follow convention and multiply the ratio by 100 to convert the statistic into a percentage.

The *CV* measures the extent to which a variable varies about its mean. Returning to the two occupational earnings distributions in Table 5.1, let us now determine if the earnings in occupation A are relatively more, or less, dispersed about their mean as compared to the earnings in occupation B.

Using equation (5.11), we calculate the following coefficients of variation for the two occupations:

$$CV_{\text{occupation A}} = \frac{\$3,535.53}{\$50,000} \times 100 = 7.07$$

$$CV_{\text{occupation B}} = \frac{\$45,961.94}{\$50,000} \times 100 = 91.92$$

Here we see that the coefficient of variation for occupation B is quite a bit larger than that for occupation A. This tells us that the earnings in occupation B are more widely dispersed than those in occupation A. Given the mathematical properties of the coefficient of variation, we can also measure the difference in that dispersion. Specifically, the dispersion of earnings in occupation B is 13 times the dispersion of earnings in occupation A.

We might wonder why we have constructed an additional statistic when we can compare the standard deviations of the two occupations and obtain the same result. Recall that standard deviation comparison are only feasible when the means of the variables are equal in value. While this condition holds for the two occupational distributions (both means are $50,000), that result rarely obtains with most economic variables. Thus we would like a statistic that does not require the equal-mean condition to hold, which is the case with the coefficient of variation.

The coefficient of variation also allows us to compare "apples and oranges," meaning we can use the *CV* to assess the relative dispersion of variables measured in different units, such as dollars and the number of people. Suppose, for example, we want to compare the variability of family income relative to the variability in the number of persons in a family. As we see in Table 5.8, we cannot, in any economically meaningful way, compare either the means or standard deviations of two variables. However, by creating a unit-free measure like the *CV*, we can assess which variable is relatively more dispersed about its mean.

According to the calculations in Table 5.8, the coefficient of variation for family incomes is 97.80 while the *CV* for family size equals 42.13. Since the *CV* for family incomes is just over twice that for family size, we can conclude that U.S. family incomes exhibited greater variability than the number of persons per family in 2008.

The coefficient of variation is also useful when the means and standard deviations of variables are far apart in value even though the variables are measured in the same units. Take, for example, the trends in daily common stock prices for Ford, Honda, and Toyota automobiles between 3 January 2000 and 31 December 2009, depicted in Chart 5.1 (p. 130). Here it appears that Toyota's stock prices exhibited greater volatility as compared with Ford and Honda.

Table 5.8 Measuring the relative dispersion of family income and number of persons per family in 2008 using the coefficient of variation

	Number of persons in family (NPF)	*Family income (FINCP)*
Mean	3.01 persons	$85,133
Standard deviation	1.27 persons	$83,262
Coefficient of variation	42.13	97.80

Source: Calculations based on data obtained from U.S. Census Bureau (2008a), American Community Survey, 2008.

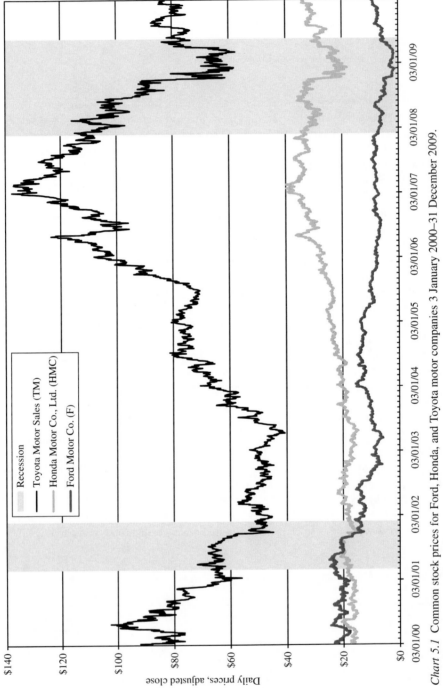

Chart 5.1 Common stock prices for Ford, Honda, and Toyota motor companies 3 January 2000–31 December 2009.

Data source: Yahoo! Finance (2010).

Table 5.9 Relative volatility of automobile common stock prices, 3 January 2000–31 December 2009

	Arithmetic mean	*Standard deviation*	*Coefficient of variation*
Ford	$11.03	$5.41	49.08
Honda	$24.37	$6.53	26.81
Toyota	$79.64	$23.64	30.97

Data source: Yahoo! Finance (2010).

Appearances can, however, be deceiving. As shown in Table 5.9, the mean (and standard deviation) of Toyota's common stock price was much larger in value than those for either Honda or Ford, with Honda's mean stock price more than double that of Ford's. Consequently, comparing the standard deviations without reference to each company's mean stock price is not economically meaningful. Instead, we must use the coefficient of variation to assess the relatively volatility of stock prices.

According to our results, the common stock prices of the Ford Motor Company were almost twice as volatile (1.8 times) as those for Honda over the ten-year period, while Toyota's stock prices were 1.2 times as volatile as Honda's. Further investigation as to *why* we observe these differences is, of course, necessary,[2] but we have established that there were noticeable differences in the relative volatility of these three companies' common stock prices during the 2000s.

Let us now examine several economic applications of the coefficient of variation statistic. One of its most popular uses is assessing the variability of incomes across time or across groups. In one recent study, Robert Dismukes and Ron L. Durst compared the coefficients of variation for income received by farm households with that for incomes received by all households in the United States between 1993 and 2003. They found that the *CV* for farm incomes ranged between 3 percent and 10 percent, while the coefficient of variation for all U.S. household income was less than 1 percent over the same period (Dismukes and Durst 2006, p. 1), results hardly surprising given the vagaries of weather, crop failures, and other issues unique to agricultural production. The authors use this statistical evidence to argue for continuing federal revenue safety net programs, which reduce the income risks inherent in farming, thereby ensuring a basic level of economic well-being for farmers.

The coefficient of variation can also be used to assess the **relative volatility** of economic variables. We might, for example, calculate the coefficients of variation for various sources of external investment in national economies to determine which source is the most stable. Economists at the World Bank undertook such an analysis to assess the relative volatility of capital inflows from net private foreign direct investment (FDI) between 1978 and 1997. Based on previous research, they expected to find less volatility for FDI capital flows as compared to non-FDI sources of capital flows such as portfolio equity, bonds, and bank- and trade-related lending. As shown in World Bank (1999, p. 55, Box 3.2, The relative volatility of FDI and other capital flows) for the 1990–97 period, the coefficients of variation for FDI capital inflows not only fell below 2 percent for all countries, but also were smaller than the coefficients of variation for non-FDI capital inflows (with the exception of Brazil). Such results have important implications for policy-makers in countries seeking stable sources for investment funds.

The coefficient of variation can also be used to assess **convergence** in an economic variable. If the coefficient of variation for an economic variable diminishes over time, this tells us that the relative dispersion of the variable across elements is growing smaller. Such convergence may be desirable if wide variation of a variable such as income or expenditures on necessities signals unequal economic outcomes.

In response to concerns about the undesirability of large differences in income levels between the U.S. states, Andrew Bernat (2001) evaluated whether the differences in per-capita income across states became smaller between 1950 and 1999. Calculating annual *CVs* for the period, he found that the coefficients of variation did decline from a value just under 0.24 in 1950 to approximately 0.15 in 1999 (Bernat 2001, p. 40),[3] leading him to conclude that convergence did, in fact, occur. (He attributed this convergence to several factors, including shifts in the distribution of high-wage industries across states as well as some states reaching their long-run steady-state growth rate.) One important reason that Bernat chose to use the coefficient of variation to measure convergence is that he did not need to adjust state incomes for price changes over the years. Thus the pure number characteristic of the *CV* finessed any issues related to inflation over time.

As these examples demonstrate, the coefficient of variation is a very useful statistic for measuring relative dispersion among two or more variables. In addition to its wide application to economic questions, the *CV* also exhibits most of the statistical properties desired in a descriptive statistic. Specifically, the *CV* is easily understood, even though it has no economic meaning in and of itself. The statistic is also a single (unique) value; it is algebraically tractable; and it utilizes all values and frequencies, although, like the mean and standard deviation, it is affected by outliers. The coefficient of variation has the added feature of enabling the comparison of economic variables whose means and standard deviations are widely different or variables that are measured in non-comparable units.

5.13 Assessing relative dispersion: index of dispersion

The composition of the coefficient of variation can be problematic when the underlying frequency distribution is not symmetric. In such cases, we might turn to a comparable measure, the **index of dispersion** (*ID*), which uses the median and interquartile range in its calculation:

$$ID = \frac{IQR}{\text{median}} \times 100 \tag{5.12}$$

Like the coefficient of variation, the index of dispersion is a pure number that allows us to compare the relative variability across variables whose (positional) measures are either different in value or measured in different units.

Since we know that neither U.S. family income nor U.S. family size are symmetrically distributed according to the Pearson coefficient of skewness, let us calculate and interpret the indices of dispersion for both variables. We first observe in Table 5.10 that the *ID* for U.S. family income equals 103.23, while the *ID* for family size equals 66.67. This tells us that family income in 2008 exhibited greater variability across elements than did the number of persons in the family. To be exact, we determine that U.S. family income was 1.55 times more variable than family size in 2008.

We might ask why we calculated the index of dispersion for family size since its skewness coefficient was very close to zero in value. The reason is that we cannot assess relative dispersion by comparing a coefficient of variation for one variable with an index of dispersion for another because the two statistics measure relative dispersion quite differently. And while here we reach the same conclusion about relative variability using either statistic, that does not necessarily obtain for all economic variables. Thus we let the shape of the underlying

Table 5.10 Variability of family income and number of persons per family in 2008 using the index of dispersion

	Number of persons in family	*Family income*
Mean	3.01 persons	$85,133
S_k	0.02	0.73
Median	3 persons	$65,000
Interquartile range	2 persons	$67,100
Index of dispersion	66.67	103.23

Source: Calculations based on data obtained from the U.S. Census Bureau (2008a), American Community Survey, 2008.

frequency distributions guide our choice between the coefficient of variation and the index of dispersion.

One popular application of the *ID* is comparing the relative variability of earnings across occupational groups. The Occupational Employment Statistics (OES) program at the U.S. Bureau of Labor Statistics produces wage estimates for over 800 occupations which are reported in the percentiles needed to construct indices of dispersion. Table 5.11 presents 2008 annual earnings statistics for the 22 major occupational groups identified by BLS and their corresponding indices of dispersion. We first observe that the mean wage for each major occupational group is greater than the median, suggesting that the underlying distributions of earnings are asymmetric. This then recommends that we use the index of dispersion to measure the relative variability of occupational earnings.

The *ID* calculations reveal noticeable earnings variability across major occupational groups. We see, for example, that the largest variability in earnings occurs in legal occupations (*ID* = 109.3), which is approximately three times as variable as earnings in food preparation and serving related occupations (*ID* = 38.8). This result is not surprising: lawyers occupy a wide range of jobs from public defenders to corporate attorneys, while many employees in food service earn at or near the minimum wage. We further observe relatively high variability in earnings for sales occupations (*ID* = 97.4), where workers' earnings may be based on commissions, and for arts and entertainment occupations (*ID* = 82.3), where some workers receive extremely high earnings while others may earn little at all. We also notice that earnings in health-care support and personal care and service occupations are among the least variable (with *ID*s of 43.6 and 51.47, respectively), which is consistent with the low wages generally earned in many of the jobs in these areas.

We could extend our analysis further since historical occupational earnings data is available by geographic area (national, state, metropolitan, and non-metropolitan areas) as well as by industry. As with the coefficient of variation, the pure number characteristic of the index of dispersion permits comparisons of nominal (non-inflation-adjusted) variables as well as those whose medians and interquartile ranges differ in magnitude. Thus we could learn a great deal about the relative variability of U.S. earnings across occupations, across geographical location, across industries, and across time using the simply constructed *ID* statistic.

5.14   Depicting relative dispersion: the Lorenz curve

Descriptive measures of relative dispersion such as the coefficient of variation and index of dispersion rely on measures that reduce a variable's distribution to two single statistics, thereby obscuring other details of the distribution. But such details can provide crucial

Table 5.11 2008 annual occupational earnings by percentiles and index of dispersion

Occupational code	Occupational title	Mean wage	25th percentile wage	Median/50th percentile wage	75th percentile wage	IQR	Index of dispersion
00-0000	All occupations	$42,270	$21,590	$32,390	$51,540	$29,950	92.47
11-0000	Management occupations	$100,310	$60,500	$87,670	$126,330	$65,830	75.09
13-0000	Business and financial operations occupations	$64,720	$42,840	$58,010	$78,450	$35,610	61.39
15-0000	Computer and mathematical science occupations	$74,500	$51,590	$71,270	$94,230	$42,640	59.83
17-0000	Architecture and engineering occupations	$71,430	$49,000	$66,750	$89,790	$40,790	61.11
19-0000	Life, physical, and social science occupations	$64,280	$40,620	$57,220	$80,560	$39,940	69.80
21-0000	Community and social services occupations	$41,790	$28,560	$38,220	$51,680	$23,120	60.49
23-0000	Legal occupations	$92,270	$45,730	$71,740	$124,120	$78,390	109.27
25-0000	Education, training, and library occupations	$48,460	$29,700	$44,230	$60,960	$31,260	70.68
27-0000	Arts, design, entertainment, sports, and media occupations	$50,670	$28,020	$41,580	$62,250	$34,230	82.32
29-0000	Health-care practitioners and technical occupations	$67,890	$40,160	$56,580	$78,390	$38,230	67.57
31-0000	Health-care support occupations	$26,340	$20,110	$24,540	$30,810	$10,700	43.60

Table 5.11 Continued

Occupational code	Occupational title	Mean wage	25th percentile wage	Median/50th percentile wage	75th percentile wage	IQR	Index of dispersion
33-0000	Protective service occupations	$40,200	$23,480	$34,630	$52,910	$29,430	84.98
35-0000	Food preparation and serving related occupations	$20,220	$15,600	$17,860	$22,530	$6,930	38.80
37-0000	Building, grounds cleaning and maintenance occupations	$24,370	$17,770	$21,880	$28,730	$10,960	50.09
39-0000	Personal care and service occupations	$24,120	$16,710	$20,420	$27,220	$10,510	51.47
41-0000	Sales and related occupations	$36,080	$17,850	$24,310	$41,520	$23,670	97.37
43-0000	Office and administrative support occupations	$32,220	$22,730	$29,780	$39,160	$16,430	55.17
45-0000	Farming, fishing, and forestry occupations	$23,560	$16,920	$19,420	$26,460	$9,540	49.12
47-0000	Construction and extraction occupations	$42,350	$28,270	$37,940	$52,930	$24,660	65.00
49-0000	Installation, maintenance, and repair occupations	$41,230	$28,560	$38,680	$51,730	$23,170	59.90
51-0000	Production occupations	$32,320	$21,880	$29,090	$39,160	$17,280	59.40
53-0000	Transportation and material moving occupations	$31,450	$19,930	$27,340	$38,060	$18,130	66.31

Source: Author's calculations, using data from U.S. Bureau of Labor Statistics (2008e), National Occupational Employment and Wage Estimates.

Table 5.12 Distribution of earnings shares in occupations A and B

Earner in occupation (element)	Cumulative percentage of all earners	Distribution of earnings						
		Perfect equal distribution of earnings				Occupation A		Occupation B
		Income received by earner	Cumulative share of income	Income received by earner	Cumulative share of income	Income received by earner	Cumulative share of income	
1	20.0		$50,000	20.0	$45,000	18.0	$10,000	4.0
2	40.0		$50,000	40.0	$50,000	38.0	$15,000	10.0
3	60.0		$50,000	60.0	$50,000	58.0	$50,000	30.0
4	80.0		$50,000	80.0	$50,000	78.0	$50,000	50.0
5	100.0		$50,000	100.0	$55,000	100.0	$125,000	100.0
Total			$250,000		$250,000		$250,000	

information about how equally the variable's observations are distributed across all elements in the sample. We now turn to two methods for assessing distributional equality, the Lorenz curve and the Gini coefficient. For both, the point of reference will be a distribution in which the variable is equally distributed across all elements in the sample.

The first method is creating a chart known as the **Lorenz curve**. Unlike the previous charts that we have constructed, Lorenz curves are used to compare the cumulative distributions for two or more heterogeneous variables against the distribution of a homogeneous variable equally distributed across a group of elements. To understand what economic questions a Lorenz curve can help us answer, let us return to the earners in hypothetical occupations A and B and outline the steps for constructing a Lorenz curve for each.

We begin by rank-ordering the elements according to their corresponding variable values. For this example, that means we want to rank-order the workers in each occupation according to their earnings, starting with the lowest-paid worker.

We next determine the percentage that each element represents among the total elements. Since occupations A and B each have five workers, each earner represents 20 percent of all elements in the occupation. From this information, we generate a **cumulative distribution of element shares**, which is presented in column two of Table 5.12. We first use these element shares to create the reference point for assessing distributional equality. Known as the **equidistribution line**, this reference point is based on the premise that the variable is equally distributed across all elements. As we see in column three of Table 5.12, under these circumstances, each of the five earners would receive $50,000, or 20 percent, of total earnings, which then yields the cumulative distribution of equally distributed earnings presented in column four.

Let us now plot the equidistribution line in a scatter diagram. Following economic convention, we plot the cumulative frequency distribution of the *elements* observed (here, the five earners) along the horizontal (X) axis of the chart, and the cumulative shares of the *variable* (here, the cumulative frequency distribution of occupational earnings) along the vertical (Y) axis.

As we see in Chart 5.2 (p. 137), the equidistribution line is a straight line at a 45° angle from the origin, reflecting the distribution of a homogeneous variable among its elements. This line will now serve as a reference against which we will compare the distributions of

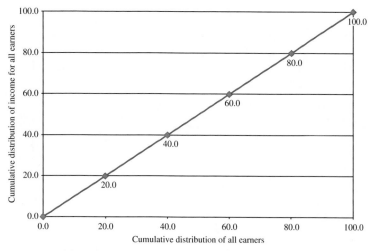

Chart 5.2 Equidistribution line five earners.

Data source: Table 5.12.

two (or more) heterogeneous variables. Thus our next step is to determine how total earnings are distributed across the workers in each occupation.

Columns five and seven of Table 5.12 present the earnings received by each worker in occupations A and B, respectively. We see, for example, that the lowest-paid worker in occupation A received $45,000, which represents 18 percent of all earnings in occupation A, while the second-highest paid worker received $50,000, or 20 percent of total earnings. In occupation B, we observe a different distribution; here the lowest-paid worker received $10,000, or 4 percent, of total occupational earnings, while the highest-paid worker received $125,000, or 50 percent of total earnings.

Once we determine the share of total earnings received by each worker, we can determine a **cumulative distribution of variable shares** for each occupation. These distributions are presented in column six of Table 5.12 for occupation A and in column eight for occupation B.

We now have the data needed to construct Lorenz curves for occupations A and B (see the two boxes "Using Excel to construct a Lorenz curve ..." at the end of the chapter for detailed directions). As with the equidistance line, we plot the cumulative shares for the element on the X-axis and the cumulative shares of the variable on the Y-axis. However, unlike the equidistance line, the Lorenz curves for both occupations are convex (bowed out), and both lie below the equidistribution line. This result is what we would expect for economic variables such as earnings and income in which those at the top receive a disproportionate share of the total. While these properties do not necessarily obtain in all cases, this is the result that economists expect to observe for most economic variables.

In Chart 5.3 (p. 138), we see that the Lorenz curve for occupation B is more convex and thus further away from the equidistribution line than is the Lorenz curve for occupation A. This tells us that the earnings in occupation B are less equally distributed across its five workers than are the earnings for occupation A's workers. Said differently, there is greater distributional equality among the workers in occupation A than in occupation B. Thus the closer a Lorenz curve is to the equidistribution line, the more equally distributed the variable is across its elements.

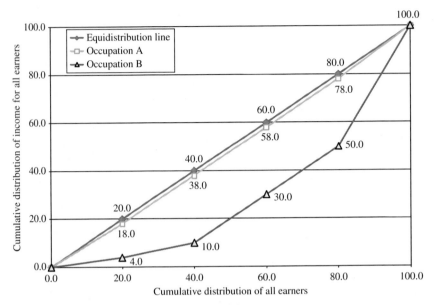

Chart 5.3 Lorenz curves for occupations A and B.

Data source: Table 5.12.

We observe in Chart 5.3 that we plotted all three curves in the same chart, which allows us to determine visually which variable's cumulative distribution is closest to the equidistribution line and which variable's distribution is farthest away. This is crucial for a meaningful comparison of Lorenz curves. Indeed, trying to compare multiple charts largely defeats the purpose behind their construction.

Let us now make one additional observation about the elements in both occupations. Since we have divided the workers into five equally sized groups, we can view the share of earnings received by each worker as the shares corresponding to five quintiles. This is particularly valuable because income distribution data is often presented by quintiles. This means we can produce Lorenz curves to assess income inequality over time for a given set of elements or across elements at a given point in time.

Let us now turn to quintile income share data from the World Bank to compare the 2005 distribution of income across four developing countries. As we observe in Table 5.13, the share of income held by a particular quintile varies noticeably across the four countries. The share of income received by the poorest 20 percent of the population in Paraguay, for example, was only 3 percent of total income, while in Bangladesh, the poorest 20 percent received 9.4 percent of all income. Similarly, the wealthiest 20 percent in Ukraine's population received over one-third of all income (37.2 percent), while that same 20 percent received almost one-half of national income (47.9 percent) in Gabon.

After calculating the cumulative distributions of variable shares for each country, we construct the Lorenz curves shown in Chart 5.4 (p. 139). As we first observe, no country's Lorenz curve is close to the equidistribution line, indicating that no country's income distribution is equal (or nearly equal) across quintiles of the population. We also see that Paraguay had the least equally distributed income by virtue of its Lorenz curve being the farthest from the

Table 5.13 Distribution of gross national income by quintile for selected countries, 2005

	Cumulative percentage of population	Bangladesh		Paraguay		Ukraine		Gabon	
		Income share	Cumulative income share	Income share	Cumulative income share	Income share	Cumulative income share	Income share	Cumulative income share
Income share held by lowest 20 percent	20.00	9.4	9.4	3.0	3.0	9.0	9.0	6.1	6.1
Income share held by second 20 percent	40.00	12.6	22.0	7.2	10.2	13.4	22.4	10.1	16.2
Income share held by third 20 percent	60.00	16.1	38.1	12.2	22.4	17.6	39.9	14.6	30.8
Income share held by fourth 20 percent	80.00	21.1	59.2	20.0	42.4	22.9	62.8	21.2	52.1
Income share held by highest 20 percent	100.00	40.8	100.0	57.6	100.0	37.2	100.0	47.9	100.0

Data source: World Bank (2010b), World Development Indicators.

equidistribution line. Gabon's income distribution is more equal than Paraguay's, but it too is considerably bowed away from the equidistribution line.

It is, however, more difficult to determine from the chart whether 2005 income was less equally distributed in Bangladesh or in the Ukraine. This illustrates one disadvantage of using Lorenz curves to assess income inequality. As we see for Bangladesh and the Ukraine in Chart 5.5 (p. 140), Lorenz curves can cross each other, rendering it impossible to determine which country exhibits greater income inequality. And even when Lorenz curves do

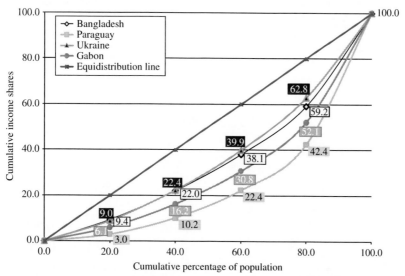

Chart 5.4 Lorenz curves for Bangladesh, Gabon, Paraguay, and Ukraine in 2005.

Data source: Table 5.13.

not cross one another, constructing more than two or three Lorenz curves in any one chart can become difficult to interpret when the income shares are close in value. Fortunately, as we shall see in the next section, we can use a variable's shares from the Lorenz curve to construct a new statistic, the Gini coefficient, which returns less ambiguous results.

Before we turn to the Gini coefficient, let us note that Lorenz curves can be constructed not just from quintile data, but also from other positional measures as well as the actual observations of a variable. While detailed directions are presented in "Using Excel to construct a Lorenz curve from ungrouped data" at the end of the chapter, Figure 5.4 summarizes the general procedures for creating a Lorenz curve from any type of data.

Chart 5.6 (p. 141) presents the Lorenz curves for monthly housing costs based on the 48,197 observations from the American Housing Survey's 2003 National Sample and the 46,716 observations from the 2004 Metropolitan Sample. Here we see that monthly housing costs in the 2004 Metropolitan Sample were distributed more equally across households than were costs in the 2003 National Sample. We also note that neither sample's costs were equally distributed across households. This empirical evidence may now lead us to investigate further *why* these differences obtain.

The Lorenz curve is an effective graphical method for illustrating how the distributions of two or more variables compare with a variable equally distributed across a group of elements. The Lorenz curve also exhibits many of the desirable properties for a descriptive statistic, including being easy to understand and using all values and frequencies in its construction. However, Lorenz curves can cross, making it impossible to determine which distribution is relatively more equal. Difficulties can also arise when comparing the distributions of many variables or when variable shares are close in value. Thus we now turn to a statistical measure that is derived from the areas associated with the Lorenz curve but which provides us with a more definitive measure of distributional inequality, the Gini coefficient of inequality.

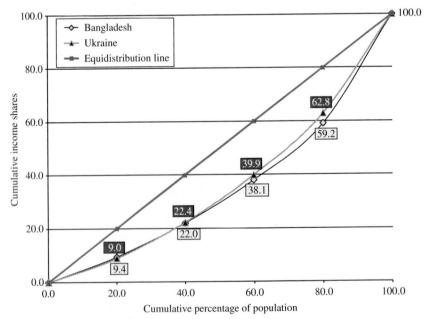

Chart 5.5 Lorenz curves for Bangladesh and Ukraine in 2005.

Data source: Table 5.13.

Step 1:	Sort the variable from the smallest to largest values.
Step 2:	Determine the share that each element represents in the sample by dividing one (1) by the total number of elements.
	Determine the share of the variable received by each element by dividing the value of the variable by the sum of the variable's values.
Step 3:	Construct a cumulative distribution of relative shares for the element *and* a cumulative distribution of relative shares for the variable.
Step 4:	Construct the equidistribution line using the *element's* cumulative distribution of relative shares for both the *X*- and *Y*-axes.
Step 5:	For the variable of interest, plot the element's cumulative share distribution on the *X*-axis and the variable's cumulative share distribution on the *Y*-axis.
Step 6:	Format the chart so it conforms to the parameters exhibited in Chart 5.6 (see the two boxes "Using Excel to construct a Lorenz curve..." at the end of this chapter for detailed instructions).

Figure 5.4 Steps for constructing a Lorenz curve.

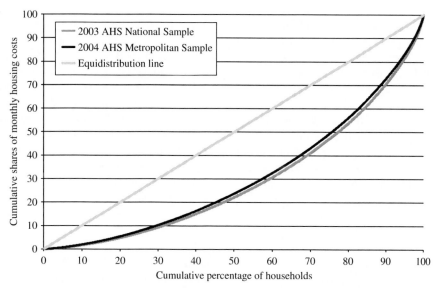

Chart 5.6 Lorenz curves for monthly housing costs, 2003 National and 2004 Metropolitan Samples, American Housing Survey.

Data sources: U.S. Census Bureau (2004b), American Housing Survey for the United States: 2003. U.S. Census Bureau (2005), American Housing Survey, 2004 Metropolitan Sample.

5.15   Assessing relative dispersion: Gini coefficient of inequality

The **Gini coefficient of inequality** is the second measure of distributional equality that we will examine. As shown in Figure 5.5, the Gini coefficient is a ratio that is equal to the area encompassed by the Lorenz curve (*A*) divided by the total area under the equidistribution line (*A* + *B*). Thus the Gini coefficient incorporates detailed variable shares data into a single statistic, summarizing the relative dispersion of a variable across its entire distribution.

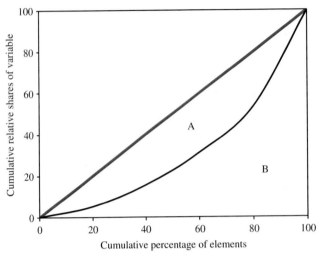

Figure 5.5 Lorenz curve and the Gini coefficient.

We can use Figure 5.5 to determine the Gini coefficient's range of values. If area *A* equals 0, the Lorenz curve is identical to the equidistribution line, indicating that all elements receive identical shares of the variable. In this case, the Gini coefficient will have a value of zero (0).

At the other extreme, when area *B* equals 0, the "wealthiest" element holds a 100 percent share of the variable, while all remaining elements have zero shares. In this case, the Lorenz curve is actually a right angle moving along the *X*-axis before turning upward to the last *X*, *Y* pair of 100, 100, which yields a Gini coefficient of one (1). Thus the Gini coefficient ranges in value from 0, indicating equally distributed shares across the elements, to 1, where only one recipient (or one group of recipients) receives the entire share of the variable.

As with many descriptive statistics, multiple formulas exist for the Gini coefficient. We will use the easily computed equation (5.13) for our calculations:

$$\text{Gini} = 1 - \left(\sum_{i=1}^{N} [f_i (Y_i + Y_{i-1})] \right) \tag{5.13}$$

where *N* is the number of groups into which the elements have been sorted, in ascending order based on unique values of the variable, f_i is the frequency of elements associated with group *i*, and Y_i is the cumulative share of the variable received by elements *through* group *i*.

Let us return to our two hypothetical occupations to demonstrate how to calculate the Gini coefficient (Table 5.14). As with the Lorenz curve, we begin by sorting the *i* elements in ascending order based on earnings received. Because we have five elements in each occupation, each element represents one-fifth of the total elements in the sample, such that f_i equals 0.20 for each element. (Note that we will follow convention and work with fractional proportions when calculating the Gini coefficient, rather than the percentages we used to construct the Lorenz curves.)

After determining the fractional proportion, or share, of total income received by each earner, we construct the cumulative frequency distribution for earners in each occupation. We will denote the value of the cumulative distribution at the *i*th earner as Y_i. The next step is to calculate the term in parentheses in equation (5.13). Here we add the value of the

Table 5.14 Gini coefficients of inequality for occupations A and B

Earner in occupation (i)	Proportion of population receiving earnings (f_i)	Occupation A				Occupation B			
		Share of income received earner	Cumulative percentage share of income (Y_i)	$Y_i + Y_{i-1}$	$f_i(Y_i + Y_{i-1})$	Share of income received earner	Cumulative percentage share of income (Y_i)	$Y_i + Y_{i-1}$	$f_i(Y_i + Y_{i-1})$
1	0.20	0.18	0.18	0.18	0.04	0.04	0.04	0.04	0.01
2	0.20	0.20	0.38	0.56	0.11	0.06	0.10	0.14	0.03
3	0.20	0.20	0.58	0.96	0.19	0.20	0.30	0.40	0.08
4	0.20	0.20	0.78	1.36	0.27	0.20	0.50	0.80	0.16
5	0.20	0.22	1.00	1.78	0.36	0.50	1.00	1.50	0.30
				Sum	0.97			Sum	0.58
				Gini coefficient	0.03			Gini coefficient	0.42

cumulative frequency distribution at the ith element to the value of the cumulative frequency distribution at the previous $(i − 1)$ element, thereby calculating the area associated with the ith element of the total area A in Figure 5.5. We next multiply the parenthetical term by the proportion of elements in group i (namely f_i) to obtain the bracketed term in equation (5.13). After summing the products across the N groups, we subtract that result from 1 to obtain the Gini coefficient.

According to the results in Table 5.14, we obtain a Gini coefficient of inequality for occupation A equal to 0.03, and 0.42 for occupation B. Both Gini coefficients are pure numbers, which means they have no units of measurement. However, because we know the range of values for the Gini coefficient, we can determine that earnings were more equally distributed across the workers in occupation A than for the workers in occupation B. Indeed, the Gini coefficient for occupation A is very close to zero, a result we would expect given the Lorenz curve we obtained for occupation A.

Figure 5.6 summarizes the steps required to calculate a variable's Gini coefficient.

Let us now return to the four countries whose Lorenz curves we constructed in the previous section. Using the quintile data from Table 5.13, we obtain the Gini coefficients shown in Table 5.15. The Gini coefficients for Paraguay and Gabon confirm what their Lorenz curves in Chart 5.4 showed. Paraguay's Gini of 0.4882 indicates that Paraguay has the least equal income distribution among the four countries, while Gabon's Gini coefficient of 0.3793 places that country second among the least equal income distributions.

Step 1:	Sort each variable from the smallest to largest values.
Step 2:	Determine the share that each element represents in the sample (f_i) *and* the share of the variable received by each element.
Step 3:	Calculate the cumulative shares of the element *and* the variable (Y_i).
Step 4:	To obtain the area between the Lorenz curve and the equidistribution line, first calculate a new measure, $(Y_i + Y_{i-1})$, and then multiply that result by the element's frequency (f_i).
Step 5:	Summarize $[f_i(Y_i + Y_{i-1})]$
Step 6:	Subtract the sum from step 5 from a value of one (1).

Figure 5.6 Steps for calculating the Gini coefficient.

Table 5.15 Gini coefficients for Bangladesh, Gabon, Paraguay, and Ukraine in 2005

	Bangladesh	*Gabon*	*Paraguay*	*Ukraine*
Gini coefficient	0.2853	0.3793	0.4882	0.2635

Source: Author's calculations based on Table 5.13.

As for Bangladesh and Ukraine, we can now determine which country's income distribution was less equal. Because Bangladesh's Gini coefficient (0.2853) is greater than Ukraine's (0.2635), we determine that, of the four countries examined, the income distribution in Ukraine is the most equally distributed across its population. Thus, unlike the Lorenz curves, the Gini coefficient provides a clear indication of the relative distribution of income among the four countries.

Gini coefficients can be used not only to compare countries (or other groups) at a given point in time, but also to assess how the income distribution of a particular country (or group) has changed over time. One popular method is to depict changes in the Gini coefficient by constructing a time-series chart of the statistic.

As Chart 5.7 (p. 145) illustrates, the Gini coefficient for U.S. household income has largely increased between 1967 and 2009, reaching its highest value in 2006. Accordingly, this tells us that the U.S. distribution of household income has, on average, become more unequal over the 41-year period. This trend is sometimes cited as empirical evidence that "the rich are getting richer and the poor are getting poorer."

Historically, the Gini coefficient has been most widely used to measure income inequality, but economists increasingly use Gini coefficients to assess such issues as earnings inequality, health-care inequalities, and education inequalities. We could, for example, calculate the Gini coefficients for monthly housing costs from the 2003 National and 2004 Metropolitan Samples of the American Housing Survey. Consistent with our previous results, we find that the Gini coefficient for the 2004 sample is smaller (0.3823) than the Gini coefficient for the 2003 sample (0.4099), indicating the more equitable distribution of housing costs for households in the Metropolitan Sample.

The Gini coefficient is also an appropriate measure to use when assessing the effects of public policies on the distribution of any economic variable. Such analyses might evaluate income distribution before and after the implementation of a redistribution policy (such as tax cuts or increased transfers), changes in the distribution of educational attainment among various race-ethnic groups resulting from changes in government funding, or changes in wealth distribution after the election of a particular political party to office.

The Gini coefficient exhibits the same desirable statistical properties as the Lorenz curve, and it has the added advantage of clearly distinguishing differences between calculated values. It is, however, not without limitations,[4] but it does provide us with a single descriptive measure that accounts for differential shares across a distribution of elements. For this reason, the Gini coefficient is still a commonly used statistical measure of inequality in economics.

Summary

We have examined several methods for describing the dispersion of a frequency distribution. The first set of measures utilized various points of reference against which that dispersion was calculated. In particular, the range and interquartile range both take positional measures at their reference points, while the average deviation and standard deviation use the frequency distribution's mean as a point of reference. Of these measures, economists most

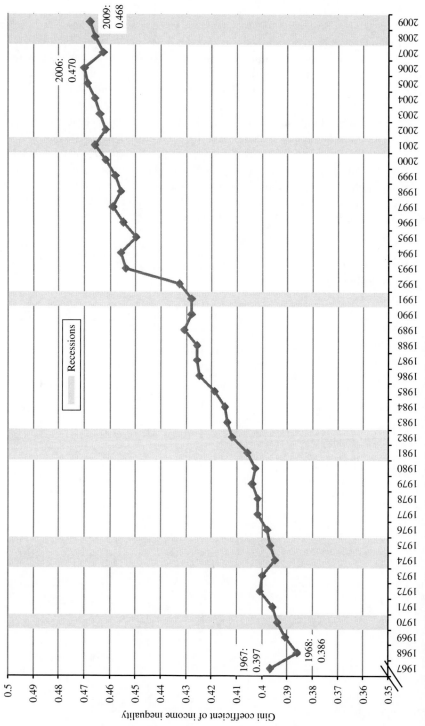

Chart 5.7 Gini coefficients of income inequality for U.S. households, 1967–2009.

Data source: Table 5.12.

frequently use the standard deviation because of its superior statistical properties and its usefulness in statistical inference.

Part of the discussion of dispersion focused on the presence of outliers among the observations of an economic variable. We examined a method for detecting them in a data series, and we considered the connection between outliers and the overall shape of a frequency distribution. This, in turn, gave us additional information about which measures of dispersion (and central tendency) are most appropriate for frequency distributions with various statistical characteristics.

We also discussed several measures of relative dispersion that invite comparisons across distributions. Among the numerical statistics used by economists, the coefficient of variation, the index of dispersion, and the Gini coefficient (along with its graphical representation, the Lorenz curve) are all popular. These statistics offer measures for comparing the relative dispersion of multiple economic variables when the units in which the variables are measured are not comparable, when the points of reference differ in value, or when we want to summarize the characteristics of a variable's distribution.

Concepts introduced

Asymmetric distribution
Average deviation
Census
Coefficient of variation
Convergence
Cumulative distribution of element shares
Cumulative distribution of variable shares
Dispersion
Equidistribution line
Extreme outlier
Gini coefficient of inequality
Index of dispersion
Interquartile range
Lorenz curve
Maximum
Mild outlier
Minimum
Negatively skewed distribution
Outliers
Pearson coefficient of skewness
Population
Positively skewed distribution
Pure number
Range
Relative volatility
Sample
Skewed distribution
Standard deviation
Survey
Symmetric distribution

Trimmed mean
Variance
Variation

Box 5.1 Using Excel to calculate measures of dispersion for ungrouped data

We can use Excel's **Descriptive Statistics** tool to calculate simultaneously many of the measures of dispersion and central tendency discussed in Chapters 4 and 5.

To begin, put a descriptive label in the first cell of the column containing the variable's observations. If you do not, Excel will label the statistical results it produces as *Column1*, which can create considerable confusion if generating descriptive statistics for multiple economic variables.

Next, under **Data Analysis** on the **Data** tab, select the **Descriptive Statistics** tool. Under **Descriptive Statistics**, fill in the following fields:

- For the **Input Range**, enter the cell references for the data you want to analyze. Be sure to include the descriptive label in this range.
- Check the **Labels in First Row** box.
- Check the **Summary Statistics** box.
- Under **Output options**, select where you want Excel to display your results.
  - If you select **Output Range**, click on the upper-left cell where you want the output table to appear in the current worksheet. Note that you will need to have columns available for each variable.
  - To insert a new worksheet in the current workbook, click **New Worksheet Ply**, and provide a name for the new sheet.
  - To create a new workbook, click **New Workbook**.

Summary statistics

Excel generates the following summary statistics in its output:

Excel statistic	Text formulas and notes
Mean	Arithmetic mean; equation (4.4)
Standard error (of the mean)	Discussion deferred until Chapter 10
Median	Equation (4.3)
Mode	Observation with greatest frequency in data
Variance	Equation (5.5)
Standard deviation	Equation (5.6)
Skewness	Do not use; calculate the Pearson coefficient of skewness (equation (5.10)) instead
Kurtosis	Not used in the text
Range	Equation (5.1)
Minimum	Lowest observation value
Maximum	Highest observation value
Sum	Total of all observations
Count	Number of observations

Excel's **Descriptive Statistics** tool does not generate the *interquartile range*. Use Excel's QUARTILE formula to determine the values for the first and third quartiles, and then apply equation (5.2).

Interpreting the descriptive statistics measures from the 2004 Metropolitan Sample of the AHS

Excel's **Descriptive Statistics** tool reports the following statistics for the 2004 Metropolitan Sample from the American Housing Survey:

2004 Metropolitan Sample, AHS		Interpretation of results
Mean	922.849	According to the 2004 American Housing Survey Metropolitan Sample, average monthly housing costs per household was $922.85 in 2004
Standard error	3.28	N/A
Median	744	Half of the monthly housing costs in the 2004 sample fell above $744, and half equaled $744 or less
Mode	400	The most frequently observed monthly housing costs in the 2004 Metropolitan Sample survey was $400
Standard deviation	708.18	The standard deviation of monthly housing costs was $708.18
Sample variance	501,525.77	The variance of monthly housing costs was 501,525.7747 ^{2}
Kurtosis	8.90	N/A
Skewness	2.23	N/A
Range	9,667	Monthly housing costs varied from a minimum value of $0 to a maximum of $9,667, yielding a range of $9,667
Minimum	0	
Maximum	9,667	
Sum	43,111,825	The total of all monthly housing costs equaled $43,111,825
Count	46,716	There were 46,716 observations in the 2004 American Housing Survey Metropolitan Sample

Note that Excel does not include the variable's units of measurement in its statistical results. It is thus our responsibility to incorporate those units in our interpretations. Excluding units of measurement from the interpretation is identical to reporting numbers without attaching any economic meaning.

Excel's QUARTILE function yields a first quartile of $445 and a third quartile of $1,200. This indicates that the middle 50 percent of monthly housing costs in the 2004 Metropolitan AHS Sample fell between $445 and $1,200. The IQR is calculated to be $755, which was the variation in monthly housing costs across the middle 50 percent of the households sampled.

Box 5.2 Using Excel to construct a Lorenz curve from quintiles

Lorenz curves can be generated using Excel's **Scatter (XY) chart** type. The example given here is based on occupation A depicted in Chart 5.3.

We begin by adding a row of zeros under the labels in row 1. This will ensure the Lorenz curve (and the equidistribution line) begins at the origin in the chart.

After indicating each quintile in column A, enter the share of earners in each quintile in column B and the share of earnings received by each quintile in column D.

We next construct the cumulative frequency distributions associated with the element (in column C) and the variable (in column E). We also include the cumulative income shares for a perfectly equal income distribution in column F:

	A	B	C	D	E	F
1	Class interval	Percent age of earners in occupation A (**element**)	Cumulative percentage of earners in occupation A	Share of income earned by quintiles (**variable**)	Cumulative share of income earned	Cumulative shares of equally distributed income earned (**equidistance line**)
2		0	0	0	0	0
3	Lowest quintile	20	20	18	18	20
4	Second quintile	20	40	20	38	40
5	Third quintile	20	60	20	58	60
6	Fourth quintile	20	80	20	78	80
7	Highest quintile	20	100	22	100	100

Let us now distinguish several features between the Excel worksheet information and the chart specifications from our discussion of Table 5.12:

- Use **Scatter with Smooth Lines and Markers** from Excel's Scatter (XY) options.
- Include a row of zeros in the first row to anchor the Lorenz curve at the origin.
- To generate the equidistribution line, plot the values in column C on the X-axis and the values in column F on the Y-axis.
- For each variable
  - Plot the cumulative percentage of the element in column C on the X-axis.
  - Plot the cumulative percentage of the variable in column E on the Y-axis.

The chart created will resemble the Lorenz curves presented in Chapter 5, but several additional modifications will be required:

- Excel will not produce the desired range of values for either axis.
  - Reset the **Maximum** value on both axes to 100. (The **Minimum** may also need to be reset to 0.)
  - Set the **Major unit** to 20 if quintile values are used to construct the Lorenz curve.
  - If we use other positional measures, set the **Major unit** (and possibly the **Minor unit**) to an appropriate number. For example, if we use percentiles, we might set the **Major unit** to 10 and the **Minor unit** to 5. The goal is to make it easy for the reader to identify specific points on the Lorenz curve.
  - Because Excel cannot generate a data table for a scatter (XY) chart, label each data point with the **Y Value** (the cumulative share of the variable).

Box 5.3 Using Excel to construct a Lorenz curve from ungrouped data

Chart 5.6 presents the Lorenz curves for the distribution of monthly housing costs using data from the 2003 National and 2004 Metropolitan American Housing Surveys. Given the large number of observations, we can use Excel to simplify our calculations, as the following worksheet excerpt for the 2004 Lorenz curve demonstrates.

- Column A: The minimum monthly housing cost in the 2004 sample was $0.
- Column B: Following the procedure outlined below this excerpt, 64 households in the sample had $0 monthly housing costs.
- Column C: The product of columns A and B, this is the total monthly housing costs for all households with the specified costs in column A.
- Column D: There were 46,716 households in the 2004 sample. Each cell's value was obtained by dividing the number in column B by 46,716 and then multiplying the result by 100.
- Column E: The sum of all monthly housing costs in 2004 was $43,111,825. Each cell's value was obtained by dividing the number in column A by $43,111,825 and then multiplying the result by 100.
- Column F: This is the cumulative distribution of element shares.
- Column G: This is the cumulative distribution of variable shares.

	A	B	C	D	E	F	G
1				2004 Metropolitan Sample			
2	Unique monthly housing costs	Frequency of households	Unique costs × frequency households	Percentage of total households	Percentage of total housing costs	Cumulative percentage of households	Cumulative percentage of total housing costs
3	0	64	0	0.1369980	0.0000000	0.1369980	0.0000000
4	1	2	2	0.0042812	0.0000046	0.1412792	0.0000046
5	2	2	4	0.0042812	0.0000093	0.1455604	0.0000139
6	4	1	4	0.0021406	0.0000093	0.1477010	0.0000232
7	5	1	5	0.0021406	0.0000116	0.1498416	0.0000348
8	6	1	6	0.0021406	0.0000139	0.1519822	0.0000487
9	8	1	8	0.0021406	0.0000186	0.1541228	0.0000673
10	9	1	9	0.0021406	0.0000209	0.1562634	0.0000881
11	10	2	20	0.0042812	0.0000464	0.1605446	0.0001345
12	13	3	39	0.0064218	0.0000905	0.1669663	0.0002250
13	14	1	14	0.0021406	0.0000325	0.1691069	0.0002575
14	15	1	15	0.0021406	0.0000348	0.1712475	0.0002923
...	...	...	...	...	...	...	...
3556	8380	1	8380	0.0021406	0.0194378	99.9978594	99.9775769
3557	9667	1	9667	0.0021406	0.0224231	100.0000000	100.0000000

As the excerpt indicates, our first step in constructing the Lorenz curve was to create an absolute frequency distribution from the unique monthly housing costs in 2004. This was necessary because Excel can only plot 32,000 data points in a scatter (X–Y) chart. Thus if we want to construct a Lorenz curve for the 2004 Metropolitan Sample, we must reduce the 46,716 observations to less than 32,000.

Fortunately, it is common for elements in a sample to share identical values of the variable being measured. Once we identify those unique values, we can then sort the observations into a frequency distribution using the following steps.

Using Excel to obtain the unique values of observations in a dataset

- Highlight all observations.
- Under **Sort & Filter** on the **Data** tab, select **Advanced**.

- Under **Action**
  - ○ Check **Copy to Another Location,** and under **Copy to**:, select where you want the results to appear.
  - ○ Check **Unique records only**.
- Click **OK**.

Using Excel to create a frequency distribution from unique values of the variable

- Create a column with all unique values of the observations.
- Use the COUNTIF function to obtain the number of elements associated with each unique value. The COUNTIF function counts the number of cells in a given range that meet the given criteria, and it has the following arguments:
  - ○ **Range** is the set of observations of the variable.
  - ○ **Criteria** tells Excel what to find in the range. To locate a value equal to the contents in a particular cell, use the following syntax: **"="&**address_of_first_cell.

The following is an excerpt of the formulas used for the 2004 Metropolitan Sample:

	A	B	C
1		Frequency of households with monthly housing costs	
2	Unique monthly housing costs	Formula	Result
3	0	=COUNTIF($U$2:$U$46717,"="&A3)	64
4	1	=COUNTIF($U$2:$U$46717,"="&A4)	2
5	2	=COUNTIF($U$2:$U$46717,"="&A5)	2
6	4	=COUNTIF($U$2:$U$46717,"="&A6)	1
7	5	=COUNTIF($U$2:$U$46717,"="&A7)	1
8	6	=COUNTIF($U$2:$U$46717,"="&A8)	1
9	8	=COUNTIF($U$2:$U$46717,"="&A9)	1
10	9	=COUNTIF($U$2:$U$46717,"="&A10)	1
11	10	=COUNTIF($U$2:$U$46717,"="&A11)	2
12	13	=COUNTIF($U$2:$U$46717,"="&A12)	3
13	14	=COUNTIF($U$2:$U$46717,"="&A13)	1

- The range of observations for the sample is in cells U2 through U46717.
- The **dollar sign ($)** is used before the row and column of the cell addresses that define the range of observations. This notation, known as an **absolute cell** (or **range**) **reference**, tells Excel to always use those cells in its calculations even when the formula is copied and pasted elsewhere.

Exercises

1. Return to Table 4.3 (size of U.S. families, 2008), and calculate the range, interquartile range, and standard deviation for the number of people per family. Explain what each statistic tells us about the dispersion of people across various family sizes in 2008.

2. In "Using Excel to calculate measures of dispersion for ungrouped data" (see below), the descriptive statistics from the 2004 Metropolitan Sample from the American Housing Survey are presented and all economically meaningful results are interpreted.

   The following table presents the same descriptive statistics from the 2003 National Sample.

Monthly housing costs, 2003 National Sample, American Housing Survey	
Mean	$892.24
Standard error	3.561545958
Median	$698.00
Mode	N/A
Standard deviation	$769.79
Sample variance	592574.223
Kurtosis	24.997
Skewness	3.163
Range	$19,765
Minimum	$0
Maximum	$19,765
Sum	$41,681,821
Count	46,716
Additional statistics from Excel's QUARTILE function	
Quartile 1	$407.00
Quartile 3	$1,163.00

   (a) Interpret the descriptive statistics for the 2003 survey.
   (b) Determine if the 2003 sample contains any outliers, and then discuss which measures of central tendency and dispersion would be more appropriate to use.
   (c) Determine for which sample (the 2003 National or the 2004 Metropolitan) monthly housing costs were more volatile.

3. Each fall, the U.S. Census Bureau publishes income data for the U.S. population from the previous year. Use the data from the "Distribution of 2009 Household Income by Race and Hispanic Origin" to answer the following questions.

   (a) Measure the absolute dispersion of U.S. household income in 2009 for the race-ethnic groups listed. Use the best statistical measure of dispersion discussed in Chapter 5, and state any assumptions made. What can you say about race-ethnic differences in household income from these results?
   (b) Table A-4 in *Income, Poverty, and Health Insurance Coverage in the United States: 2009* (DeNavas-Walt *et al.* 2010) is entitled "Number and real median earnings of total workers and full-time, year-round workers by sex and female-to-male earnings ratio: 1960 to 2009." Explain why the particular measure of central tendency mentioned was selected for comparing the full-time, year-round earnings of males and females.

Race and Hispanic origin of householder and year	Number (thousands)	Total	Under $5,000	$5,000 to $9,999	$10,000 to $14,999	$15,000 to $24,999	$25,000 to $34,999	$35,000 to $49,999	$50,000 to $74,999	$75,000 to $99,999	$100,000 and over
White alone, not Hispanic	83,158	100	10.6	11.0	10.3	14.0	18.8	12.4	13.4	5.1	4.4
Black alone	14,730	100	23.5	15.4	13.4	14.6	15.1	8.7	6.3	1.8	1.2
Asian alone	4,687	100	11.7	7.9	8.2	11.1	16.9	11.8	16.9	7.8	7.7
Hispanic (any race)	13,298	100	16.5	15.2	14.3	15.4	17.6	9.1	7.8	2.2	1.7

Income in 2009 CPI-U-RS adjusted dollars. Households as of March of the following year.

Data source: U.S. Census Bureau (2010c), Income. Table H17, Households by total money income, race, and Hispanic origin of householder.

(c) Table A-2 in that publication (DeNavas-Walt *et al.* 2010), entitled "Selected measures of household income dispersion: 1967 to 2009," reports several statistics that can be used to assess changes in the dispersion of U.S. household income. Interpret the following household income ratios of selected percentiles for 1974, 1991, and 2009: P90/P10, P95/P50, and P20/P50.

Household income ratios of selected percentiles	1974	1991	2009
P90/P10	8.58	10.22	11.36
P95/P20	6.30	7.66	8.80
P95/P50	2.73	3.20	3.62
P80/P50	1.73	1.88	2.01
P80/P20	3.98	4.51	4.89
P20/P50	0.43	0.42	0.41

(d) The Census report also contains income data by quintiles. Use the following data to construct Lorenz curves for household income in 1974, 1991, and 2009.

Shares of household income in quintiles	1974	1991	2009
Lowest quintile	4.3	3.8	3.4
Second quintile	10.6	9.6	8.6
Third quintile	17.0	15.9	14.6
Fourth quintile	24.6	24.2	23.2
Highest quintile	43.5	46.5	50.3

(e) Explain what the results in part (d) indicate about the distribution of household income across the three years, and discuss if they are consistent with the statistical results in part (c).

4. Table 5.15 presented the 2005 Gini coefficients for Bangladesh, Gabon, Paraguay, and Ukraine. Use the quintile data in Table 5.13 to verify that the results in Table 5.15 are correct.

Part II

Temporal descriptive statistics

Change over time

The descriptive statistics discussed in Part I allow us to describe certain characteristics of economic data primarily at a given point of time. While we could compare these static measures of central tendency or dispersion at different points in time, more sophisticated statistical methods are required to fully capture the time dimension.

In Chapter 1's discussion of a country's economic goals, we mentioned price stability and economic growth as two key objectives among policy-makers. Price instability in the form of inflation, for example, may threaten an economy operating close to its capacity (full employment) since it destabilizes planning for the future. Such concerns have led economists to create an aggregate measure of how the overall price level changes over time. Similarly, to evaluate the goal of economic growth, economists have constructed numerous measures of an economy's development over time. These statistics allow policy-makers to examine short-run and long-run patterns of growth and to identify any anomalies in temporal trends.

Part II presents statistical methods for measuring such changes in economic variables over time. In Chapter 6, we examine how to measure dynamic changes in economic variables like the overall price level and industry production levels using the construction known as an index number. In Chapter 7, we discuss how economists explain short-term growth in economic variables and introduce the very important concept of compounding. Chapter 8 presents basic methods for describing long-run growth trends in economic variables, and concludes with an introduction to how economists account for temporal instability in those trends.

6 Measuring changes in price and quantity

6.1 Important index numbers in empirical economics

Surveys of any economy, past or present, quickly reveal concerns about rising price levels. As discussed in Chapter 1, contemporary discussions about such changes usually rely upon a measure known as the Consumer Price Index (CPI) for empirical evidence. This statistic is designed to describe relative changes in the general level of prices over time. But price indices are not confined to tracking the overall price level; they can also be constructed to measure relative changes in the prices of the goods and services that comprise consumer budgets, selling prices received by domestic producers, and the costs of labor compensation incurred by employers, to name just a few.

Index numbers can also be constructed to track changes in quantities of gross domestic product (GDP) and its components. Part of the National Income and Product Accounts (NIPA) in most nations, these quantity indices provide a consistent method for assessing short-term economic growth and productivity. In addition, we can create index numbers to assess changes in a composite of economic variables. Two popular composite indices are the Conference Board's Index of Leading Economic Indicators (LEI) and the United Nations Development Programme's Human Development Index (HDI). The LEI aggregates into an index the relative changes in ten economic variables that economists believe foreshadow changes, such as downturns, in the overall economy. The HDI, on the other hand, combines variables that measure economic growth, health, and education, and is used to assess changes in the quality of life across countries as well as over time.

As this brief introduction suggests, index numbers have wide application in economics. However, all index numbers share basic features that help to explain their popularity in empirical research.

6.2 Why economists use index numbers

One desirable feature of an index number is that it efficiently communicates information about changes in the economy. As we have noted, empirical economics often employs variables whose observations are expressed in very large units of measurement. In 2007, for example, real U.S. GDP equaled $13,843,000,000,000, an increase from its 2006 value of $13,194,700,000,000. While subtraction tells us real GDP was $648,300,000,000 higher in 2007, that difference is difficult for most of us to grasp. If, however, we reported that the annual increase in real GDP was 4.9 percent between 2006 and 2007, most of us would understand the magnitude of the relative change in GDP.

Calculating percentage changes, too, can be cumbersome, particularly when the variable is measured in trillions of dollars. What if, instead, we used an alternative approach in which

Table 6.1 Economic variables pertaining to the health of the French economy, 2004 and 2005

Variable	Units of measurement	2004	2005	2004 base year	
				2004	2005
GDP per capita	US dollars, current prices and PPPs*	29,006.19	30,266.43	100.0	104.3
CPI: all items	(year 2000 = 100)	108.0209	109.8957	100.0	101.7
Employment rates: total	Share of persons of working age (15 to 64 years) in employment	62.3707	62.3066	100.0	99.9

*Statistics reported in "U.S. dollars with purchasing power parity (PPP)" means one currency unit in a particular nation has the same purchasing power over that nation's domestic GDP as the U.S. dollar has over U.S. GDP. This allows for a standard comparison of actual price levels between nations.

Data source: OECD (2010).

we calculate the ratio of a year's GDP relative to a point of reference such as the value of GDP in a specific year? This method has the effect of **normalizing** a variable's value in the specified year, denoted as the **base period**, to a value of 100, with all subsequent values of the variable measured relative to that base.

Let us apply this methodology to determine the change in real GDP between 2006 and 2007. We begin by normalizing 2006 real GDP, which means we divide $13,194,700,000,000 by itself, and then multiply the quotient by 100, yielding a value of 100.0 for 2006 GDP. We next divide 2007 GDP by 2006 real GDP (and multiply the result by 100), to obtain an index of 104.9. Lastly, we subtract the 2006 result from that for 2007, producing a value of 4.9 (percent), which is exactly the same result as we obtained from the less wieldy process of retaining the variable's original units.

This approach, frequently used by economists to calculate **relative change**, is a simple example of how to construct an **index number**. That is, by establishing a base period to serve as a point of reference, we can quickly determine how another observation of the variable differs from the base. Thus constructing an index number leads to easy comprehension of relative change in a variable.

Index numbers can also be used to compare unlike series, thereby resolving the so-called "apples and oranges" problem. Suppose, for example, we determine that per-capita gross domestic product, the Consumer Price Index, and employment rates are indicators of the overall health of the French economy. As we note in Table 6.1, the first variable, per-capita GDP, is measured in current U.S. dollars with purchasing power parity (PPP) per person. The second variable, the Consumer Price Index, is an index number with a base year of 2000, while the third variable, the employment rate, is the percentage of the working age population classified as employed. Given the disparate units of measurement, we cannot compare the actual changes in the three variables. For example, how do we compare a $1,260.24 change in per-capita GDP to a 1.875-point change in the CPI?

We could calculate the percentage change of each variable to compare those values, but a more efficient method is to construct an index for each variable using 2004 as the base year. As shown in the last column of Table 6.1, we can quickly determine that French per-capita GDP experienced the greatest relative change (4.3 percent) between 2004 and 2005. This was followed by a 1.7 percent increase in the CPI, while the French employment rate actually declined by 0.1 percent.

Another advantage of working with index numbers derives from their property of always being a positive (non-zero) value. That is, given their construction, index numbers will never have a value of zero, nor will they take on a negative value, both of which can occur when we calculate percentage changes. This feature makes index numbers particularly attractive when we want to use (natural) logarithms to assess time-series data, since we can always take the natural logarithm of an index number, but we cannot take that of a negative or zero percentage change value.

These are several of the reasons why economists often utilize index numbers in empirical analysis. Let us now turn our attention to how index numbers are constructed, beginning with the simple price index. As we shall see, the basic idea behind a price index is fairly intuitive, although the formulas used for statistically "good" indices can be quite complex and remain a subject of debate among statisticians and economists.[1]

6.3 Constructing a simple price index

The principal purpose of a **price index** is to measure relative changes in prices over time for some relevant group of goods and services (more generally, a group of variables) and group of purchasers (more generally, a group of elements), such as the prices paid by college students for higher education. To illustrate the general concepts associated with a price index, we will use the formula for constructing a **simple index number**:

$$\text{simple index number } (P) = \frac{P_t}{P_0} \times 100 \tag{6.1}$$

Let us apply equation (6.1) to construct a simple price index for total tuition, room, and board for all U.S. four-year institutions between the 2007–08 academic year ($19,323) and 2008–09 ($20,435):

$$\text{single index number for tuition and fees} = \frac{P_{2009}}{P_{2008}} \times 100 = \frac{\$20,435}{\$19,323} \times 100 = 105.8$$

The single index number result indicates that the total costs of four-year institutions in the U.S. increased by 5.8 percent between the 2007–08 and 2008–09 academic years.

A simple price index can also be constructed to analyze changes in the price of a single good over time. Table 6.2 presents data for average undergraduate tuition, room, and board at four-year institutions in the U.S. between 1976–77 and 2008–09. As we observe, the average costs for U.S. undergraduates at four-year colleges and universities rose from $2,577 in the 1976–77 academic year to $20,435 in 2008–09.

Let us now convert the total costs of attending four-year post-secondary institutions into a price index. As with the real GDP calculations above, we first establish a base period. Here we will select the 1976–77 academic year as the base, the point of reference against which the costs from all other years are compared. We note that, again, the base-period price index equals 100, which will always be any index's value in the base period.

The remaining price index values are calculated by dividing each year's tuition by the dollar value of tuition in 1976–77, and then multiplying the quotient by 100 to convert the reported value into percentage terms. Once we have calculated a price index for each year, we can determine, through simple subtraction, that the average costs of an undergraduate education at four-year institutions in the U.S. increased 693.0 percent between 1976–77 and 2008–09.

Table 6.2 Average undergraduate tuition, room, and board in the United States

Academic year	Average tuition and fees ($)	Price index (1976–77 = 100)	Academic year	Average tuition and fees ($)	Price index (1976–77 = 100)
1976–77	2,577	100.0	1993–94	9,296	360.8
1977–78	2,725	105.7	1994–95	9,728	377.5
1978–79	2,917	113.2	1995–96	10,330	400.9
1979–80	3,167	122.9	1996–97	10,841	420.7
1980–81	3,499	135.8	1997–98	11,277	437.6
1981–82	3,951	153.3	1998–99	11,888	461.3
1982–83	4,406	171.0	1999–2000	12,352	479.3
1983–84	4,747	184.2	2000–01	12,922	501.5
1984–85	5,160	200.3	2001–02	13,639	529.3
1985–86*	5,504	213.6	2002–03	14,439	560.3
1986–87	5,964	231.4	2003–04	15,505	601.7
1987–88	6,272	243.4	2004–05	16,509	640.6
1988–89	6,725	261.0	2005–06	17,447	677.0
1989–90	7,212	279.9	2006–07	18,471	716.8
1990–91	7,602	295.0	2007–08	19,323	749.8
1991–92	8,238	319.7	2008–09†	20,435	793.0
1992–93	8,758	339.8			

*Room and board data are estimated.

†Preliminary data based on fall 2007 enrollment weights.

Data source: National Center for Education Statistics (2010), Digest of Education Statistics: 2009, Table 334, Average undergraduate tuition and fees and room and board rates charged for full-time students in degree-granting institutions, by type and control of institution: 1964–65 through 2008–09.

What if we are now interested in determining the increase in the cost of an undergraduate education between 2004–05 and 2008–09? It is important to recognize that we can no longer use subtraction alone to obtain the answer. That is, if the index for the initial period is *not* the base period (that is, if the initial period's index value does *not* equal 100.0), we must use the following formula to calculate the percentage change:

$$\%\Delta\text{index} = \left[\left(\frac{\text{final year index}}{\text{initial year index}}\right) - 1\right] \times 100 \tag{6.2}$$

Let us now apply equation (6.2) to determine the average percentage change in the cost of an undergraduate education between 2004–05 and 2008–09:

$$\%\Delta P = \left[\left(\frac{793.0}{640.0}\right) - 1\right] \times 100 = 23.8\%$$

We see that, over the five-year period, the cost of an undergraduate education in the U.S. increased by a total of 23.8 percent.

Determining how a variable changes using index numbers introduces an important distinction – the difference between a **percentage change** and a **percentage point change**. In the above example, we used equation (6.2) to calculate the *percentage change* in the college cost index between two academic years.

What if we had simply subtracted the 2004–05 index of 640.0 from the 2008–09 index of 793.0? Here we obtain a change of 152.0 *percentage points*. This result has no statistical

Table 6.3 Price index of college costs for two base periods

Academic year	Price index (1976–77 = 100)	Price index (2004–05 = 100)
2004–05	640.6	100.0
2005–06	677.0	105.7
2006–07	716.8	111.9
2007–08	749.8	117.0
2008–09	793.0	123.8

Source: Author's calculations from Table 6.2.

or economic meaning because we lack a point of reference. It is important to recognize this distinction between the percentage change in an index and the percentage point change of an index, because, if we do not, we may incorrectly report how a variable changes.

Another way we could have calculated the change in college costs over the 2004–05 to 2008–09 time period is to rebase the index to 2004–05. This would be particularly desirable if we wanted to compare increases in the subsequent costs of college following the 2004–05 academic year.

Table 6.3 reports the results for the rebased college cost index. Here we can quickly determine by how much the cost of undergraduate education changed after 2004–05. For example, costs increased by 11.9 percent between 2004–05 and 2006–07. As previously noted, subtracting one index value from another to obtain the percentage change is only feasible when we want to determine change relative to the base period of 2004–05. If we are interested in a different change, then we must use equation (6.2).

Before we turn to more complex price indices, let us note one additional feature shared by all index numbers. Index numbers are *unit-free* measures; that is, they are pure numbers. As we previously established in Chapter 5, such statistics are particularly valuable for making comparisons between economic variables.

6.4 Constructing a weighted price index

The simple price index from Section 6.3 is sometimes referred to as an **unweighted price index**. Unweighted price indices are appropriate to use when we want to determine the relative change in the price of a single item, such as we did for four-year college tuition, room, and board above. However, college students incur additional expenses when obtaining a college education, such as those on textbooks. Thus, to obtain a complete picture of the cost of higher education, we must also account for these additional items.

Let us now assume that the cost of a four-year university undergraduate education includes just two items, tuition and fees, and textbooks. (Economists consider this is a reasonable assumption, since students would need to pay for room and board whether or not they attended university.) Table 6.4 presents the average prices for the two expenditure categories for 2004–05 and 2005–06 as well as the percentage change in the price for each item.

By how much did the overall cost of higher education change between 2005 and 2006? Our first impulse might be to average the relative price changes in the two components using the geometric mean, which yields 6.256 percent increase in the price of higher education.

But is it appropriate to assign equal weights to textbook expenses and to the expense of tuition and fee charges in our calculations? As we see, the purchase of one textbook (or even ten) will have far less impact on a college student's total expenditures than paying for one

Table 6.4 Average tuition and required fees and average textbook prices for U.S. undergraduate students: 2004–05 and 2005–06

Item	Average price		Relative change in price (%)
	2004–05	*2005–06*	
Tuition and fees	$10,051.00	$10,666.00	6.117
College textbooks	$86.62	$92.16	6.401

Data source: Tuition and fees: National Center for Education Statistics (2010) Digest of Education Statistics: 2009, Table 334, average undergraduate tuition and fees and room and board rates charged for full-time students in degree-granting institutions, by type and control of institution: 1964–65 through 2008–09.

Textbook prices: calculated from estimates in U.S. Government Accountability Office (2005), College Textbooks: Enhanced Offerings Appear to Drive Recent Price Increases.

year's tuition and fees. Thus, to construct a meaningful measure of relative change in the price of college, we must account for the different weights that each item contributes to total college expenditures.

Equation (6.1) can be easily modified to include the weights of items that comprise a price index. Doing so yields the **weighted price index** (I_t) presented in equation (6.3):

$$\text{weighted price index} = I_t = \frac{\sum p_t q_0}{\sum p_0 q_0} \times 100 \tag{6.3}$$

Let us now apply equation (6.3) to construct a weighted price index for higher education using the data in Table 6.4. We will assume that the quantity of tuition and fees was one, and that ten textbooks were purchased during both academic years. The resulting price index for a year of college in 2005–06 was

$$I_{2005-06} = \frac{\sum p_t q_0}{\sum p_0 q_0} \times 100 = \frac{(1 \times \$10,666) + (10 \times \$92.16)}{(1 \times \$10,051) + (10 \times \$86.62)} \times 100 = 106.140$$

Accordingly, the overall increase in the price paid by students for a four-year university undergraduate education between 2004–05 and 2005–06 was 6.140 percent. This value is less than the result we obtained from the (geometric) mean percentage change in tuition and textbooks because the cost of the more heavily weighted tuition and fees rose by a smaller percentage than the cost of ten textbooks. Like the weighted averages in Chapter 4, weighted price indices will generate a more accurate description of how a group of prices change since it balances the relative importance of each item in overall expenditures.

Before we discuss the appropriate weights for constructing a price index, let us comment further on some general features of a weighted price index. The numerator and the denominator in equation (6.3) both calculate the total money value of expenditures on the various items in the index. That is, both components aggregate the product of prices (p), the variable whose relative change we wish to measure, and quantities (q), which serve as the weights. Aggregation is a necessary characteristic if we want the index to describe how prices for all items change from one period to another.

Equation (6.3) also shows subscripts on both prices and quantities. The zero subscript designates the base period, which provides the benchmark, or reference point, for measuring price changes. The t subscript designates the period for which the relative change in prices is to be measured. Since we want to measure how the price of college changed between 2004–05 and 2005–06, 2004–05 serves as the base period, while t corresponds to the 2005–06 academic year in our example.

We also observe that equation (6.3) uses base-period quantities in both the numerator and denominator of the price index. (Selecting the base-period quantities was somewhat arbitrary on our part; we could have instead used quantities in period t as the weights, as we shall discuss further in the next section.) We recognize that the denominator ($\sum p_0 q_0$) is the product of base-period prices and base-period quantities, thus measuring the total value of all expenditures in the base year. The numerator ($\sum p_t q_0$) is also a price-quantity product, but here the quantities purchased in the base period are valued at the prices existing in the tth year.

The q_0 term in equation (6.3) acts as the **weight** for the price index. Because it appears in both the numerator and denominator of the ratio of expenditure values, q_0 cannot be the source of any difference between values in the numerator and the denominator. Thus any change in a weighted price index is entirely attributable to price differences between time periods.

As with the simple price index, the last step in constructing a weighted price index is to multiply the expenditure ratio by 100, thereby converting the index from a proportion to a percentage. However the percent sign is typically omitted from reported values of an index.

Index numbers are typically reported using one to three decimal places. Prior to January 2007, for example, the U.S. Consumer Price Index was reported to one decimal place, but is now reported using three decimal places. The Organization of Economic Cooperation and Development (OECD), a major source for statistics on developed economies, reports some countries' price indices to three decimal places (as for Germany) and others to four (such as France). In the remainder of this chapter, we will report three decimal places across the time periods being examined.

6.5   Selecting appropriate weights for an index number

Selecting base-year weights for equation (6.3) is not without controversy. Economists and statisticians have debated long and hard over what set of quantity weights are most appropriate to use in a price index. While we will not review the literature here, we will examine in this section three popular price index formulas that suggest some of the issues associated with the appropriate weighting of prices.

The price index presented in equation (6.3) is known as a **Laspeyres price index** (L_P). (We use the subscript "P" to denote that it is a price index.) This method uses base-period quantity weights in the price index's construction:

$$\text{fixed-weight } L_P = \frac{\sum p_t q_0}{\sum p_0 q_0} \times 100 \tag{6.4}$$

One advantage of a Laspeyres price index is that it only requires quantity data from the base period. This, in turn, means that, in a Laspeyres index, the base-period market basket of expenditures is held constant. Consequently, any changes in the index value can be attributed solely to changes in the price.

Holding the base-period market basket constant has led to criticism. As we know, economists predict that consumers will purchase relatively cheaper substitutes if the relative price of a good increases. If this substitution occurs between the base period and the end period, the quantity weights in the Laspeyres index will not accurately reflect the end-period market basket. This, in turn, has the effect of overweighting goods whose prices have increased over the time period.

What if, instead, we used end-year quantities as weights? Such a price index is known as the **Paasche price index** (P_P):

$$\text{fixed-weight } P_P = \frac{\sum p_t q_t}{\sum p_0 q_t} \times 100 \tag{6.5}$$

The Paasche index's advantage is that, by using current-year weights, it will reflect the current buying habits by consumers. Unfortunately, it too suffers from similar limitations as the Laspeyres index in that it also does not account for substitution among commodities. For the Paasche price index, goods whose prices have declined will be overweighted. As a result, both indices may not accurately reflect price changes, especially when the base-year and current-year periods are far apart.

In 1922, economist Irving Fisher developed an index that incorporated both base-period and current-period quantity weights. The **Fisher ideal price index** (F_P) is the geometric mean of the Laspeyres price index and the Paasche price index:

$$\text{fixed-weight } F_P = \sqrt{P_P \times L_P} \tag{6.6}$$

The Fisher ideal price index is considered a "superlative" index compared to the other two because each year's index accounts for buying patterns in both the base period as well as the current period. While it still does not account for quantity changes between the base and current periods, economists prefer the Fisher ideal price index to the Laspeyres and Paasche indices.

The three price indices presented thus far are typically designated as **fixed-weight price indices**, since each index uses the same quantity of each item as the weights. We now turn to how fixed-weight price indices are constructed using the price and quantity data for two cruciferous vegetables, broccoli and cabbage, over a five-year period.

As we see in the final column of Table 6.5, U.S. expenditures on broccoli and cabbage increased from just over $2 million in 1996 to just over $2.3 million in 1997, an increase of

Table 6.5 U.S. expenditures on broccoli and cabbage, 1996–2000

	Broccoli			Cabbage			
	Utilization (millions of pounds)	*Average annual price per pound*	*Total expenditures (millions of dollars)*	*Utilization (millions of pounds)*	*Average annual per pound*	*Total expenditures millions of dollars*	*Total expenditures on broccoli and cabbage (millions of dollars)*
1996	1,217.833	0.910	1,108.330	2,226.903	0.403	897.813	2,006.143
1997	1,355.083	0.980	1,327.981	2,459.847	0.398	979.429	2,307.410
1998	1,392.994	1.096	1,527.186	2,327.161	0.450	1,046.447	2,573.632
1999	1,720.249	1.005	1,728.850	2,109.700	0.419	884.140	2,612.990
2000	1,662.809	1.138	1,892.415	2,513.180	0.437	1,097.506	2,989.921

Data source: Utilization: Economic Research Service, U.S. Department of Agriculture (2007), 2007 Vegetables and Melons Yearbook, Table 60, U.S. fresh broccoli: supply, utilization, and price, farm weight, 1979–2007; and Table 65, U.S. fresh cabbage: supply, utilization, and price, farm weight, 1979–2007.

Prices: U.S. Bureau of Labor Statistics (2010b), Consumer Price Index, U.S. city average annual price.

Table 6.6 Constructing the 1997 Laspeyres composite price index for broccoli and cabbage (constant 1996 quantities)

1996	Utilization (millions of pounds)	Average annual price per pound	Total expenditures (dollars)	1997	Utilization (millions of pounds)	Average annual price per pound	Total expenditures (dollars)
Broccoli	1,217.833	0.910	1108329516.08	Broccoli	1,217.833	0.980	1,193,476,340.00
Cabbage	2,226.903	0.403	897813059.50	Cabbage	2,226.903	0.398	886,678,544.50
		Total	2,006,142,575.58			Total	2,080,154,884.50
						Laspeyres	103.689

Source: Author's calculations from Table 6.5.

Table 6.7 Constructing the 1997 Paasche composite price index for broccoli and cabbage (constant 1997 quantities)

1996	Utilization (millions of pounds)	Average annual price per pound	Total expenditures (dollars)	1997	Utilization (millions of pounds)	Average annual price per pound	Total expenditures (dollars)
Broccoli	1,355.083	0.910	1,233,238,453.58	Broccoli	1,355.083	0.980	1,327,981,340.00
Cabbage	2,459.847	0.403	991,728,315.50	Cabbage	2,459.847	0.398	979,429,080.50
		Total	2,224,966,769.08			Total	2,307,410,420.50
						Paasche	103.705

Source: Author's calculations from Table 6.5.

15.017 percent. However, we need to determine how much of that increase is due to changes in the composite price of broccoli and cabbage versus changes in the quantities of broccoli and cabbage consumed.

To distinguish the two sources of change, we will begin by constructing a Laspeyres composite price index for broccoli and cabbage. As shown in Table 6.6, we hold the 1996 quantities of broccoli and cabbage consumed constant. The Laspeyres price index indicates that the composite price of broccoli and cabbage increased by only 3.689 percent between 1996 and 1997, a very different result from the previous total expenditure calculation. The reason is that the Laspeyres price index assumes that the amount of broccoli and cabbage consumed between the two years does not change.

Now let us calculate how prices for broccoli and cabbage change if we utilize the current-year quantities consumed to construct a Paasche price index. As shown in Table 6.7, here we used the pounds of cruciferous vegetables consumed from 1997. According to the calculations in Table 6.7, if we hold 1997 quantities constant, the composite price of broccoli and cabbage increased by 3.705 percent between 1996 and 1997. We also note that this result is slightly higher than the price increase obtained from the Laspeyres index.

We now have the necessary statistics to calculate the 1997 Fisher price index for cruciferous vegetables. Applying equation (6.6), we obtain a value of 103.697. This result, as we would expect, lies between the Laspeyres and Paasche price index values. According to the 1997 Fisher price index, the composite price of broccoli and cabbage increased by 3.697 percent between 1996 and 1997.

We might wonder why all three price indices generate changes so much smaller than the change in total expenditures on broccoli and cabbage. As we see in the data from Table 6.5, the consumption of both vegetables, especially broccoli, increased quite noticeably between

Table 6.8 Fixed-weight cruciferous vegetable composite price indices, 1996–2000

	Laspeyres price index (1996 = 100)	Paasche price index (current year = 100)	Geometric mean price index
1996	100.000	100.000	100.000
1997	103.689	103.705	103.697
1998	116.468	116.666	116.567
1999	107.529	108.148	107.838
2000	117.563	118.341	117.952

Source: Author's calculations based on data in Table 6.5.

1996 and 1997. This suggests that changes in quantities consumed were the main driving force behind the increase in expenditures on cruciferous vegetables between the two years.

The procedures just outlined can be applied to construct composite price indices for cruciferous vegetable for the remaining three years. To calculate the fixed-weight Laspeyres price indices, we use 1996 quantities as the weights. For the fixed-weight Paasche price indices, we use the current-year quantities as the weights, while the Fisher price index for each year uses, in a sense, a combination of base- and current-year quantities in its calculation. The results are presented in Table 6.8.

As we observe, the composite price of cruciferous vegetables increased relative to that for 1996 in each of the subsequent years. However, composite prices did not increase each year. We observe, for example, that the composite price index for cruciferous vegetables was approximately 8 percent higher in 1999 when compared to 1996, but actually declined by approximately 7.5 percent between 1998 and 1999.

Let us now examine a visual representation of the three fixed-weight price indices for cruciferous vegetables. We see in Chart 6.1 (p. 167) that the three price indices are initially very close in value, and they actually equal each other in the first year after the base period. However, as we move farther away from the base period, the indices begin to diverge. This is not surprising, since we expect consumers to adjust their purchases of each item in the index in response to price changes. Thus, because base-year and current-year quantities may differ by considerable amounts, economists consider the Fisher fixed-weight price index to provide the most accurate measure of changes among the three fixed-weight price indices presented here.

6.6   Chained price indices

The statistical superiority of the Fisher fixed-weight price index has been appreciated by economists and statisticians for many years. Because it captures changes both in relative prices and in the composition of expenditures over time, it more accurately tracks price changes than either the Laspeyres or Paasche fixed-weight price indices. However, this fixed-weight index does not account for quantity changes from one time period to the next.

Recent advancements in data collection and computing capabilities have allowed economists to construct price indices that *do* incorporate changes in the quantity weights across adjacent time periods. If we have both per-period price and per-period quantity data available, we can construct an index that "chains" the relative changes in prices and the quantity weights from one period to the next. This system is referred to as **chain weighting**.

As with the fixed-weighted price indices, we distinguish between chain-weighted Laspeyres, chain-weighted Paasche, and chain-weighted Fisher price indices. The formulas

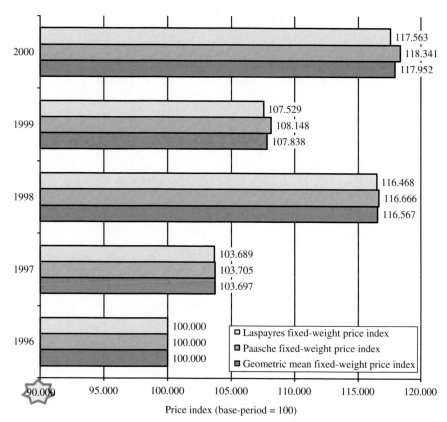

Chart 6.1 Comparison of fixed-weight price indices for cruciferous vegetable, 1996–2000.

for calculating the indices are presented in equations (6.7), (6.8), and (6.9), respectively:

$$\text{chain-weighted } L_P \text{ for time period } t = \frac{\sum P_t \times Q_{t-1}}{\sum P_{t-1} \times Q_{t-1}} \tag{6.7}$$

$$\text{chain-weighted } P_P \text{ for time period } t = \frac{\sum P_t \times Q_t}{\sum P_{t-1} \times Q_t} \tag{6.8}$$

$$\text{chain-weighted } F_P \text{ for time period } t = \sqrt{\text{chain-weighted } L_P \times \text{chain-weighted } P_P} \tag{6.9}$$

Here we observe that the base-period (or following-period) quantities change for each pro-gressive chain-weighted price index. By constantly adjusting the quantity weights to reflect price-initiated changes in consumer buying habits, we obtain a more accurate picture of how prices change over time.

Chain weighting is perhaps best understood through an example of its application, so let us return to the cruciferous vegetable data in Table 6.5. We first note that the chain-weight price indices for 1996 and 1997 will be identical to their three fixed-weight price index counterparts. Thus, let us now calculate and then interpret the chain-weighted indices for 1998.

Table 6.9 The 1998 chain-weighted price indices for cruciferous vegetables

1997	Utilization (millions of pounds)	Average annual price per pound	Total expenditures (millions of dollars)	1998	Utilization (millions of pounds)	Average annual price per pound	Total expenditures (millions of dollars)
			Laspeyres chain-weighted price index				
Broccoli	1,355.08	0.98	1,327,981,340.00	Broccoli	1,355.08	1.096	1,485,622,662.33
Cabbage	2,459.85	0.398	979,429,080.50	Cabbage	2,459.85	0.45	1,106,111,201.00
		Total	2,307,410,420.50			Total	2,591,733,863.33
						Laspeyres	112.322
			Paasche chain-weighted price index				
Broccoli	1,392.99	0.98	$1,365,134,120	Broccoli	1,392.99	1.096	$1,527,185,755
Cabbage	2,327.16	0.398	$926,597,938	Cabbage	2,327.16	0.45	$1,046,446,730
		Total	$2,291,732,058			Total	$2,573,632,485
						Paasche	112.301
				Geometric mean chain-weighted price index			112.311

Data source: Author's calculations based on data in Table 6.5.

Table 6.9 presents the components needed to construct the 1998 chain-weighted Laspeyres, Paasche, and Fisher price indices for cruciferous vegetables. As we see, the chain-weighted Laspeyres price index utilizes 1997 quantity weights for broccoli and cabbage, while the 1998 Paasche price index uses 1998 quantity weights. Taking the appropriate total expenditures ratios yields a chain-weighted Laspeyres price index of 112.322 and a chain-weighted Paasche composite price index of 112.301, while the chain-weighted Fisher price index equals 112.311.

Let us now compare the 1998 chain-weighted results with the 1998 fixed-rate results presented in Table 6.8. We observe that each fixed-weight price index is higher than its chain-weighted counterpart, which indicates a greater increase in the composite price of cruciferous vegetables between 1997 and 1998 when we use base-period quantity weights. Such results are not surprising. If we assume that cabbage consumers behave rationally, the increase in cabbage prices from 38.9 cents per pound in 1997 to 45 cents per pound in 1998 should trigger a decline in the consumption of cabbage, which is what we observe. Thus if we do not adjust for the 132.69 million pound decline in cabbage consumption, we will overestimate the increase in the composite price for cruciferous vegetables.

Let us now interpret the chain-weighted values for all three price indices between 1996 and 2000 and the percentage change in each index across successive periods in Table 6.10. We observe, for example, that, according to the chain-weighted (1996) Fisher price index, the price of cruciferous vegetables decreased by 7.747 percent (92.236–100.000) over the three-year period between 1996 and 1999, while the price for the one-year period between 1998 and 1999 fell by considerably more, declining by 17.875 percent.

Note that we have included "(1996)" when presenting the chained price indices. While a chained index does not have a base year *per se*, it does have a beginning reference point. In the above results, for example, we began our calculations in 1996. Thus we always want to indicate the initial time period from which our chained price indices indicate relative change.

The current discussion might suggest that constructing chained price indices is a straightforward process once the appropriate data series have been collected. Such a conclusion

Table 6.10 U.S. chained (1996) composite price indices for cruciferous vegetables, 1997–2000

	Chained (1996) Laspeyres	Percentage change in chained L_P between consecutive years	Chained (1996) Paasche	Percentage change in chained P_P between consecutive years	Chained (1996) geometric mean	Percentage change in chained GM_P between consecutive years
1996	100.000		100.000		100.000	
1997	103.689	3.689	103.705	3.705	103.697	3.697
1998	112.322	8.326	112.301	8.288	112.311	8.307
1999	92.291	−17.834	92.181	−17.916	92.236	−17.875
2000	110.184	19.387	109.748	19.057	109.966	19.222

Source: Author's calculations based on Table 6.9.

ignores the many technical details involved with index number design and construction. Thus while chained price indices are increasingly utilized, their calculation is still limited by such considerations. As a result, most statistical agencies publish chained price indices as a supplement to fixed-weight price indices. In the United States, for example, the monthly chained Consumer Price Index for All Urban Consumers (designated as the C-CPI-U) is available beginning with December 1999, while the monthly fixed-weight Consumer Price Index for All Urban Consumers (designated as the CPI-U) is available from January 1947.

6.7 Price index applications

Let us now move from the technical details for constructing a weighted price index to the ways in which economists utilize these very useful statistics. As we have already discussed, weighted price indices can be used to determine how the composite price of a market basket of goods and services changes over time. We also saw in Chapter 1 that economists use price indices to measure inflation and deflation in the economy, enabling them to assess the macroeconomic goal of price stability. Price indices are also popularly used as a deflator of nominal data, as a mechanism for determining the purchasing power of a currency unit, and as a cost-of-living adjustor for income payments.

Let us now consider these different functions of a price index. In our discussion, we will use the most commonly used price index for the United States, the Consumer Price Index for All Urban Consumers (CPI-U). The CPI-U is constructed from the prices of all expenditure items in over 200 categories of goods and services, approximately 80,000 items each month, purchased by urban consumers, who represent approximately 87 percent of the U.S. population. (For more detail, see the U.S. Bureau of Labor Statistics (2011), Consumer Price Index, Frequently Asked Questions.) Because of the large number of items and consumers included in the construction of the CPI-U, this price index is used to assess changes in the overall price level, and thus price stability, in the U.S. economy.

Chart 6.2 (p. 170) presents the trend and annual percentage change in the CPI-U between 1960 and 2009. As we see, the CPI-U exhibits an overall upward trend across the 49-year period. However, we also observe that the annual percentage change in the CPI-U varied noticeably over the period, ranging from a maximum annual percentage change of 13.5 percent between 1979 and 1980 to a minimum 0.4 percent decrease between 2008 and 2009. Thus if we strictly define price stability as there being no change in the CPI-U between years, the United States has failed to achieve that particular macroeconomic goal. But since economists and policy-makers typically expect some inflation to obtain, they are more likely

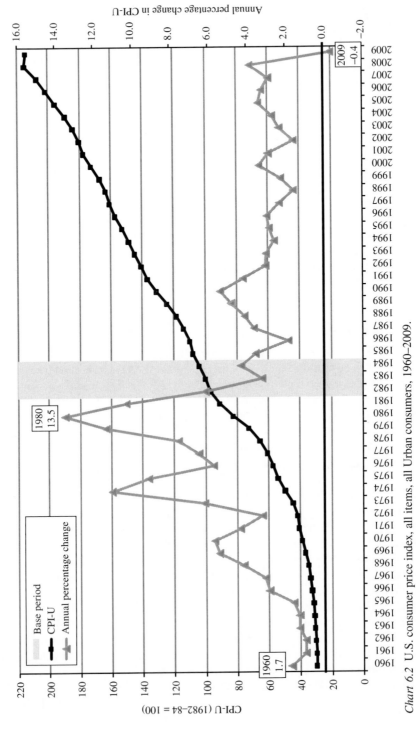

Chart 6.2 U.S. consumer price index, all items, all Urban consumers, 1960–2009.

Data source: U.S. Bureau of Labor Statistics (2010a), Consumer Price Index, All Urban Consumers (Current Series).

to compare the annual percentage change in the CPI-U against a low threshold value (which will vary depending on economic conditions and administrative priorities) when assessing the goal of price stability.

Economists also use price indices to deflate economic variables measured in nominal, or current dollar, terms to reflect real changes in the variable. By factoring out the effects of changes in the overall price level over time, we can determine how a variable such as income has changed in terms of its purchasing power, a desirable feature when assessing economic well-being.

Suppose, for example, that we decide to assess economic well-being using per-capita income. If per-capita income increases over time, this would suggest that economic well-being has improved. However, if per-capita income increases from one year to the next *solely* because the overall level of prices has increased, per-capita income cannot purchase more goods and services this year than it could in the previous year. Thus economists argue that **real income**, not **nominal income**, is the more appropriate measure of economic well-being.

Real income, however, is not directly observable because income data is reported in the dollars in which it is received; that is, it is reported in nominal, or current, dollars. To convert these current dollars into real, or constant, dollars, we can apply equation (6.10):

$$\text{real income} = \frac{\text{current dollar income}}{\text{price index}} \times 100 \qquad (6.10)$$

in which the denominator is a specified price index.

Let us apply equation (6.10) to evaluate how both nominal and real per-capita income changed each decade from 1960 to 2000. As we see in Table 6.11, nominal per-capita disposable personal income (DPI) in the U.S. equaled $3587 in 1970, while the 1970 CPI-U (1960 = 100) equaled 131.1. Thus real per-capita DPI in 1970 equals

$$\text{real 1970 income} = \frac{\$3{,}587}{131.1} \times 100 = \$2{,}736.47 \text{ in 1960 dollars}$$

What this tells us is that in 1970, real per-capita DPI could purchase $2,736.47 of goods and services valued at 1960 prices.

Table 6.11 U.S. nominal and real per-capita disposable personal income by decade

	Per-capita disposable personal income (current dollars)	Change in nominal per-capita DPI from previous decade	CPI-U (1960 dollars)	Per-capita disposable personal income (1960 dollars)	Change in real per-capita DPI from previous decade
1960	$2,022.00		100.0	$2,022.00	
1970	$3,587.00	$1,565.00	131.1	$2,736.47	$714.47
1980	$8,822.00	$5,235.00	278.4	$3,169.07	$432.59
1990	$17,131.00	$8,309.00	441.6	$3,879.71	$710.64
2000	$25,467.00	$8,336.00	581.8	$4,377.60	$497.90
	Total change	$23,445.00		Total change	$2,355.60

Source: Author's calculations of CPI-U, using data from *Economic Report of the President* (2004), Table B-31, Total and per capita disposable personal income and personal consumption expenditures, and per capita gross domestic product, in current and real dollars, 1959–2003.

We further observe in Table 6.11 that real per-capita DPI increased by considerably fewer dollars ($2,355.60) than did nominal per-capita DPI ($23,445.00) over the 40-year time period. This reflects the fact that much of the increase in nominal income was due solely to a rise in the overall price level in the economy. Thus consumers can purchase only $2,355.60 of additional goods and services with their income in 2000 as compared to purchases in 1960, and not $23,445 more as suggested by the increase in nominal income.

Deflating nominal (current) dollars using a price index is thus akin to determining the purchasing power of the dollar. The **purchasing power of a dollar** (or any currency unit) indicates the dollar value of the market basket of goods and services that the dollar will buy at a specific date, and it is calculated using equation (6.11):

$$\text{purchasing power of the dollar in period } t = \left(\frac{\$1.00}{\text{CPI-U}_t} \right) \times 100 \tag{6.11}$$

Suppose, for example, we want to know the purchasing power of the U.S. dollar in 1980 relative to its 1960 value. We see in Table 6.11 that the CPI-U (1960 = 100) equals 278.4 in 1980. Applying equation (6.11), we determine that the purchasing power of the 1980 dollar equals $0.36 relative to its value in 1960. That is, one dollar ($1.00) in 1980 could have purchased only $0.36 worth of goods and services in 1960.

The differences between real and nominal values can be substantial, particularly over long time periods or periods of price instability. Thus we must always consider which category of values, real or nominal, is more appropriate for our analysis. In cases where we want purchasing power to be held constant, we select, or construct, real values of the variable. In a few cases, such as projecting future costs and benefits of an infrastructure project or when working with financial variables, we may choose to use nominal values. In either case, it is important that we make a conscious decision about which type of variable to use.

6.8   Shifting an index's reference period

All price indices include a reference to a particular base period, information that is critical, since all index numbers are relative measures. Indeed, if the base-period reference is omitted, statistical results using the index number are essentially meaningless.

We may have also noticed that not all indices use the same base period. We can, for example, obtain two fixed-weight price index series for the All Items CPI-U from the Bureau of Labor Statistics, one that uses 1967 as the base (known as the Old Series) and one that uses a 1982–84 average base (referred to as the Current Series). BLS also uses other base periods for some items, as we shall see in Table 6.13.

We cannot, however, compare results derived from indices with different reference points. Fortunately, it is straightforward to rebase (technically, re-reference) price indices to have the same reference period. (Note that we cannot rebase a fixed-weight index to compare with a chained index (or vice versa), as the two indices are constructed differently.) Let us now turn to the procedures for shifting the base period of a price index.

Table 6.12 presents data for two indices – the All-Items CPI-U index, with a base period of 1982–84 (which we will discuss below), and the CPI-U for Information Technology, Hardware and Services, with a base period of December 1988. Our first task is to determine which series to set as the base period. Here we will select a base period of 1989, the first period for which an annual technology index could be calculated. This choice was somewhat arbitrary, as is any base-period choice. In some cases, the indices available may suggest a particular

Table 6.12 Converting the base period of an index

	All Items CPI-U (1982–84 = 100)	Information Technology, Hardware and Services CPI-U (December 1988 = 100)	All Items CPI-U (1989 = 100)	Information Technology, Hardware and Services CPI-U (1989 = 100)
1989	124.0	92.7	100.0	100.0
1990	130.7	90.3	105.4	97.4
1991	136.2	86.6	109.8	93.4
1992	140.3	81.3	113.1	87.7
1993	144.5	75.1	116.5	81.0
1994	148.2	68.3	119.5	73.7
1995	152.4	61.0	122.9	65.8
1996	156.9	53.9	126.5	58.1
1997	160.5	47.4	129.4	51.1
1998	163.0	34.8	131.5	37.5
1999	166.6	28.2	134.4	30.4
2000	172.2	23.8	138.9	25.7
2001	177.1	19.8	142.8	21.4
2002	179.9	17.2	145.1	18.6
2003	184.0	15.3	148.4	16.5
2004	188.9	14.2	152.3	15.3
2005	195.3	13.1	157.5	14.1
2006	201.6	11.2	162.6	12.1
2007	207.342	10.215	167.211	11.019
2008	215.303	9.906	173.631	10.686

Data source: U.S. Bureau of Labor Statistics (2010b), Consumer Price Index.

base period, but in others, the choice is up to the researcher. We next want to re-reference each series to a 1989 base period by dividing each year's index value by its 1989 value. Following convention, we then multiply each quotient by 100, obtaining the results presented in the last two columns of Table 6.12.

We can now easily compare the relative change in U.S. information technology prices between 1989 and any subsequent year with that for the overall U.S. price level over the same period. The most striking result is seen in Chart 6.3 (p. 174). While the overall U.S. price level increased by 73.631 percent between 1989 and 2008, prices for information technology, hardware and services fell by 89.314 percent. It also appears that the biggest annual decline in information technology, hardware and services prices occurred between 1998 and 2001, which we can confirm by applying equation (6.2) to the data. While further investigation would be needed, one contributing factor may be the rush by businesses and government worldwide to upgrade their information technology to avoid problems at the transition to the year 2000 (colloquially known as Y2K problems).[2]

Returning to Table 6.12, let us comment further on the unusual base period, 1982–84, for the All-Items CPI-U. The U.S. Bureau of Labor Statistics uses the three-year average of the CPI-U for 1982, 1983, and 1984 as the base period for indices in its Current Series. As a result, we will not observe a value of 100.0 in either the CPI-U monthly or annual series. This makes it even more important to include the reference period when using the CPI-U, since it cannot be determined by examining the index numbers themselves.

The index numbers presented thus far have been constructed to measure price changes in individual items, groups of items, and the overall price level in the economy. We have also seen how these indices can be used to measure inflationary tendencies in an economy and to

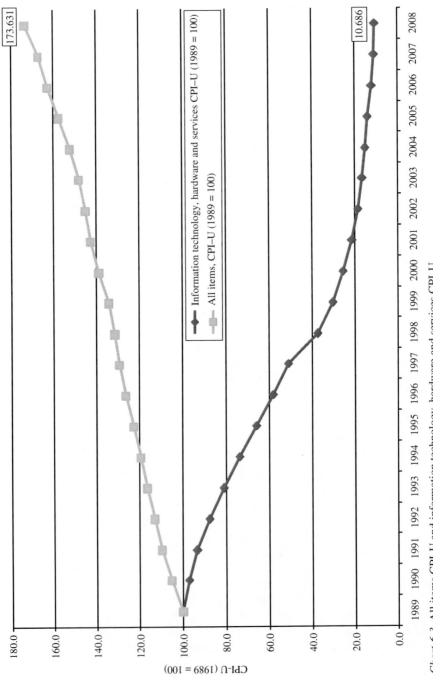

Chart 6.3 All items CPI-U and information technology, hardware and services CPI-U.

convert nominal series into real values. Regardless of how they are constructed, price indices are unit-free, non-negative, and non-zero statistics that offer economists greater flexibility for conducting certain types of empirical analysis. These valuable properties apply to other indices used in economics, so we now turn to a brief discussion of some of the more popular non-price indices.

6.9 Quantity indices

The basic construct of an index number allows us to track the movement of one economic variable weighted by another, closely associated variable. As we have seen, a price index weighs price by the amount of the goods or services consumed at that price. But what if we wanted to measure how quantities change over time? Not surprisingly, we can use the same method to construct a **quantity index**, in which the amount purchased is weighted by the cost, or price, of purchase.

Suppose, for example, that we wanted to determine how the quantity (utilization) of cruciferous vegetables changed between 1996 and 2000. We could return to the expenditure data in Table 6.5 and apply one of the formulas in Figure 6.1 to obtain the series of quantity indices presented in Table 6.13.

We observe, for example, that between 1996 and 1999, the relative quantity of cruciferous vegetables consumed increased. According to the fixed-weight Fisher quantity index, that increase equaled 20.783 percent, but we know this index does not account for the relatively higher prices observed for broccoli and cabbage after 1996. Thus the fixed-weight quantity index overstates the quantity by which cruciferous vegetables increased.

	Fixed-weight quantity index		Chain-weighted quantity index	
Laspeyres	$L_Q = \dfrac{\sum P_0 Q_t}{\sum P_0 Q_0}$	(6.12)	$L_Q = \dfrac{\sum P_{t-1} Q_t}{\sum P_{t-1} Q_{t-1}}$	(6.15)
Paasche	$P_Q = \dfrac{\sum P_t Q_t}{\sum P_t Q_0}$	(6.13)	$P_Q = \dfrac{\sum P_t Q_t}{\sum P_t Q_{t-1}}$	(6.16)
Geometric mean	$GM_Q = \sqrt{\text{fxd-wt } L_Q \times \text{fxd-wt } P_Q}$	(6.14)	$GM_Q = \sqrt{\text{chn-wt } L_Q \times \text{chn-wt } P_Q}$	(6.17)

Figure 6.1 Formulas for quantity indices.

Table 6.13 Quantity indices for cruciferous vegetables, 1996–2000

	Fixed-weight Laspeyres quantity index (1996 = 100)	Fixed-weight Paasche quantity index (1996 = 100)	Fixed-weight geometric mean quantity index (1996 = 100)	Chained (1996) Laspeyres quantity index	Chained (1996) Paasche quantity index	Chained (1996) Paasche geometric mean index
1996	100.000	100.000	100.000	100.000	100.000	100.000
1997	110.908	110.925	110.916	110.908	110.925	110.916
1998	109.961	110.148	110.055	99.321	99.302	99.311
1999	120.437	121.130	120.783	110.141	110.010	110.075
2000	125.939	126.773	126.356	104.262	103.849	104.055

Source: Author's calculations based on data in Table 6.5.

Table 6.14 Real gross domestic product, quantity indices (index numbers, 2005 = 100)

	2005	*2006*	*2007*	*2008*	*2009*
Personal consumption expenditures	100.000	102.886	105.335	105.057	103.797
Durable goods	100.000	104.064	108.418	102.798	99.011
Motor vehicles and parts	100.000	96.823	98.589	85.000	79.093
Furnishings and durable household equipment	100.000	105.326	106.659	102.827	96.222
Recreational goods and vehicles	100.000	113.145	125.670	129.771	131.643
Other durable goods	100.000	103.712	104.899	101.041	98.669

Data source: U.S. Bureau of Economic Analysis (2010c), National Economic Accounts, Table 1.5.3, Real gross domestic product, expanded detail, quantity indexes.

The chained Fisher quantity index does factor in the changes in the price weights over the three-year period, and it indicates that cruciferous vegetable consumption increased by only 10.075 percent. As with price indices, the chained Fisher quantity index is the "best" statistical measure of quantity changes among the six quantity indices presented here.

Quantity indices are constructed for many economic variables. The U.S. Bureau of Economic Analysis, for example, produces quantity indices for variables in the National Income and Product Accounts (NIPA) as well as for those associated with state and industry-level output.

Table 6.14 presents quantity indices for real personal consumption expenditures (PCE) and several of its components in the United States between 2005 and 2009. We immediately observe the effects of the 2008–09 recession on PCE, since its quantity index fell each successive year after 2007. We also see that real expenditures on motor vehicles and parts fell the most among the major durable goods categories, and that this decline began prior to the beginning of the current recession. Interestingly, the quantity of recreational goods and vehicles increased each successive year between 2005 and 2009, a result economists typically do not expect with durable goods (whose purchases can be deferred until economic conditions improve), suggesting an interesting question for further analysis.

6.10   Composite indices

The indices discussed thus far combine the price and quantity of multiple items to measure how either prices or quantities change over time. An alternative interpretation of these indices is that they track the time-series movement, or trend, of one variable determined in product markets (either the prices or quantities of multiple items) or one component of the macroeconomy (aggregate price level or aggregate national product).

Index numbers can also be used to assess trends in a heterogeneous group of variables. Such a so-called **composite index** combines often disparate variables into a summary measure. With composite indices, individual indices are constructed for each variable relative to a benchmark and then combined using weights that reflect each variable's relative importance to the overall trend being measured.

The methods used to construct composite indices are as varied as the number of indices themselves. However, we can turn to one composite index to examine the basic steps involved in their construction. We will also discuss how this index is used, and interpret its empirical results.

The **Human Development Index** (HDI), constructed by the United Nations Development Programme (UNDP), "measures the average achievements in a country in three basic dimensions of human development: a long and healthy life, knowledge and a decent standard of living" (United Nations Development Programme 2007). It was developed in response to concerns among economists and policy-makers about the limitations of using GDP per capita as the sole indicator of human development. Acknowledging that changes in real GDP per capita may indicate short-term changes in a region's standard of living and provide a base for future economic growth, development professionals argued that longer-term, or permanent, progress also depends on the overall health and knowledge of the region's population. Thus the HDI was created as a more complete measure of a country's economic and social progress.

The first step in creating any composite index is to identify variables that are deemed relevant to what the composite index is attempting to measure *and* that can be measured quantitatively. For the HDI, UNDP selected four variables to capture three dimensions – believed to enhance a country's long-term development.

The first dimension, the health dimension, relates to the general health of a country's population and is measured by life expectancy at birth. As we would expect, not only is there intrinsic value in a long life, longevity also enables individuals to pursue various economic goals, and it is closely correlated to other desirable characteristics such as good health and nutrition.

The second dimension captured in the HDI, the knowledge dimension, relates to current and potential levels of education in a country and is measured by two variables – the mean years of schooling, and the expected years of schooling. The mean years of schooling for adults 25 years and over is a key indicator of the adult population's ability to learn and build knowledge, making it crucial to both short-term and long-term economic progress. The expected years of schooling measures the years of formal education that "a child of school entrance age can expect to receive if prevailing patterns of age-specific enrolment rates were to stay the same throughout the child's life" (United Nations Development Programme 2010c, p. 223). This measure thus serves as an indicator of potential productivity for the country's future labor force.

The final dimension, a decent standard of living, is measured by a country's gross national income (GNI) per capita to capture each resident's access to income from both domestic and foreign sources. Here, GNI per capita is measured in purchasing power parity (PPP) in constant U.S. dollars to account for price differences between countries, which more accurately reflects living standards across countries than would GNI per capita valued in each country's own currency.

These four variables not only have theoretical connections to human development, but also are quantifiable. While initial statistics for many developing countries were not always available and their reliability was sometimes suspect, UNDP and other international agencies have developed detailed protocols and provided financial and technical support to these countries to ensure the quality of the data used to construct the HDI. In addition, UNDP continues to refine the HDI itself, introducing major changes, most recently in 2010. This demonstrates the dynamic nature of economic statistics, which must evolve to obtain an accurate picture of economic phenomena in light of changing economic conditions as well as improved data collection methods and more refined knowledge of how economies work.

Once the dimensions of a composite index are determined, the next step is to convert each of the index's components into its own index number, or sub-index. In the process, we typically establish some benchmark, or standard, against which we compare the actual observations of the variable. These benchmarks differ from the reference periods used for price and quantity indices in that they use specific values of the variable based on underlying economic conditions.

The individual indices for the three dimensions of the HDI are constructed by establishing what UNDP calls "goalposts." As shown in equation (6.18), the actual value of the variable is compared to minimum and maximum values for the variable as established by UNDP to create so-called dimension indices for health, education, and living standards:

$$\text{dimension index} = \frac{\text{actual value} - \text{minimum value}}{\text{maximum value} - \text{minimum value}} \tag{6.18}$$

Let us now turn to the Life Expectancy Dimension Index to demonstrate how a dimension index is constructed. As of 2010, UNDP established 20 years of age as the minimum value of the Life Expectancy Dimension Index (based on historical evidence for the minimum life expectancy at birth) and 83.2 years of age as the maximum (again, based on historical evidence). This yields the following formula for the Life Expectancy Dimension Index:

$$\text{Life Expectancy Dimension Index} = \frac{\text{actual value} - 20}{83.2 - 20}$$

Once the average annual life expectancy at birth is measured for a country, its Life Expectancy Dimension Index can be quickly calculated.

The process used to construct a dimension index is another form of *normalizing* the data. Because the three dimensions of human development are measured in different units, we must create a way to aggregate the indicators. As we can see for the Life Expectancy Dimension Index, this statistic is unit-free, and its values, because of the minimum and maximum values chosen, will fall between 0 and 1.

Indices for the other two dimensions are similarly calculated. Constructing the Education Dimension Index, however, is a two-step process. First sub-indices must be constructed for both the mean years of schooling (MYS) and the expected years of schooling (EYS). The index for mean years applies equation (6.18) using a minimum value of 0, and a maximum value of 13.2 years, while the index for expected years also uses a minimum value of 0 but a maximum value of 20.6 years. Next, the geometric mean of the two indices is calculated to yield the Education Dimension Index:

$$\text{mean years of schooling (MYS)} = \frac{\text{mean years of schooling} - 0}{13.2 - 0}$$

$$\text{expected years of schooling (EYS)} = \frac{\text{expected years of schooling} - 0}{20.6 - 0}$$

$$\text{Education Dimension Index} = \sqrt{\text{MYS} \times \text{EYS}}$$

The GNI Dimension Index uses a modified form of equation (6.18). Here logarithms of the actual, minimum ($163), and maximum ($108,211) values are taken to reflect economists' belief that increasing income offers diminishing returns to one's standard of living:

$$\text{GNI Dimension Index} = \frac{\ln(\text{GNI per capita}) - \ln(163)}{\ln(108,211) - \ln(163)}$$

Once the three dimension indices have been calculated, they are combined using the geometric mean to generate the Human Development Index (HDI). Let us now turn to data for several countries and examine their HDIs for 2010.

Table 6.15 2010 Human Development Index and components for selected countries

	Life expectancy at birth (years)	Mean years of schooling (years)	Expected years of schooling (years)	GNI per capita (PPP US$ 2008)	Life Expectancy Index	Education Index	GDP Index	Human Development Index
Norway	81.006	12.631	17.322	58,809.525	0.965	0.943	0.906	0.938
United States	79.584	12.445	15.745	47,093.853	0.943	0.893	0.872	0.902
United Kingdom	79.775	9.469	15.941	35,087.162	0.946	0.783	0.827	0.849
China	73.474	7.548	11.385	7,258.474	0.846	0.591	0.584	0.664
India	64.352	4.399	10.307	3,337.366	0.702	0.429	0.465	0.519
Côte d'Ivoire	58.365	3.307	6.300	1,624.858	0.607	0.291	0.354	0.397
Sierra Leone	48.246	2.879	7.206	808.722	0.447	0.290	0.246	0.317

Data source: United Nations Development Programme (2010a), Calculating the Human Development Indices.

We first observe in Table 6.15 that, in 2010, among the seven countries, Norway had the highest HDI, and thus the highest level of human development, while Sierra Leone was ranked the lowest. We also note that each dimension index preserves the rank-order of the variable across countries, a very desirable statistical property. Thus we see that the Health Dimension Index for the United Kingdom is larger than that of the United States, reflecting the slightly higher life expectancy in the U.K.

The HDI is used not only to compare countries (or regions) at a given time period, but also to assess human development progress over time and across socio-economic and demographic groups. While developers of the HDI acknowledge its limitations (see, for example, United Nations Development Programme 2007, pp. 36–37), policy-makers have found the HDI useful for tracking progress in human development beyond what can be gleaned simply from changes in GNI per capita. Additional composite indices have also been constructed to measure progress in such areas as gender equality and poverty reduction, thus demonstrating the value of such indices to economic analysis and policy-making.

Summary

Index numbers are frequently used constructions that facilitate the calculation of relative changes in a variable over time. They are also unit-free measures, thus enabling comparisons of heterogeneous variables, and they are always positive in value, so they can be transformed, or rescaled, by logarithms.

Any index of relative change, regardless of its purpose, must be aggregative in form, and it must contain weights that measure the relative importance of the items included. It is also vital that the point of reference be noted when using index numbers, since all indices measure relative change.

Index numbers can also be used to combine a variety of economic and social indicators into a summary measure of economic activity. These composite indices are particularly useful for reducing complex information about the economy into a single measure whose trends can be tracked over time or across groups such as countries.

Concepts introduced

Base period
Chain weighting
Composite index

Fisher ideal price index
Fixed-weight price index
Human Development Index
Index number
Laspeyres price index
Nominal income
Normalizing
Paasche price index
Percentage change
Percentage point change
Price index
Purchasing power of a dollar
Quantity index
Real income
Relative change
Simple index number
Unweighted price index
Weight
Weighted price index

Box 6.1 Where can I find...? Index numbers frequently used by economists

Economists have developed index numbers for numerous purposes in empirical economics. The following table identifies just some of the more frequently used price and quantity indices as well as several composite indices.

A key feature of these indices is the wide variety of uses to which they can be applied, but the basic concepts and characteristics of index numbers underlie them all. Because index construction requires considerable data and expertise, always use published indices in empirical analyses whenever possible.

Frequently used U.S. price indices

Index	Brief description and usage	Publisher and web address
Consumer Price Index (CPI)	The consumer price indices program produces monthly data on changes in the prices paid by urban consumers for a representative basket of goods and services. See Chapter 6 for more details	U.S. Bureau of Labor Statistics available from <www.bls.gov/cpi/>
Producer Price Index (PPI)	The Producer Price Index is a family of indices that measures the average monthly change over time in the selling prices received by domestic producers of goods and services. It differs from the CPI in	U.S. Bureau of Labor Statistics Available from <www.bls.gov/ppi/>

Index	Brief description and usage	Publisher and web address
	perspective of the seller rather than the purchaser. (Sellers' and purchasers' prices may differ due to government subsidies, sales and excise taxes, and distribution costs.) Conceptually, the PPI is calculated using a fixed-weight Laspeyres formula	
	PPIs are available for the products of virtually every industry in the mining and manufacturing sectors of the U.S. economy. New PPIs are gradually being introduced for the products of industries in the transportation, utilities, trade, finance, and services sectors of the economy	
Employment Cost Index (ECI)	Part of the National Compensation Survey, the Employment Cost Index (ECI) is a quarterly measure of changes in the cost of labor compensation and includes changes in wages and salaries as well as changes in employer costs for employee benefits. Its construction uses fixed weights to control for shifts among occupations and industries using a Laspeyres formula	U.S. Bureau of Labor Statistics Available from <http://www.bls.gov/ncs/ect/>
Import/Export Price Indices	The International Price Program (IPP) produces Import/Export Price Indices (MXP) of the average monthly changes in the prices of non-military goods and services traded between the U.S. and the rest of the world	U.S. Bureau of Labor Statistics Available from <http://www.bls.gov/mxp/>
GDP Chained Price Indices	Current-dollar GDP is estimated by BEA and then separated into a price-change component and a quantity-change component. The GDP chained price indices measure the price components of changes in current-dollar GDP data using a chained Fisher price index formulation. These statistics are found in NIPA table's ending with the number 4	U.S. Bureau of Economic Analysis Available from <http://www.bea.gov/national/nipaweb/Index.asp>
Implicit (GDP) Price Deflator	An implicit price deflator (IPD) is the ratio of the current-dollar value of a series divided by its corresponding chained-dollar value, multiplied by 100. It is very close in value to the chained-dollar price indices. These statistics are found in NIPA table's ending with the number 9	U.S. Bureau of Economic Analysis Available from <http://www.bea.gov/national/nipaweb/Index.asp>.

Index	Brief description and usage	Publisher and web address
	• The implicit price deflator was used to calculate real estimates in the National Income and Product Accounts prior to 1996. The introduction of chained indices in 1996 largely supplanted the IPD	
	• BEA continues to use the IPD for estimating terms of trade	

A note on non-U.S. price indices

International agencies such as the United Nations (UN), the International Labour Organization (ILO), the Organisation for Economic Co-operation and Development (OECD), the World Bank (WB), and the International Monetary Fund (IMF) apply a different methodology from U.S. agencies in constructing price indices. In 2007, these non-governmental agencies created the *Knowledge Base on Economic Statistics* (United Nations 2011), available online at <http://unstats.un.org/unsd/EconStatKB/Knowledgebase.aspx> for use by developing countries, academics, government, and business.

The following links direct the interested reader to further details on some of the methodologies used:

- *Consumer Price Index Manual: Theory and Practice.* 2004. Available online from <http://unstats.un.org/unsd/EconStatKB/KnowledgebaseArticle10130.aspx> (accessed 16 August 2010)
- *Producer Price Index Manual: Theory and Practice.* 2004. Available online from <http://unstats.un.org/unsd/EconStatKB/Attachment118.aspx> (accessed 16 August 2010)
- *Export and Import Price Index Manual Theory and Practice.* Available online from <http://www.imf.org/external/np/sta/xipim/pdf/xipim.pdf> (accessed 16 August 2010)
- United Nations Statistics. *System of National Accounts; 2008.* Available online from <http://unstats.un.org/unsd/EconStatKB/Attachment132.aspx> (accessed 16 August 2010)

Frequently used U.S. quantity indices

Index	Brief description and usage	Compiler of index
GDP Quantity Indices	GDP Quantity Indices are the second change component into which current-dollar GDP is estimated by the Bureau of Economic Analysis and measure the level of quantity produced, assuming the price does not change due to inflation or deflation. These statistics are found in NIPA table's ending with the number 3, and are constructed using a chained Fisher quantity index formulation	U.S. Bureau of Economic Analysis Available online from <http://www.bea.gov/national/nipaweb/Index.asp>
Index of Industrial Production (IIP)	The industrial production index (IIP) measures the real output of the manufacturing, mining, and electric and gas utilities industries. Constructed using a Fisher index formula, the IIP measures how the volume of industrial production activity, weighted by value added, changes over time	Federal Reserve Board of Governors Available online from <www.federalreserve.gov/releases/g17/>

Index	Brief description and usage	Compiler of index
Index of Industrial Capacity (IIC)	The IIC measures how the volume of potential production activity, weighted by value added, changes over time for industries in manufacturing, mining, and electric and gas utilities. It is constructed using a complex six-step procedure not covered in this text	Federal Reserve Board of Governors Available online from <www.federalreserve.gov/releases/g17/>
Capacity Utilization Rate	This is calculated using the IIP and IIC indices: Capacity Utilization Rate $$= \frac{\text{Index of Industrial Production}}{\text{Index of Industrial Capacity}}$$ It measures the ratio of actual production in industry to potential production. It is used by the Fed as an indicator of future inflation. The closer actual production is to its capacity, the more likely that production bottlenecks will occur, which in turn could reduce the supply of output, thereby raising prices	
Labor Productivity Index	Indices of labor productivity measure changes in the ratio of output to hours of labor input. Indices are published for the business, non-farm business, and manufacturing (total, durable, and non-durable) sectors, for non-financial corporations, and for selected 2-, 3-, 4-, 5-, and 6-digit NAICS industries	Bureau of Labor Statistics Available online from <http://www.bls.gov/lpc/>
Multifactor Productivity Indices	Multifactor Productivity Indices measure output per unit of some combined set of inputs that cannot be accounted for by the change in combined inputs. As a result, multifactor productivity measures reflect the joint effects of many factors, including new technologies, economies of scale, managerial skill, and changes in the organization of production.	Bureau of Labor Statistics Available online from <http://www.bls.gov/mfp/home.htm>

Frequently used composite indices

Index	Brief description and usage	Compiler of index
	BLS publishes multifactor productivity for major U.S. sectors and for total manufacturing and 20 2-digit Standard Industrial Classification manufacturing industries. These indices are constructed using an annual-weighted (Fisher ideal) output index and annually chained (Tornqvist) indices for inputs	
U.S. Index of Leading Economic Indicators (LEI)	Composite index that measures trends in ten economic variables (Average weekly hours in manufacturing; Average weekly initial claims for unemployment insurance; Manufacturers' new orders, consumer goods and materials; Index of supplier deliveries – vendor performance; Manufacturers' new orders, non-defense capital goods; Building permits for new private housing units; Stock prices of 500 common stocks; Money supply (M2); Interest rate spread = 10-year Treasury bonds less federal funds; and Index of consumer expectations) that are assumed to precede by some months changes in aggregate economic activity. Note that the Conference Board charges a fee for time-series LEI data. Recent statistics are available in each month's press release	The Conference Board Available online from <http://www. conference-board.org/data/ bci.cfm> See "Brief description and usage" column for complete explanations for each component's inclusion in the LEI, CEI, and LAG Available online from <http://www. conference-board.org/data/bci/ index.cfm?id=2160> A more detailed description of these indices can be found in Conference Board (2001)
U.S. Coincident Economic Index (CEI)	Composite index that measures current economic activity by tracking trends in four economic variables (Employees on non-agricultural payrolls; Real personal income less transfer payments; Index of industrial production; and Sales at the manufacturing, wholesale and retail trade levels)	
U.S. Lagging Economic Index (LAG)	Composite index that measures by tracking trends in seven economic variables (Average duration of unemployment; Inventories to sales ratio, manufacturing and trade; Labor	

Index	Brief description and usage	Compiler of index
	cost per unit of output, manufacturing; Average prime rate; Commercial and industrial loans; Consumer installment credit to personal income ratio; and Consumer price index for services)	
International Business Cycle Indicators	The Conference Board also produces LEI and CEI indices for the following countries and regions: Australia, China, Euro Area, France, Germany, Japan, Korea, Mexico, Spain, and the United Kingdom. Links to these indices for each country/region are available	The Conference Board Available online from <http://www.conference-board.org/data/bci.cfm>
OECD Composite Leading Indicators (CLI)	The OECD Composite Leading Indicators (CLI) are calculated for 29 OECD countries (Iceland is not included), six non-member economies and nine zone aggregates. The full list of component series used in calculating each country's CLI is available. According to the metadata from OECD's StatExtracts (at <http://stats.oecd.org/OECDStat_Metadata/ShowMetadata.ashx?Dataset=MEI_CLI&Lang=en&>): The OECD CLI is designed to provide qualitative information on short-term economic movements, especially at the turning points, rather than quantitative measures. Therefore, the main message of CLI movements over time is the increase or decrease, rather than the amplitude of the changes	Organisation for Economic Co-Operation and Development Available online from <www.oecd.org/std/cli>
Index of Consumer Sentiment (ICS)	Measures how consumers feel their personal finances have changed from the past year, and how they think their personal finances, overall business conditions, and buying conditions for consumer durables will change in the near (one year) and long-term (five years) future. Note that the Conference Board publishes a similar index, the Index of Consumer Expectations, as part of the LEI	University of Michigan, Surveys of Consumers Available online from <http://www.sca.isr.umich.edu/main.php> To access the ICS, follow the link, Time-Series Archive, All Households

Index	Brief description and usage	Compiler of index
UNDP Human Development Indices	UNDP constructs four composite human development indices in addition to the Human Development Index (HDI): • Gender Development Index (GDI). This composite index contains "equally distributed" indices based on female and male indices for each of the three dimensions in the HDI. It adjusts the HDI for inequalities between men and women • Gender Empowerment Index (GEM). This composite index measures gender inequality in the following three areas: political participation and decision-making power as measured by women's and men's percentage shares of parliamentary seats; economic participation and decision-making power as measured by women's and men's percentage shares in public and private leadership positions and in professional and technical professions; and power over economic resources as measured by women's and men's estimated earned income (PPP \$US) • Human Poverty Index for developing countries (HPI-1). This composite index measures deprivations in the three basic dimensions captured in the HDI. Its components include the probability at birth of not surviving to age 40; the adult illiteracy rate; and an unweighted average of the percentage of the population not using an improved water source and the percentage of children underweight for age • Human Poverty Index for selected OECD countries (HPI-2). This composite index measures deprivations in the same dimensions as the HPI-1 plus social exclusion. Its components include the probability at birth of not surviving to age 60; the percentage of adults (ages 16–65) lacking functional literacy skills; the percentage of the population living below the income poverty line (as defined by 50 percent of median-adjusted household disposable income); and the rate of long-term (12 months or more) unemployment	United Nations Development Programme Available online from <http://hdr.undp.org/en/statistics/data/> See Technical Note 1 for details on calculating all five UNDP Human Development Indices Available online from <http://hdr.undp.org/en/media/HDR_20072008_Tech_Note_1.pdf>

Box 6.2 Using Excel to construct index numbers

Calculating price indices, or any index number, can be quite involved and occasionally confusing. Excel can help us systematically organize the data and perform the various calculations required. Here we will use the cruciferous vegetable data from Table 6.5 and Excel to construct the six types of price indices discussed in Chapter 6.

	U.S. expenditures on broccoli and cabbage, 1996–2000						
	Broccoli			*Cabbage*			
	Utilization (millions of pounds)	*Average annual price per pound*	*Total expenditures (millions of dollars)*	*Utilization (millions of pounds)*	*Average annual price per pound*	*Total expenditures (millions of dollars)*	*Total expenditures on broccoli and cabbage (millions of dollars)*
1996	1,217.833	0.910	1,108.330	2,226.903	0.403	897.813	2,006.143
1997	1,355.083	0.980	1,327.981	2,459.847	0.398	979.429	2,307.410
1998	1,392.994	1.096	1,527.186	2,327.161	0.450	1,046.447	2,573.632
1999	1,720.249	1.005	1,728.850	2,109.700	0.419	884.140	2,612.990
2000	1,662.809	1.138	1,892.415	2,513.180	0.437	1,097.506	2,989.921

Data source: Table 6.5.

- First, place the price and quantity data for each item in the same worksheet. This allows us to calculate the total expenditures on each item for each year and the total expenditures on all items per year.
- Next, construct a matrix of total expenditures on broccoli and cabbage. Note that the values in the diagonal of the expenditure matrix equal the total expenditures in the last column of the table above.

Expenditure matrix for cruciferous vegetables (millions of dollars)					
Total (both vegetables)	*Quantity*				
	1996	*1997*	*1998*	*1999*	*2000*
Price					
1996	**2006.14**	2224.97	2205.97	2416.13	2526.53
1997	2080.15	**2307.41**	2291.73	2525.86	2630.22
1998	2336.51	2591.73	**2573.63**	2834.63	2953.09
1999	2157.18	2392.74	2375.23	**2612.99**	2724.36
2000	2358.48	2616.41	2601.61	2879.09	**2989.92**

Fixed-weight Laspeyres price index

The fixed-weight Laspeyres price index holds the quantity weights fixed at their base-period values and the base-period price constant.

In terms of the expenditure matrix:

- The first cell (total expenditures = 2006.14) is the denominator for all fixed-weight Laspeyres price index calculations.
- The numerator of each fixed-weight Laspeyres price index is the value in the first column of the expenditure matrix.

	Laspeyres fixed-weight price index for cruciferous vegetables (1996 = 100)	
	Excel formula	*Price index value*
1996	= 2006.14/2006.14	100.000
1997	= 2080.15/2006.14	103.689
1998	= 2336.51/2006.14	116.468
1999	= 2157.18/2006.14	107.529
2000	= 2358.48/2006.14	117.563

Fixed-weight Paasche price index

The fixed-weight Paasche price index holds the base price constant in the denominator, while the numerator is the product of the current-period price and quantity.

In terms of the expenditure matrix:

- Each value in the first row of the expenditure matrix serves as the denominator for that year.
- The numerator for a given year is the value in the expenditure matrix's diagonal for that year.

	Paasche fixed-weight price index for cruciferous vegetables (1996 = 100)	
	Excel formula	*Price index value*
1996	= 2006.14/2006.14	100.000
1997	= 2307.41/2224.97	103.705
1998	= 2573.63/2205.97	116.666
1999	= 2612.99/2416.13	108.148
2000	= 2989.92/2526.53	118.341

Fixed-weight Fisher price index

The fixed-weight Fisher price index is the square root of the product of the fixed-weight Laspeyres and Paasche price indices:

	Fixed-weight Fisher price index for cruciferous vegetables (1996 = 100)	
	Excel formula	*Price index value*
1996	= SQRT(100.000 * 100.000)	100.000
1997	= SQRT(103.689 * 103.705)	103.697
1998	= SQRT(116.486 * 116.666)	116.567
1999	= SQRT(107.529 * 108148)	107.838
2000	= SQRT(117.563 * 118.341)	117.952

We can also use the expenditure matrix to calculate the chained price indices, but now we use its adjacent cells in a stepwise fashion.

Chained Laspeyres price index

The chained Laspeyres price index uses the previous year's quantities as weights.

- The denominator for a given year is the diagonal value from the previous year.
- The numerator's value is in the cell below the denominator.

Expenditure matrix for cruciferous vegetables (millions of dollars)							Laspeyreschained (1996) price index for cruciferous vegetables		
	Total	Quantity						Excel formula	Price index value
		1996	1997	1998	1999	2000			
Price 1996		**2006.14**	2224.97	2205.97	2416.13	2526.53	1996	=2006.14/2006.14	100.000
1997		2080.15	**2307.41**	2291.73	2525.86	2630.22	1997	=2080.15/2006.14	103.689
1998		2336.51	2591.73	**2573.63**	2834.63	2953.09	1998	=2591.73/2307.41	112.322
1999		2157.18	2392.74	2375.23	**2612.99**	2724.36	1999	=2375.23/2573.63	92.291
2000		2358.48	2616.41	2601.61	2879.09	**2989.92**	2000	=2879.09/2612.99	110.184

Chained Paasche price index

The chained Paasche price index uses the current year's quantities as weights.

- The index's numerator uses the diagonal value for the given year.
- The denominator uses the value immediately above the numerator.

Expenditure matrix for cruciferous vegetables (millions of dollars)							Paasche chained (1996) price index for cruciferous vegetables		
	Total	Quantity						Excel formula	Price index value
		1996	1997	1998	1999	2000			
Price 1996		**2006.14**	2224.97	2205.97	2416.13	2526.53	1996	=2006.14/2006.14	100.000
1997		2080.15	**2307.41**	2291.73	2525.86	2630.22	1997	=2307.41/2224.97	103.705
1998		2336.51	2591.73	**2573.63**	2834.63	2953.09	1998	=2573.63/2291.73	112.301
1999		2157.18	2392.74	2375.23	**2612.99**	2724.36	1999	=2612.99/2834.63	92.181
2000		2358.48	2616.41	2601.61	2879.09	**2989.92**	2000	=2989.92/2724.36	109.748

The chained Fisher price index is the square root of the product of the chained Laspeyres and Paasche price indices:

	Chained (1996) Fisher price index for cruciferous vegetables	
	Excel formula	*Price index value*
1996	= SQRT(100.000 * 100.000)	100.000
1997	= SQRT(103.689 * 103.705)	103.697
1998	= SQRT(112.322 * 122.301)	112.311
1999	= SQRT(92.291 * 92.181)	92.236
2000	= SQRT(110.184 * 109.748)	109.966

Exercises

1. Verify the index values in Table 6.8.
2. Verify and comment on the *quantity* indices for cruciferous vegetables in Table 6.14.
3. Growing health concerns have likely affected both the prices and consumption of fruits and vegetables since 2000. Update the cruciferous vegetable data in Table 6.5 through the

most current year, and construct all six price indices for those years. Discuss how your results compare with those presented in Chapter 6.

4. Use the attached data for gasoline prices and quantities to construct a monthly composite price index for all grades of gasoline:

  (a) fixed-weight Laspeyres price index (2000 = 100) between 2000 and 2009,
  (b) fixed-weight Fisher price index (2000 = 100) between 2000 and 2009,
  (c) chain-weighted Laspeyres price index between 2000 and 2009,
  (d) chain-weighted Fisher price index between 2000 and 2009.

	U.S. conventional retail gasoline prices (cents per gallon)			U.S. gasoline volume through company outlets (thousand gallons per day)		
Date	Regular	Midgrade	Premium	Regular	Midgrade	Premium
1994	107.2	115.6	124.6	34344.3	8451.9	9968.7
1995	110.3	119.6	128.7	34331.3	9560.7	9821.9
1996	119.2	128.0	137.1	37152.5	9545.3	8413.3
1997	118.9	127.7	136.8	40533.1	9692.1	8426.3
1998	101.7	110.6	119.9	41711.9	10115.4	9613.9
1999	111.6	120.7	129.7	41594.5	9900.7	8677.4
2000	146.2	154.9	163.9	43637.4	8576.1	6911.1
2001	138.4	146.8	156.1	45036.1	8267.7	7096.2
2002	131.3	139.9	149.6	46228.3	8341.0	7521.3
2003	151.6	160.1	169.7	47333.0	7761.6	7124.2
2004	181.2	189.8	199.7	44505.7	6430.9	5914.5
2005	224.0	233.0	243.6	46360.1	5945.1	5241.6
2006	253.3	262.5	273.5	48159.5	5471.5	4932.5
2007	276.7	286.0	297.5	46981.7	4950.2	4727.5

5. Go to the BLS Consumer Price Index home page (http://www.bls.gov/cpi/) to answer the following questions.

  (a) According to the CPI's Frequently Asked Questions, reading or interpreting a price index requires that we distinguish between a percentage point change and a percentage change. Since we will follow BLS's conventions, use the following data to calculate both statistics for the change in the overall price level for the Midwest region between 2007 and 2008.

      Midwest CPI-U 2007 :   198.123

      Midwest CPI-U 2008 :   205.382

  (b) As we have noted, BLS constructs price indices for many consumer goods and services. We can use these indices to compare changes in the relative price levels. To retrieve the data for such comparisons, on the home page mentioned, go to the "Create Customized Tables" (multiple screens) under **Get Detailed CPI Statistics** to retrieve seasonally adjusted Consumer Price Index-All Urban Consumers for the following:

      All items from 1978 to 2010
      College tuition and fees from 1978 to 2010

    i. Calculate the annual average index for each year.

    ii. Discuss the relative price level changes for the two indices. You may find constructing a chart useful for the comparisons.

6. UNDP has recently made available 40 years of data from their *Human Development Reports* free of charge online. Go to the *Indicators* link (http://hdr.stats.undp.org/en/indicators/default.html), and construct the three dimension indices and the Human Development Index for the following countries: Djibouti, Japan, Mali, and Zimbabwe. Compare your results with those in Table 6.15, and comment on the relative ranking of the countries.

7 Descriptions of stability

Short-run changes

7.1 Measuring economic change over time

Observers of economies have long been interested in how economic variables change over time. Stock market analysts, policy-makers, business owners, and consumers all carefully watch the daily, weekly, or monthly changes in gross domestic product, employment levels, and interest rates for signs of economic health or illness. Annual changes in variables also garner considerable attention, often serving as yardsticks of economic progress, while cyclical changes worry policy-makers and frustrate economic forecasters. And since the rise of modern economics, economists have tried to understand those factors that account for the long-run trends in economic growth which characterize most economies. Given these interests, we now consider how economists measure change over various time periods.

We will first examine quantitative methods used to empirically estimate short-term changes in economic variables. As previously discussed in Chapter 6, index numbers provide one approach for assessing the change in an economic variable relative to some base period. We shall now examine additional methods used to measure short-run changes in a variable.

Table 7.1 contains seven short-run growth calculations for real GDP (columns (3) through (9)). These values, obtained from the St. Louis Federal Reserve's FRED® (Federal Reserve Economic Data) database, will comprise the range of short-run growth rates that we might compute for a particular economic variable. (See Box 7.1 at the end of this chapter for details about where to find FRED®.) As we explain how each rate is calculated, we will also discuss under what circumstances the particular growth rate is most appropriate for understanding short-run movement in a variable.

As we see in column (2) of Table 7.1, fourth-quarter real GDP in 2009 equaled $13,019,010,000,000. When compared to the previous quarter's real GDP, we observe an increase of $158,212,000,000, as reported in column (3). We further observe that in the fourth quarter of 2009, real GDP was $25,347,000,000 lower than it was in the fourth quarter of 2008, a change reported in column (4).

We have previously discussed how, for most of us, units in the trillions of dollars are not immediately meaningful values, and, even if they were, they are cumbersome to manipulate mathematically. Consequently, economists typically rely on rates, rather than units, of change when assessing how an economic variable varies over time.

7.2 Calculating percentage growth

Let us first re-examine the most easily understood and frequently presented type of change in an economic variable, the **per-period percentage change**, also known as **percentage growth**. As illustrated in equation (7.1), we calculate this growth rate by constructing a ratio

Table 7.1 Growth rates for real (chained 2005 dollars) gross domestic product first quarter 2008–second quarter 2010

Year, quarter (1)	Billions of chained 2005 dollars (2)	Change, billions of chained 2005 dollars (3)	Change from year ago, billions of chained 2005 dollars (4)	Percent change (5)	Percent change from year ago (6)	Compounded annual rate of change (7)	Continuously compounded rate of change (8)	Continuously compounded annual chained 2005 change (9)	Natural log of billions of dollars (10)
2007-10-01	13,363.488	–	–	–	–	–	–	–	9.5003
2008-01-01	13,339.175	−24.313	249.859	−0.182	1.909	−0.726	−0.182	−0.728	9.4985
2008-04-01	13,359.046	19.871	164.898	0.149	1.250	0.597	0.149	0.595	9.4999
2008-07-01	13,223.507	−135.539	−44.951	−1.015	−0.339	−3.997	−1.020	−4.079	9.4898
2008-10-01	12,993.665	−229.842	−369.823	−1.738	−2.767	−6.773	−1.753	−7.014	9.4722
2009-01-01	12,832.619	−161.046	−506.556	−1.239	−3.798	−4.866	−1.247	−4.989	9.4597
2009-04-01	12,810.012	−22.607	−549.034	−0.176	−4.110	−0.703	−0.176	−0.705	9.4580
2009-07-01	12,860.800	50.788	−362.707	0.396	−2.743	1.595	0.396	1.583	9.4619
2009-10-01	13,019.012	158.212	25.347	1.230	0.195	5.012	1.223	4.891	9.4742
2010-01-01	13,138.832	119.820	306.213	0.920	2.386	3.733	0.916	3.665	9.4833
2010-04-01	13,194.862	77.706	406.526	0.591	3.174	2.387	0.590	2.359	9.4892

Data source: FRED, Federal Reserve Economic Data, Federal Reserve Bank of St. Louis (2010a), Real Gross Domestic Product, 3 Decimal (GDPC96).

of a variable's final-period value divided by the variable's initial-period value:

$$\text{percentage growth} = \left[\left(\frac{\text{final} - \text{year value}}{\text{initial} - \text{year value}} \right) - 1 \right] \times 100 \tag{7.1}$$

We again note the conventions of multiplying the bracketed term by 100 and omitting the percentage sign (%) when reporting percentage growth.

Let us now calculate the percentage growth in real GDP, or **quarter-to-quarter growth rate**, between the third and fourth quarters of 2009 using data from column (2) in Table 7.1:

$$\text{quarter-to-quarter growth rate} = \left[\left(\frac{13,019.01}{12,860.80} \right) - 1 \right] \times 100 = 1.230$$

or 1.23 percent. Accordingly, we determine that real GDP rose by 1.23 percent between the third and fourth quarters of 2009. As we see, this result equals the **Percent Change** value from FRED® presented in column (5) of Table 7.1. To express the result as a rate, we say that real U.S. GDP grew at an average *quarterly* rate of 1.23 percent between the third and fourth quarters of 2009.

It is important that we specify both the type of periodic rate (here, quarterly) and the time frame (here, between the third and fourth quarters of 2009) when reporting results, because the growth rate is a relative measure. That is, we determine the growth in the final-period value *relative* to the initial-period value.

Table 7.1 suggests another percentage change of interest to economists, that is, how a variable differs in value from the same period a year earlier, what FRED® labels as the **Percent Change from Year Ago**. Agencies such as the U.S. Bureau of Economic Analysis use this statistic to eliminate seasonal patterns from non-seasonally adjusted data (which we will discuss further in Chapter 8). The Percent Change from Year Ago statistic may also be preferred when analyzing volatile economic series.

Let us consider one application of the Percent Change from Year Ago. According to the National Bureau of Economic Research (NBER), the U.S. economy officially moved into a recession in December 2007, leading us to suspect that real GDP declined between the fourth quarter of 2007 and the fourth quarter of 2008. Using equation (7.1) and the initial-period value for 1 October 2007 (the first day of the fourth quarter), we find that real GDP did decline, by 2.77 percent, over the 12-month period between the fourth quarter of 2007 and the fourth quarter of 2008:

$$\text{Percent Change from Year Ago} = \left[\left(\frac{12,993.67}{13,363.49} \right) - 1 \right] \times 100 = -2.767$$

We can now generalize the above formula to determine what we will call the **year-to-year growth rate**. When working with quarterly data, denoted by Q_t, the year-to-year growth rate is calculated using equation (7.2):

$$\text{year-to-year growth rate} = \left[\left(\frac{Q_t}{Q_{t-4}} \right) - 1 \right] \times 100 \tag{7.2}$$

The year-to-year growth rate using monthly data, denoted by M_t, is calculated by equation (7.3):

$$\text{year-to-year growth rate} = \left[\left(\frac{M_t}{M_{t-12}}\right) - 1\right] \times 100 \tag{7.3}$$

Economists often utilize percentage growth rates from sub-annual periods such as quarters or months to determine annualized rates of growth. **Annualized growth rates** indicate the value of an economic variable that would obtain if the quarter-to-quarter (or month-to-month) growth rate was maintained over a full year. These rates are particularly valuable when we want to compare growth in variables that are measured across different time periods (for example, months and quarters).

Let us now demonstrate how annualized growth rates are calculated. Suppose we know only the quarter-to-quarter growth rate in real GDP between the third and fourth quarters of 2008, but we want to know the change in real GDP between the fourth quarter of 2008 and the fourth quarter of 2009. Our first impulse might be to take the quarterly result, and multiply it by four. However, doing this ignores a fundamental feature inherent in most economic variables, including real GDP. That is, the value of GDP in the fourth quarter of 2009 depends not only on the value of GDP in the previous quarter but on the values of all quarters before the last one as well. Consequently, we must calculate the annualized growth rate as shown in Table 7.2.

In Table 7.1, we have already determined that real GDP grew at a -1.738 percent quarterly rate between the third and fourth quarters of 2008. Let us now apply that rate to the value of real GDP at the beginning of the fourth quarter, $12,993.665 billion. Here we find that GDP will equal $12,832.619 billion at the beginning of the first quarter of 2009 (which we will denote as 2009-Q1). (Note that we must divide the percentage change rate presented in Table 7.1 by 100 to convert it back into a ratio.)

We next apply the -1.738 percent rate of growth to the newly calculated (not actual) 2009-Q1 value of real GDP. This yields a 2009-Q2 value for real GDP of $12,810.012 billion. If we continue to apply the -1.738 constant quarterly growth rate to the calculated value of real GDP until we reach the fourth quarter of 2009, we obtain a 2009-Q4 value of real GDP of $12,113,558 billion.

We can now use this 2009-Q4 value of real GDP to determine the annualized rate of growth in real GDP between 2008-Q4 and 2009-Q4:

$$\text{annualized growth rate} = \left[\left(\frac{\$12,113,558}{\$12,993.665}\right) - 1\right] \times 100 = -6.773$$

Table 7.2 Annualized growth in 2008 real GDP

	Real GDP (billions of chained 2005 dollars)	Quarterly percentage change (r_q)	Real GDP growing at a constant discrete quarterly rate of -1.738 percent
2008-07-01	13,223.507 ⎫		
2008-10-01	12,993.665 ⎭	-1.738	
2009-01-01			12,767.818
2009-04-01			12,545.896
2009-07-01			12,327.832
2009-10-01			12,113.558

Data source: Author's calculations based on Table 7.1.

According to this result, real GDP grew at an annualized rate of -6.773 percent between the fourth quarters of 2008 and 2009 based on the growth in real GDP between the third and fourth quarters of 2008.

Determining the annualized rate of growth acknowledges that a variable's value depends on more than its value in the previous period, reflecting the **multiplicative process** by which many economic variables change over time. In slightly different terms, this means that the value of such economic variables is a compounded result of its past values, or that the variable's growth is compounded over time. To explore this important concept of **compound growth**, let us shift our focus temporarily from macroeconomic indicators to financial instruments and a lump-sum investment in an interest-bearing account.

7.3   Compound growth

Most of us have a basic understanding of how an investment accrues interest over time. Suppose, for example, we have $1,200 that we want to deposit into an interest-bearing account with an annual interest rate of 3 percent. Because the investment accrues interest over that first year, we know we will have a balance greater than $1,200 by the end of the year (which is, of course, why we place our money into the account rather than in a mattress!). But how do we determine how much interest our investment will earn during the first year?

Based on the 3 percent annual interest rate, we know the value of our investment at the end of year one will equal the initial $1,200 deposit plus the 3 percent accrued interest on that amount. That is,

$$P_1 = \$1,200 + (0.03 \times \$1,200) = \$1,236.00$$

Accordingly, at the beginning of year two, our investment is worth $1,236, which now becomes the amount on which the interest accrued in year two is based:

$$P_2 = \$1,236 + (0.03 \times \$1,236) = \$1,273.08$$

By the end of year two, then, the investment will equal $1,273.08.

Instead of calculating the investment's value at the end of each year, we can also determine its value as follows:

$$P_2 = \$1,200 \times (1+0.03) \times (1+0.03) = \$1,200 \times (1+0.03)^2 = \$1,273.08$$

As this second approach suggests, we can derive an efficient formula for calculating the value of the investment at the end of a specified future time period, n:

$$P_n = P_0 \times (1+r)^n \tag{7.4}$$

where P_n is the value of the investment at the end of n years (and is also known as the investment's **future value**), P_0 is the original amount invested (also known as its **present value**), r is the rate of interest earned (also known as the investment's **rate of return**), and n is the number of time periods over which the investment accrues interest.

Let us now apply equation (7.4) to determine the future value of our investment in, say, five years:

$$P_n = \$1,200 \times (1+0.03)^5 = \$1,200 \times 1.159 = \$1,391.13$$

As we see, the initial $1,200 investment will have earned $191.13 after five years if it grows at a 3 percent interest rate compounded annually at the end of each year.

Economists use the concept of compound growth, along with the current, or present, observed value of a variable and its observed final, or future, value to determine the *constant per-period rate* at which the economic variable changes. To derive a formula for calculating this **compound rate of growth** (r), we solve equation (7.4) for r by first dividing both sides of equation (7.4) by P_0:

$$(P_n/P_0) = (1+r)^n$$

To isolate r from the exponent n, we take the natural logarithm of both sides,

$$\ln(P_n/P_0) = n\ln(1+r)$$

and then divide each side by n, to give

$$[\ln(P_n/P_0)]/n = \ln(1+r)$$

Solving for r requires that we take the antilog of both sides of the above equation[1] such that

$$e^{\{[\ln(P_n/P_0)]/n\}} = 1+r$$

Once we rearrange terms, we obtain the **per-period compound growth rate** formula presented as equation (7.5):

$$r = [e^{\{[\ln(P_n/P_0)]/n\}} - 1] \times 100 \qquad (7.5)$$

Since growth rates are typically reported as percentages, we multiply the calculated value by 100.

We can now use equation (7.5) to determine the compounded annual rate of interest required for an initial investment of $1,200 to be worth a particular value, say $1,500, at the end of five years:

$$r = [e^{\{[\ln(1,500/1,200)]/5\}} - 1] \times 100 = [e^{0.045} - 1] \times 100 = 0.0456 \times 100 = 4.56.$$

To have $1,500 at the end of five years on an initial investment of $1,200, we would need to earn a compounded annual rate of 4.56 percent.

What happens to the rate of return needed for a specific future value if we change the period over which the investment is compounded? Using the above example, suppose we find a financial instrument that offers an interest rate that is compounded quarterly rather than annually. In other words, how does more frequent compounding affect the interest rate we must earn on $1,200 to have $1,500 at the end of five years?

Before we do the actual calculations, we can intuit that the quarterly rate will be less than the annual rate given the principles of compounding. We first remember that each period's value builds on the values in all previous periods. Thus, here, interest will now be added to the investment at the end of each quarter rather each year. As a result, our investment will reach the target $1,500 more quickly when interest compounds quarterly than when it compounds annually, making the required quarterly rate of return lower than the required annual rate.

How much lower than the annual rate will that quarterly rate be? Because of the compounding process, we cannot simply divide the annual rate by 4 to obtain a quarterly rate. In fact, the quarterly rate will be *less than one-quarter* the value of the annual rate. To demonstrate this result, let us determine the quarterly rate required to take $1,200 to a future value of $1,500 over the 20 quarters in five years:

$$r = [e^{\{[\ln(1,500/1,200)]/20\}} - 1] \times 100 = [e^{0.0112} - 1] \times 100 = 0.0112 \times 100 = 1.12.$$

Because our investment accrues interest each quarter rather than each year, we determine that the necessary quarterly rate is 1.12 percent. We also realize that this rate is less than the 1.14 percent that is one-fourth of the annual 4.56 percent rate.

Before we apply the concept of compound growth further, let us note two key assumptions about the compound growth rate formula defined by equation (7.5). The first is that the formula generates a constant, or fixed, rate (r) of growth over time. While this may not be a realistic assumption about how an economic variable changes over the long run, economists consider it to be reasonable for calculating change over short periods of time.

The second assumption relates to how frequently the variable is compounded. When we apply equation (7.5), we are assuming that the variable grows discretely. This means we assume that the variable is compounded at fixed points in time – in other words, it is subject to **discrete compounding** – rather than being continuously or instantaneously compounded. In the investment example, we first assumed that the interest earned on the $1,200 investment compounded at the end of the year, and then we assumed interest was compounded each quarter. We will reconsider if this assumption of discrete compounding is reasonable for all economic variables later in the chapter.

7.4   Annualized growth rates from sub-annual rates

Let us now consider the process of compound growth as it relates to the real GDP data presented in Table 7.1. We begin by applying equation (7.5) to the 2008 third and fourth quarter values of real GDP to determine the quarter-to-quarter compound growth rate for the fourth quarter of 2008:

$$r = [e^{\{[\ln(12,993.67/13,223.51)]/1\}} - 1] \times 100$$
$$= [e^{-0.0175} - 1] \times 100 = [0.9826 - 1] \times 100 = -1.738$$

We see that this result is identical in value to the Percent Change calculation in column (5) of Table 7.1. In fact, it can be shown that the formula for the percentage change over one period (equation (7.1)) is the same as the per-period compound growth rate formula (equation (7.4)) when $n = 1$. In less technical terms, we see that the percentage change in a variable equals the per-period compound rate of change as we move from one period to the next.

Now suppose we want to annualize this quarterly growth rate to account for the compound growth process. As shown in Table 7.2, applying the -1.738 percent quarter-to-quarter growth rate to the value of real GDP in the fourth quarter of 2008 over the next four quarters yields an annualized rate of change equal to -6.773 percent. That is,

$$\text{fourth-quarter 2009 real GDP} = \text{fourth-quarter 2008 real GDP} \times \{[1 + (-0.01738)]^4\}$$

which tells us that the annualized quarterly growth rate is $[1 + (-0.01738)]^4$. Thus we can calculate the annualized rate by applying equation (7.6):

$$APR = [(1 + r_q)^4 - 1] \times 100 \qquad (7.6)$$

where r_q is the quarterly growth rate obtained from either equation (7.1) or equation (7.5). This annualized quarterly growth rate is more commonly known as the **annualized percentage (growth) rate**, or **APR**, an acronym familiar to anyone with a credit card or car loan. This rate is called the **Compounded Annual Rate of Change** by FRED®.

Let us now apply the APR formula using real GDP's fourth-quarter 2008 quarterly growth rate. Here we obtain the same result as before:

$$APR = \{[1 + (-0.01738)]^4 - 1\} \times 100 = -6.773$$

We further observe that this rate equals that found in column (7) of Table 7.1 for 2008-Q4.

Annualized growth rates can be calculated from any sub-annual rate such as monthly, weekly, or daily rates. We can thus generalize equation (7.6) for any sub-annual rate:

$$APR = [(1 + r_p)^{\text{number of periods per year } (p)} - 1] \times 100 \qquad (7.7)$$

where r_p is the sub-annual periodic rate and "p" refers to the periodicity of the data. If, for example, we wanted to annualize monthly rates (r_m) that compound 12 times over the course of a year, the corresponding APR formula would be

$$APR = [(1 + r_m)^{12} - 1] \times 100$$

while the weekly APR formula is

$$APR = [(1 + r_w)^{52} - 1] \times 100$$

When annualizing sub-annual rates, we are, in essence, rescaling those sub-annual growth rates to a yearly base. As we know, rescaling a variable to a common reference point, or base, facilitates comparisons of variables that are measured in different units or over different time periods. Since not all economic variables are measured over the same period of time, annualizing their sub-annual rates of change allows economists to compare changes in the variables. For example, if we wanted to compare trends in the Consumer Price Index, which is calculated monthly, and real GDP, which is measured quarterly, annualizing the rates of change for both variables will allow us to discuss comparable changes in the two variables.

Let us now note several important characteristics of the compound growth rates calculated from equations (7.5)–(7.7):

- Implicit in the calculations is that the variable grows discretely and at a constant per-period rate.
- All sub-annual and annualized rates are *average* growth rates for the specified time period.
- All annualized growth rates depend solely on the per-period growth rates used in their calculation. This means that we can obtain widely varying APRs for any given year, and that irregularities in the data will be magnified, as we see in Chart 7.1 (p. 200).

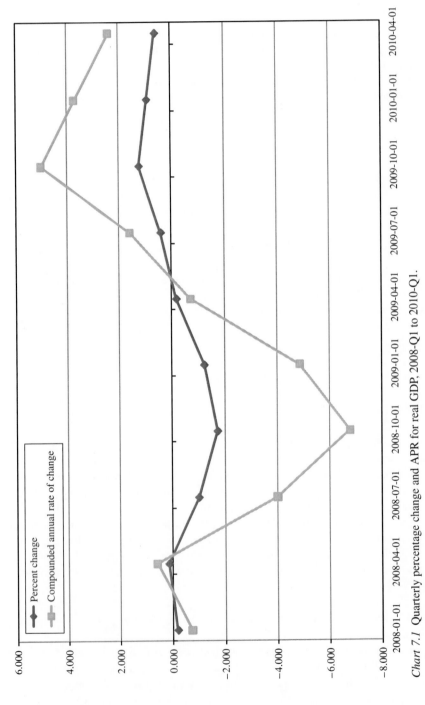

Chart 7.1 Quarterly percentage change and APR for real GDP, 2008-Q1 to 2010-Q1.

Data source: Table 7.1.

- Not all economic series lend themselves to annualization. The U.S. Bureau of Economic Analysis (2010b), for example, does not calculate annualized growth rates for corporate profits (measured quarterly) or personal income and outlays (measured monthly) because of the considerable volatility associated with these variables. Instead BEA only publishes quarter-to-quarter and month-to-month rates for these series.
- The difference between the per-period growth rate and the corresponding APR value becomes larger as the sub-annual time period gets smaller. If we move from quarterly to monthly per-period rates, for example, we would observe a larger difference between the monthly rate and its corresponding annualized rate than we would observe between a quarterly rate and its corresponding APR.
- None of the discrete growth rates consider intermediate values of the variable in their calculation. As we established in Figure 4.1, this is not a desirable property for a descriptive statistic, which helps to explain why these growth rates are not typically applied to calculating change over long time periods. (We will discuss more appropriate statistics for determining long-run change in Chapter 8.)

7.5 Annualized growth rates from supra-annual rates

Let us now examine one additional use of annualized growth rates – converting growth rates for periods *longer than a year*, or **supra-annual growth rates**, into annualized rates. Economists sometimes must work with variables that are measured only every five or ten years. In such cases, it can be useful to annualize such series, particularly for comparison purposes. Fortunately, we can modify equation (7.5) to determine annualized growth rates from periods longer than one year.

Table 7.3 and Chart 7.2 (p. 202) present decennial census data of the number of people living in the United States since the first Census was conducted in 1790. As we see, the U.S. population has increased, usually by increasing amounts. While these increases have not been at a constant rate (factors such as economic depressions, waves of immigration, and wars, as well as social attitudes toward population growth, have all contributed to a more complex process), a major portion of the change in the U.S. population between 1790 and 2000 results from a steady multiplicative process by which each generation is larger than the preceding one. It is thus reasonable to describe that growth using a compounding process.

Suppose we are interested in the average annual rate of population growth over the 210-year period. Recognizing that *n* now equals 210, we can apply equation (7.5) to determine that annual rate:

$$r = \left[e^{[\ln(281,422,509/3,929,214)/210]} - 1 \right] \times 100 = 0.02055 \times 100 = 2.06.$$

We find that the U.S. population grew at an average annual rate of 2.06 percent between 1790 and 2000.

Determining the average rate over the entire time frame might, however, obscure differences in the population's growth between sub-periods in U.S. history. As we see in Table 7.3, decennial increases in the U.S. population dropped below 25 percent starting in 1900. This suggests we might calculate the population growth rate for the period between 1790 and 1900, when it appears the U.S. population grew more rapidly, and the apparently slower post-1900 rate.

Applying equation (7.5) to obtain the average annual growth rates for the 110-year and 100-year periods, we determine that the U.S. population grew at an average annual growth

Table 7.3 Decennial growth in the U.S. population since 1790

Census year	Total population	Increase over preceding census		Census year	Total population	Increase over preceding census	
		Number	Percentage			Number	Percentage
1790	3,929,214			1900	76,212,168	13,232,402	21.0
1800	5,308,483	1,379,269	35.1	1910	92,228,496	16,016,328	21.0
1810	7,239,881	1,931,398	36.4	1920	106,021,537	13,793,041	15.0
1820	9,638,453	2,398,572	33.1	1930	123,202,624	17,181,087	16.2
1830	12,866,020	3,227,567	33.5	1940	132,164,569	8,961,945	7.3
1840	17,069,453	4,203,433	32.7	1950	151,325,798	19,161,229	14.5
1850	23,191,876	6,122,423	35.9	1960	179,323,175	27,997,377	18.5
1860	31,443,321	8,251,445	35.6	1970	203,302,031	23,978,856	13.4
1870	39,818,449	8,375,128	26.6	1980	226,542,199	23,240,168	11.4
1880	50,189,209	10,370,760	26.0	1990	248,718,302	22,176,103	9.8
1890	62,979,766	12,790,557	25.5	2000	281,422,509	32,704,207	13.1

Data source: U.S. Census Bureau (2010f), Statistical Abstract of the United States: 2009, No. 1 – Population and Area: 1790 to 2000.

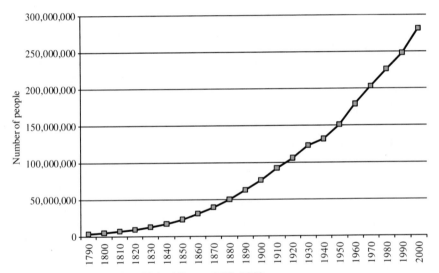

Chart 7.2 Population of the United States, 1790–2000.

Data source: Table 7.3.

rate of 2.73 percent between 1790 and 1900, while the population grew at a slower rate of 1.32 percent between 1900 and 2000. Such results may then lead us to consider the possible reasons for the decline, including changes in U.S. immigration policy and global conditions. In fact, we might want to examine even smaller sub-periods to identify other interesting population trends. Regardless of how we define the time frame, we can use equation (7.5) to calculate discrete compound growth rates for any economic variable.

7.6 Continuous compound growth

But is it realistic to think that a population grows discretely? Every second of every day children are born, the elderly die, and people emigrate out of or immigrate into a geographical region. This suggests that population growth may be better described by a process of continuous, rather than discrete, compounding.

To explain the process of **continuous compounding**, let us return to equation (7.4), where we calculated the future value of an investment earning a discretely compounded interest rate over the life of the investment:

$$P_n = P_0 \times (1+r)^n \tag{7.4}$$

Suppose we assume that the per-period interest rate, now written as r_p, is the inverse of the number of compounding periods in a given year, p, that is,

$$r_p = \frac{1}{p}$$

and that r_p is compounded at the end of each period, p. We can now rewrite equation (7.4) as equation (7.4*):

$$P_n = P_0 \times \left(1+\frac{1}{p}\right)^p \tag{7.4*}$$

Let us apply equation (7.4*) to determine the value of an initial $1,000 investment as we increase the number of compounding periods in a year, beginning with the investment's principal compounded once, at the end of the year:

$$P_n = \$1,000 \times \left(1+\frac{1}{1}\right)^1 = \$2,000.00$$

The investment will have a value of $2,000 at the end of the year.

Now suppose that interest on the investment is compounded monthly, such that p equals 12:

$$P_{12} = \$1,000 \times \left(1+\frac{1}{12}\right)^{12} = \$2,613.04$$

Monthly compounding at a rate of 1/12 yields an end-of-year value of $2,613.04, which is larger than the investment's end-of-year value with annual compounding.

As shown in Table 7.4, we find that, if we continue to increase the frequency at which interest is compounded, the final period value of the investment grows larger, but it does so at a decreasing rate. This suggests that eventually the final period value of the investment will converge to a specific value. We can actually determine the exact amount by recognizing that

$$\lim_{p \to \infty} \left(1+\frac{1}{p}\right)^p = e$$

where e is a mathematical constant (the base of natural logarithms, with a truncated value equal to 2.71828). Given this property, we can view e as the base rate of growth shared by all continuously growing variables.

Table 7.4 End-of-year value of $1,000 investment compounded over different sub-annual periods

Compounding periods	Period (p) equation (7.4*)	End-of-year value of a $1,000 investment
Year	$1 = \$1{,}000 \times \left(1 + \dfrac{1}{1}\right)^{1}$	$2,000.0000
Month	$12 = \$1{,}000 \times \left(1 + \dfrac{1}{12}\right)^{12}$	$2,613.0353
Week	$52 = \$1{,}000 \times \left(1 + \dfrac{1}{52}\right)^{52}$	$2,692.5970
Day	$365 = \$1{,}000 \times \left(1 + \dfrac{1}{365}\right)^{365}$	$2,714.5675
Hour	$8{,}760 = \$1{,}000 \times \left(1 + \dfrac{1}{8{,}760}\right)^{8{,}760}$	$2,718.1267
Minute	$525{,}600 = \$1{,}000 \times \left(1 + \dfrac{1}{525{,}600}\right)^{525{,}600}$	$2,718.2792
Second	$31{,}536{,}000 = \$1{,}000 \times \left(1 + \dfrac{1}{31{,}536{,}000}\right)^{31{,}536{,}000}$	$2,718.2818

The exponential e essentially rescales all continuously growing variables to a common annual rate. Thus we can combine equation (7.4*) with the above property to determine that our investment will approach an end-of-one-year value equal to its initial value times e raised to the interest rate earned:

$$P_1 = P_0 \times e^r$$

Because most investments are not cashed in at the end of only one year, let us also add a time dimension to our formula such that:

$$P_t = P_0 \times e^{rt} \tag{7.8}$$

where t equals the time period over which the investment accrues interest. Thus equation (7.8) calculates the future value of an investment with an initial value of P_0 if the interest rate r accrues *continuously* over t time periods.

As we did for the discrete compounded growth rate, we can solve equation (7.8) for the continuous rate of growth, r. Doing so yields the formula for the **continuously compounded growth rate** (or **exponential growth rate**), what FRED® calls the **Continuously Compounded Rate of Change**:

$$r = \left(\frac{\ln(P_t / P_0)}{t}\right) \times 100 \tag{7.9}$$

Equation (7.9)[2] looks quite similar to equation (7.5), but there is a key difference – the frequency with which compounding occurs. Given that compounding here is instantaneous, the growth rate we calculate using equation (7.9) will *always* be smaller than the discrete compounded growth rate from equation (7.5).

Let us now return to the U.S. population data from Table 7.3 to demonstrate this result. We previously determined that the population grew discretely at a constant 2.06 percent rate of growth between 1790 and 2000. If we now apply equation (7.9), we find that the U.S. population grew at a continuously compounded growth rate of

$$r = \left(\frac{\ln(281,422,509/3,929,214)}{210} \right) \times 100 = 2.034 \text{ percent}$$

between 1790 and 2000. As we see, the continuously compounded rate is smaller than the discretely compounded rate by 0.021 percentage points.

We can similarly calculate a continuously compounded quarterly rate for real GDP in 2008-Q4 using data from Table 7.1:

$$r = \left(\frac{\ln(12,993.67/13,223.51)}{1} \right) \times 100 = -1.753$$

Real GDP thus grew at a continuously compounded rate of -1.753 percent between the third and fourth quarters of 2008, which corresponds to the result obtained from FRED® in column (8) of Table 7.1. We also observe that the exponential growth rate of -1.753 percent is, in fact, smaller than the discretely compounded growth rate of -1.738 percent obtained from equation (7.5).

We must now ask which compound growth rate, discrete or continuous, better describes the growth process underlying the population and real GDP variables. The answer depends on two considerations – which type more realistically describes how the variable actually changes over time, and which type is consistent with how the variable is measured.

Theoretically, most economic variables, including population and real GDP, change continuously. In the case of real GDP, for example, the multiplier process ensures constant change in the dollar value of goods and services in the economy, suggesting a continuous compounding process. However, most economic variables are measured only at intervals, such as months or quarters, which suggests that the discrete compound growth rate is more appropriate. In practice, organizations like the World Bank and the Organisation for Economic Co-operation and Development (OECD) use the continuous (per-period) compound rate to describe indicators such as the labor force and population, and the discrete rate to describe all other economic variables, while most U.S. federal agencies publish discrete (per-period) compound growth rates for all variables. Regardless of which conventions we follow, it is important to recognize that these growth rates are calculated using only two values of the variable, those from the initial and final periods. Thus neither the discrete nor the continuous growth rates presented in Chapter 7 account for intermediate values of the variable, an undesirable property for descriptive statistics. However, when examining change over short periods of time, these statistics are generally considered by economists to be sufficient for our purposes.

7.7 Continuously compounded annual growth and logarithms

Let us now consider one final short-run growth rate, an annualized rate based on the continuously compounded rate in equation (7.9). We derive its formula in Table 7.5, since the underlying process of change may appear counterintuitive.

As in Table 7.2, we first determine the continuous quarterly rate of change between 2008-Q3 and 2008-Q4, which we calculated to be -1.753 percent. We next assume that this growth

Table 7.5 Annualized continuous growth in 2008 real GDP

	Real GDP (billions of chained 2005 dollars)	Quarterly continuous rate (r_q)	Real GDP growing at constant continuous quarterly rate of -1.753 percent	ln(GDP)
2008-07-01	13,223.51 ⎫			
2008-10-01	12,993.67 ⎭	-1.753	12,993.665	9.47222
2009-01-01			12,767.818	9.45468
2009-04-01			12,545.896	9.43715
2009-07-01			12,327.832	9.41961
2009-10-01			12,113.558	9.40208

Data source: Author's calculations based on Table 7.1.

rate holds constant over the next four quarters and apply it to real GDP. As we see in column (4), 2009-Q4 real GDP will equal $12,113.558 billion. Because we assume that GDP grows continuously, we then take the natural logarithm of each quarterly value of GDP. The last step is to apply equation (7.9) using 2009-Q4 (P_t) and 2008-Q4 GDP (P_0) to obtain an annualized continuous growth rate of -7.014 percent. (The relevant time frame, t, in this case is one, since we are determining an annualized rate.)

We could have obtained this annualized rate more efficiently by recognizing that it equals the quarterly rate, -1.753, times the four quarters between 2008-Q4 and 2009-Q4. Thus to obtain the **continuously compounded annual growth rate** (CCAGR), we can apply equation (7.10):

$$CCAGR = \{[\ln(X_t) - \ln(X_{t-1})] \times 100\} \times p \qquad (7.10)$$

where p is the periodicity of observations in a year. FRED® calls this rate the **Continuously Compounded Annual Rate of Change**.

We should not be surprised to learn that the CCAGR is smaller in value than the APR reported in column (7) of Table 7.1. Thus both per-period and annualized continuously compounding growth rates will, by virtue of their more frequent compounding, be smaller than their discretely compounded counterparts.

The final column presented in Table 7.1 reports the natural logarithm of the real GDP values in column (2). As we have seen, several of the short-run growth rate formulas utilize natural logarithms in their calculation, so having these values readily available can facilitate our calculations.

The natural logarithm values in column (10) of Table 7.1 serve another useful purpose. Natural logarithms rescale the actual values of a variable to approximate the percentage change in the variable from one period to the next. If we then construct a time-series chart of the natural log values and fit a straight line to those values, the *slope* of the resulting line equals the average per-period percentage growth rate in the variable.

Chart 7.3 (p. 207) displays time series of the natural log of real GDP for the longest period of economic expansion in the United States from April 1991 through February 2001. We observe that the natural log of real GDP increased fairly constantly over the time frame. As a result, it appears that the assumption of constant growth implicit in the short-run growth rate formulas presented in Chapter 7 may, in fact, be appropriate for describing the overall change in real GDP during the expansion.

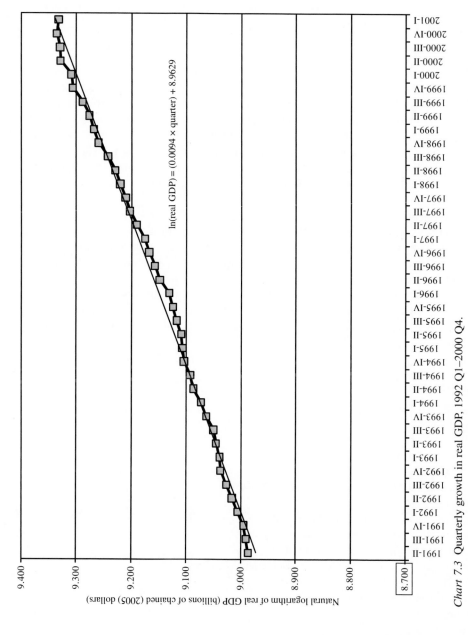

Chart 7.3 Quarterly growth in real GDP, 1992 Q1–2000 Q4.

Data source: U.S. Bureau of Economic Analysis (2010c), National Economic Accounts, Table 1.1.6, Real gross domestic product, chained dollars.

ln(real GDP) = (0.0094 × quarter) + 8.9629

Chart 7.3 also includes a linear trendline that has been fitted to the time-series data using the accompanying equation. According to these results, real GDP grew at a constant average quarterly rate of 0.94 percent between the second quarter of 1991 and the first quarter of 2001. As we shall explain in the following chapter, this method for calculating economic growth is more appropriate for the longer time period presented here. Thus the natural logarithm of a variable's values are not only useful for determining short-run growth rates, but also can be used to calculate constant long-run growth rates.

Summary

Assessing the change in economic variables over time is the focus of much economic analysis, so economists have developed several methods for describing the different types of change we might observe. The basic percentage growth measurement provides us with a starting point for assessing change, but we also recognize that this procedure may not most accurately capture the process by which an economic variable actually changes. Even a cursory examination of economic activity suggests that many economic variables change through a multiplicative process, which requires us to use statistics that account for the compounding inherent in such processes.

Economists also frequently want to compare growth rates across economic variables that are measured over different periods of time. The annualization method introduced here facilitates such comparisons, although volatile economic series may not lend themselves to this technique. Finally, we have examined how the frequency with which an economic variable grows affects our calculations of growth, highlighting the distinction between variables that grow at discrete rates versus those that grow at continuous rates over time. As we shall see in Chapter 8, this distinction, along with assumptions made about change being by constant amounts or constant rates, will inform which methods we use to measure the long-run growth of economic variables.

Concepts introduced

Annualized growth rate
Annualized percentage (growth) rate
APR
Compound growth
Compound rate of growth
Compounded Annual Rate of Change (FRED®)
Continuous compounding
Continuously Compounded Annual Rate of Change (FRED®)
Continuously compounded annual growth rate
Continuously compounded growth rate
Continuously Compounded Rate of Change (FRED®)
Discrete compounding
Exponential growth rate
Future value
Multiplicative process
Per-period compound growth rate
Per-period percentage change
Percent Change (FRED®)

Percent Change from Year Ago (FRED®)
Percentage growth
Present value
Quarter-to-quarter growth rate
Rate of return
Supra-annual growth rates
Year-to-year growth rate

Box 7.1 Where can I find...? FRED®

Macroeconomic time-series data is among the most readily available economic data. Most nations, particularly in the industrialized world, have governmental statistical agencies that measure and publish various indicators of economic activity, which are then used to assess the health of that nation's economy.

In the United States, the St. Louis Federal Reserve Bank collects 25,176 datasets generated by federal agencies and other institutions and makes them available online through the Federal Reserve Economic Data, or FRED® (at <http://research.stlouisfed.org/fred2/>). FRED® offers a wide variety of time-series data that can be quickly downloaded into Microsoft Excel.

FRED® also generates a range of growth rate calculations for most series. In addition, FRED® can generate time-series charts for each data series across all periods or for specified sub-periods.

FRED® classifies its data into the following 15 categories (numbers in parentheses indicate the data series available in each category):

- Banking (432)
- Business/Fiscal (453)
- Consumer Price Indexes (CPI) (28)
- Employment & Population (494)
- Exchange Rates* (95)
- Foreign Exchange Intervention* (21)
- Gross Domestic Product (GDP) and Components (232)
- Interest Rates (260)
- Monetary Aggregates (128)
- Producer Price Indexes (PPI) (18)
- Reserves and Monetary Base (148)
- U.S. Trade & International Transactions (1083)
- U.S. Financial Data (341)
- Regional Data (21,575)
- International Data (509)

Retrieving data from FRED®

Select the desired data series. This will lead you to a webpage like the following for Real Gross Domestic Product:

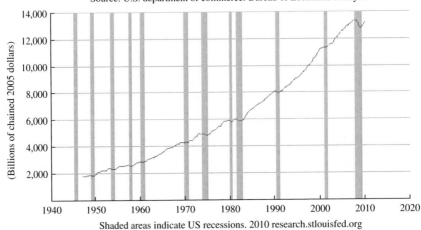

Home > Economic Data - FRED® > Categories > Gross Domestic Product (GDP) and Components > GDP/GNP > Series: GDPC96, Real Gross Domestic Product, 3 Decimal

Series: GDPC96, Real Gross Domestic Product, 3 Decimal

View Data | Download Data | Notify Me of Updates | Add to My Data List

Real gross domestic product, 3 decimal (GDPC96)
Source: U.S. department of commerce: Bureau of Economic Analysis

Shaded areas indicate US recessions. 2010 research.stlouisfed.org

Graph: Edit | Print | PDF | Save

Type: Line | Bar | Pie | Scatter   **Size:** Medium | Large | X-Large

Range: 1yr 5yrs 10yrs Max   **Recession Bars:** On | Off   **Log Scale:** Left

Units: Levels | Chg. | Chg. from Yr. Ago| % Chg. | % Chg. from Yr. Ago| Comp. Annual Rate of Chg. | Cont. Comp. Rate of Chg. | Cont. Comp. Annual Rate of Chg.

   Notes: Growth Rate Calculations | US recession dates

Latest Observations:

Date	2009:Q2	2009:Q3	2009:Q4	2010:Q1	2010:Q2
Value	12810.012	12860.800	13019.012	13138.832	13216.538

Series Properties:

Series ID:	GDPC96
Source(s):	U.S. Department of Commerce: Bureau of Economic Analysis
Release:	Gross Domestic Product
Units:	Billions of Chained 2005 Dollars
Frequency:	Quarterly
Seasonal Adjustment:	Seasonally Adjusted Annual Rate
Observation Range:	1947:Q1 to 2010:Q2

Last Updated:	2010-07-30 3:16 PM CDT
Notes:	A Guide to the National Income and Product Accounts of the United States (NIPA) - (http://www.bea.gov/national/pdf/nipaguid.pdf)

Related Categories:

Gross Domestic Product (GDP) and Components > GDP/GNP

- Under the Series: **GDPC96, Real Gross Domestic Product, 3 Decimal**, select the **Download Data** option. You will be taken to the following screen:

Download Data for Real Gross Do mestic Product, 3 Decimal (GDPC96)

Download a data file for the following series:

Series Properties:

Series ID:	GDPC96
Title:	Real Gross Domestic Product, 3 Decimal
Source(s):	U.S. Department of Commerce: Bureau of Economic Analysis
Release:	Gross Domestic Product
Units:	Billions of Chained 2005 Dollars
	Change, Billions of Chained 2005 Dollars
	Change from Year Ago, Billions of Chained 2005 Dollars
	Percent Change
	Percent Change from Year Ago
	Compounded Annual Rate of Change
	Continuously Compounded Rate of Change
	Description of growth rate formulas
Frequency:	Quarterly
Seasonal Adjustment:	Seasonally Adjusted Annual Rate
Date Range:	1947-01-01 to 2010-07-01
Last Updated:	2010-11-23 11:16 AM CST
Notes:	A Guide to the National Income and Product Accounts of the United States (NIPA) - (http://www.bea.gov/national/pdf/nipaguid.pdf)
File Format:	Excel
	Text, Space Delimited
	Text, Comma Separated

Download Data

Note: CSV files do not contain header information.

- Select **Data Range** and **File Format** (select the Excel option) you want, and click the **Download Data** button. You will get a pop-up window asking "Do you want to open or save this file?" If you select **Open**, a new worksheet containing the data will open in Excel. If you select **Save**, you will be asked for where you want the data file to be saved.

FRED® is only one source of empirical information available at the St. Louis Federal Reserve. Go to EconDISC® (at <http://econdisc.stlouisfed.org/>) for other data and materials that are available.

Box 7.2 Using Excel to calculate short-run growth rates

Excel can facilitate our calculations of the various short-run growth rates presented in Chapter 7. Using the real GDP data from Table 7.1, this section presents the formulas needed by Excel to calculate these various rates.

Calculating percentage growth

Calculating percentage growth from either the previous period or one year earlier requires two values for real GDP, at the initial and final time periods. As the following worksheet excerpt shows, once the change between the appropriate periods has been determined in columns C and D, calculating the percentage growth rates can be completed using equation (7.1):

$$\text{percentage growth} = \left[\left(\frac{\text{final} - \text{year value}}{\text{initial} - \text{year value}} \right) - 1 \right] \times 100$$

	A	B	C	D	E	F
	Year, quarter (1)	Billions of chained 2005 dollars (2)	Change, billions of chained 2005 dollars (3)	Change from year ago, billions of chained 2005 dollars (4)	Percentage growth [Percent Change in FRED] (5)	Percentage growth from one year ago [Percent Change from Year Ago in FRED] (6)
1						
2	2007-01-01	13,099.901				
3	2007-04-01	13,203.977	=B3-B2		=(C3/B2)*100	
4	2007-07-01	13,321.109	=B4-B3		=(C4/B3)*100	
5	2007-10-01	13,391.249	=B5-B4		=(C5/B4)*100	
6	2008-01-01	13,366.865	=B6-B5	=B6-B2	=(C6/B5)*100	=(D6/B2)*100
7	2008-04-01	13,415.266	=B7-B6	=B7-B3	=(C7/B6)*100	=(D7/B3)*100
8	2008-07-01	13,324.600	=B8-B7	=B8-B4	=(C8/B7)*100	=(D8/B4)*100
9	2008-10-01	13,141.920	=B9-B8	=B9-B5	=(C9/B8)*100	=(D9/B5)*100
Formula in text				equation (7.1)	equation (7.2)

Note: Do *not* include spaces in the above formulas when entering them into Excel.

Calculating discrete compound growth rates

In Section 7.4, we discussed how sub-annual growth rates can be converted into annualized rates of growth. After using Excel to calculate the percentage growth values, we can apply equation (7.6)

$$\text{APR} = [(1 + r_p)^{\text{number of periods per year}} - 1] \times 100$$

to calculate the annualized percentage (growth) rate.

	A	B	C	D
1	Year, quarter (1)	Billions of chained 2005 dollars (2)	Percentage growth [Percent Change in FRED] (5)	Annualized percentage rate [Compounded Annual Rate of Change in FRED] (7)
2	2007-01-01	13,099.901		
3	2007-04-01	13,203.977	0.7945	=(((1+(C3/100))^4)-1)*100
4	2007-07-01	13,321.109	0.8871	=(((1+(C4/100))^4)-1)*100
5	2007-10-01	13,391.249	0.5265	=(((1+(C5/100))^4)-1)*100
6	2008-01-01	13,366.865	-0.1821	=(((1+(C6/100))^4)-1)*100
7	2008-04-01	13,415.266	0.3621	=(((1+(C7/100))^4)-1)*100
8	2008-07-01	13,324.600	-0.6758	=(((1+(C8/100))^4)-1)*100
9	2008-10-01	13,141.920	-1.3710	=(((1+(C9/100))^4)-1)*100
Formula in text			equation (7.1)	equation (7.6)

Note that the exponent used in equation (7.6) and column D depends on the per-period percentage change calculated in column C. In this example, it is 4, corresponding to quarterly percentage changes. If column C's values were monthly percentage changes, the exponent in equation (7.6) and column D would equal 12.

Calculating continuously compounding growth rates

In Section 7.6, we saw how logarithms can be used to calculate continuously compounding growth rates. To determine the per-period continuously compounded rate of growth, we apply equation (7.9)

$$r = \left(\frac{\ln(P_t / P_0)}{t} \right) \times 100$$

and to determine the annual continuously compounded growth rate, we apply equation (7.10)

$$CCAGR = \{[\ln(X_t) - \ln(X_{t-1})] \times 100\} \times p$$

	A	B	C	D	E
1	Year, quarter (1)	Billions of chained 2005 dollars (2)	Continuously compounded quarterly growth rate [Continuously Compounded Rate of Change in FRED] (8)	Continuously compounded annual growth rate [Continuously Compounded Annual Rate of Change in FRED] (9)	Natural log of billions of chained 2005 dollars (10)
2	2007-01-01	13,099.901			=LN(B2)
3	2007-04-01	13,203.977	=(E3-E2)*100	=C3*4	=LN(B3)
4	2007-07-01	13,321.109	=(E4-E3)*100	=C4*4	=LN(B4)
5	2007-10-01	13,391.249	=(E5-E4)*100	=C5*4	=LN(B5)
6	2008-01-01	13,366.865	=(E6-E5)*100	=C6*4	=LN(B6)
7	2008-04-01	13,415.266	=(E7-E6)*100	=C7*4	=LN(B7)
8	2008-07-01	13,324.600	=(E8-E7)*100	=C8*4	=LN(B8)
9	2008-10-01	13,141.920	=(E9-E8)*100	=C9*4	=LN(B9)
	Formula in text		equation (7.9)	equation (7.10)	

After first calculating the natural log of each value of real GDP, we can then generate the continuously compounded quarterly growth rate in column (8) and, from that, the continuously compounded annualized growth rate based on the quarterly rates in column (9).

Exercises

1. Calculate the following growth rates for the U.S. CPI-U data given in the table **to six decimal places**:

    (a) per month percentage change;
    (b) per month compound growth rate;
    (c) per month continuously compounded growth rate.
    (d) Comment on your findings.

    U.S. CPI-U, December 2006–January 2008

	CPI-U (1982–84 = 100)		CPI-U (1982–84 = 100)
Dec-06	203.3	Jul-07	207.338
Jan-07	203.574	Aug-07	207.520
Feb-07	204.357	Sep-07	208.382
Mar-07	205.348	Oct-07	209.133
Apr-07	205.920	Nov-07	211.166
May-07	206.682	Dec-07	211.737
Jun-07	207.023	Jan-08	212.495

    Source: U.S. Bureau of Labor Statistics (2010c), Consumer Price Index (at <www.bls.gov/cpi/>). The values presented are seasonally adjusted.

2. Suppose that your first credit card charges 12.49 percent interest to its customers, compounding that interest monthly. Within one day of getting your first credit card, you "max out" the credit limit by spending $1,200. If you do not buy anything else on the card and

you do not make any payments, calculate how much money you would owe the company after six months.

3. India's large and growing population has garnered considerable attention from policymakers and economists. In 1985, India's population stood at 765,147,000 people, and the annual growth rate was calculated to be 2.043 percent.

   (a) Determine the expected number of India's population in 2009 based on the 1985 results. Use the most appropriate growth rate for determining changes in population.
   (b) The World Bank estimated India's population at 1,155,347,678 people in 2009. Determine the actual annual rate of population growth between 1985 and 2009, and compare it to the projected 1985 rate.

4. Calculate what price you should pay for a financial instrument that guarantees you $1,000 three years from now if you expect the interest rate to be 8 percent per year for this three-year period.

5. Use data on real motor vehicle output from the table below to answer the following.

   (a) Calculate the quarterly percentage growth in:

      i. final sales of motor vehicles to domestic purchasers,
      ii. private fixed investment in new autos and new light trucks, and
      iii. domestic output of new autos, and sales of imported new autos between the first quarter of 2006 and the fourth quarter of 2009.

   (b) Plot and analyze the growth rates.

U.S. real motor vehicle output (billions of chained (2005) dollars)

Year, quarter	Final sales of motor vehicles to domestic purchasers	Private fixed investment in new autos and new light trucks	Domestic output of new autos	Sales of imported new new autos
2006-I	528.9	198.4	114.5	93.7
2006-II	524.1	186.2	107.9	93.6
2006-III	529.9	185.8	105.7	96.2
2006-IV	530.3	190.7	105.0	97.7
2007-I	519.2	191.0	104.3	99.2
2007-II	510.6	189.1	102.3	100.3
2007-III	506.7	194.3	103.5	98.4
2007-IV	509.0	191.6	109.0	98.6
2008-I	482.9	182.8	108.1	94.0
2008-II	438.7	164.3	98.4	102.1
2008-III	402.1	151.0	106.8	87.5
2008-IV	337.4	114.7	86.5	71.5
2009-I	315.7	89.6	41.6	67.4
2009-II	309.3	100.2	48.2	67.5

Source: U.S. Bureau of Economic Analysis (2010e), National Economic Accounts, Interactive Access..., Table 7.2.6B, Real motor vehicle output, chained dollars. Seasonally adjusted at annual rates.

6. Use the "Final sales of motor vehicles to domestic purchasers" data in the previous table to calculate the following growth rates:

    (a) annualized percentage (growth) rate,

    (b) continuously compound growth rate, and

    (c) continuously compounded annualized growth rate.

    (d) Discuss what the three different rates indicate about short-run changes in automobile sales.

8 Patterns of long-term change

8.1 Economic growth over time

The economic growth rates discussed in Chapter 7 are used to describe short-term changes in economic variables. For all calculations, we assumed that the variable changed constantly by either amount or rate, through the process of compounding or not, and periodically or continuously. However, we have little *a priori* reason to expect that all economic variables necessarily change constantly over time.

The short-run growth calculations in Chapter 7 also only utilized two observations, those from the initial period and the final period, but, as we have previously established, a superior descriptive statistic incorporates all observations in its calculations. Because an economic variable's value may vary widely between the first observation and the last, such methods may omit important information about how the variable actually changes between those two periods, particularly as we expand the time horizon.

We now turn to statistical methods that incorporate all observations of the variable when calculating its growth over more than a few time periods. Generating a measure known as the variable's **long-run secular trend**, these methods estimate the underlying direction of a time series by smoothing out, to the extent possible, all seasonal, cyclical, and irregular effects in the time series. In other words, a variable's secular trend captures the overall tendency of the variable's growth over time.

Methods for estimating a variable's secular trend can generate statistics that measure not only the constant average rate at which the variable changes over time, but also the constant amount by which a variable grows. With these latter methods, we will relax the assumption that a variable grows at a constant rate. And because a variable's secular trend is rarely uninterrupted, we will conclude Chapter 8 by briefly discussing how seasonal effects, business cycles, and other influences may cause a variable's pattern of growth to deviate from its long-run secular trend.

8.2 Constant long-run rates of growth

Suppose we are interested in measuring the average rate at which real quarterly personal consumption expenditures on services ($PCE_{services}$) grew over the longest period of expansion in the U.S. economy's history, April 1991 through February 2001. While we could apply one of the growth rate formulas from Chapter 7, we recognize that we might omit important information about the growth in PCE if we only utilize its value in the initial (1991-Q2) and final (2001-Q1) quarters.

How, then, might we estimate the quarterly growth rate from all observations of $PCE_{services}$? As we noted in Section 7.7, if we rescale a time-series variable by taking its

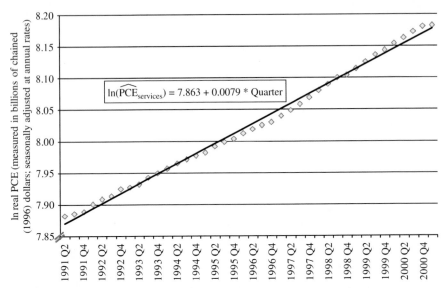

Chart 8.1 Natural log of real personal consumption expenditures on services.

Data source: U.S. Bureau of Economic Analysis (2010c), National Economic Accounts, Table 1.2, Real gross domestic product.

natural log and then fit a linear trendline to the logged values, the slope of that trendline will estimate the variable's per-period growth rate. That is, as shown in Chart 8.1 (p. 218), we can plot the natural log of real PCE$_{services}$ and then fit a linear trendline across the time period of interest.

Chart 8.1 also includes the equation that generated the fitted trendline. According to the estimated slope, real PCE on services grew at an average per-quarter rate of 0.79 percent, slightly higher than the continuously compounded growth rate of 0.77 percent calculated using equation (7.9). But how did we fit the linear trendline to the data? That is, what method did we use to obtain the equation that produces the constant average per-quarter growth rate for personal consumption expenditures on services?

We might have started simply by drawing straight lines through as many of the data points as possible, noting the equation that corresponds to each line. As we recall from high-school algebra, each of those equations will take the general form of

$$y = a + bx \qquad (8.1)$$

In the terminology of the algebra of linear relationships, y is called the **dependent variable** and x is the **independent variable**, because the values of y depend on, or can be explained by, values of x. Because we know the paired values for (x, y), we want to determine the values of a, the intercept term, which measures the value of y when x equals zero, and b, the line's slope, which measures the average change in y for a given change in x over all pairs of (x, y) that generate each line.

If equation (8.1) describes a perfect linear relationship between x and y, we could draw a straight line through every pair of (x, y) values to determine the values of a and b. However, x and y are rarely perfectly linearly related, so any line we estimate will not intersect all pairs of (x, y). This means that we can expect to observe a difference between the actual value of

y and its value estimated by the linear equation for the straight line, which we will denote as $\hat{y}$ (called "y-hat").

Once we have drawn several straight lines through the paired observations, we next want to select the line that best fits the data. But how do we know which line is "best"? Statisticians employ the **criterion of least squares**, which states that the "best" fitting line is the one that minimizes the sum of the squared differences between each observation of the dependent variable, y, and its value estimated by the line, $\hat{y}$. Thus the least-squares criterion requires that we minimize the following sum:

$$\min \sum (y_t - \hat{y}_t)^2 \tag{8.2}$$

where $\hat{y}_t = \hat{a} + \hat{b}x_t$ and the hats ($^\wedge$) signify the least-squares estimated values.

Let us now define the difference between the actual and estimated values of the dependent variable as the **residual** $\hat{e}_t$ (technically, the residual associated with the tth observation) such that

$$\hat{e}_t = y_t - \hat{y}_t \tag{8.3}$$

Restating the least-squares criterion as

$$\min \sum \hat{e}_t^2 = \min \sum (y_t - [\hat{a} + \hat{b}x_t])^2 \tag{8.4}$$

we see that the least-squares procedure will determine the values of $\hat{a}$ and $\hat{b}$ that minimize $\sum \hat{e}_t^2$.

Those familiar with differential calculus will recognize that we can derive the equations for estimating a and b by first differentiating the right-hand side of equation (8.4) with respect to a and b, and then setting each result equal to zero. The resulting equations, called the **normal equations**,[1] are solved first for b and then for a, yielding

$$\hat{b} = \frac{\sum x_t y_t - (\bar{x} \sum x_t)}{\sum x_t^2 - (\bar{x} \sum x_t)} \tag{8.5}$$

$$\hat{a} = \bar{y} - \hat{b}\bar{x} \tag{8.6}$$

where $\bar{x}$ and $\bar{y}$ are the arithmetic means of the independent and dependent variables, respectively. Thus equations (8.5) and (8.6) provide the formulas for the least-squares estimators, $\hat{b}$ and $\hat{a}$.

Let us now apply the least-squares method to estimate the average quarterly growth rate of real personal consumer expenditures on services depicted in Chart 8.1. Table 8.1 presents the first ten observations of actual quarterly expenditures (Y_t) and their natural log ($\ln(Y_t)$) and the corresponding independent variable of time (x_t in equations (8.5) and (8.6)). It also includes the components required to solve equations (8.5) and (8.6).

We first solve equation (8.5) to give

$$\hat{b} = \frac{6,622.510 - (8.024 \times 820)}{22,140 - (21 \times 820)} = 0.00787$$

Table 8.1 Personal consumption expenditures on services and expenditures estimated using the least-squares principle

Year-quarter	Monthly PCE on services	$\ln(Y_t)$	t	$\ln(Y_t) \times t$	t^2
1991-Q2	2651.1	8.3299	1	8.330	1
1991-Q3	2658.3	8.3338	2	16.668	4
1991-Q4	2668.3	8.3354	3	25.006	9
1992-Q1	2698.9	8.3379	4	33.352	16
1992-Q2	2720.9	8.3419	5	41.710	25
1992-Q3	2734.1	8.3464	6	50.078	36
1992-Q4	2765.0	8.3464	7	58.425	49
1993-Q1	2772.7	8.3514	8	66.811	64
1993-Q2	2784.9	8.3496	9	75.146	81
1993-Q3	2816.0	8.3509	10	83.509	100
		$\vdots$	$\vdots$	$\vdots$	$\vdots$
Sum		320.955	820	6,622.510	22,140
Mean		8.024	21		

Source: Author's calculations based on data from Chart 8.1.

and then equation (8.6) to give

$$\hat{a} = 8.024 - [(0.00787) \times (21)] = 7.85882$$

Thus the linear function that minimizes the sum of the squared deviations of the given values of y is

$$\hat{y} = 7.859 + 0.0079x$$

or, for our example,

$$\widehat{PCE}_{services} = 7.859 + 0.0079t_Q$$

What does the estimated equation tell us about the long-run growth of personal consumption expenditures on services? The literal interpretation of the intercept term ($\hat{a}$) is the value of real personal consumption expenditures on services in "quarter 0", which is logically the first quarter of 1991. However, we cannot assume that $\hat{a}$ is the value of U.S. expenditures on services in March 1991 because the estimated intercept is a statistical artifact of the least-squares method that did not use data from the first quarter of 1991 in its calculations. As a result, the intercept is neither statistically reliable nor economically meaningful, so we will follow convention and focus our attention solely on the slope estimated by the least-squares method.

The slope estimated by equation (8.5) equals 0.0079. After multiplying it by 100, we obtain the same value as that associated with the linear trendline in Chart 8.1. Again, we have determined that U.S. expenditures on services grew at an average per-quarter rate of 0.79 percent between the second quarter of 1991 and the first quarter of 2001.

Let us now make several observations about using the least-squares method to estimate long-run growth. First, it can be shown that the linear trendline estimated by the least-squares method derives from the formula for the continuously compounded per-period growth rate for the variable, Y:

$$r = \left(\frac{\ln(Y_t/Y_0)}{t} \right)$$

which we recognize as equation (7.9) without the conversion to percentages. Rearranging the equation to isolate $\ln(Y_t)$, which is the dependent variable in Chart 8.1, yields

$$\ln(Y_t) = \ln(Y_0) + rt \tag{8.7}$$

Now let us consider equation (8.7) from a different perspective. Suppose we create a new variable, y, that equals $\ln(Y_t)$, and we define $a = \ln(Y_0)$ such that equation (8.7) can be rewritten as

$$y = a + rt$$

As we recognize from our earlier discussion, this equation has the same general form as the straight line from equation (8.1). Thus we can now apply least squares to equation (8.7) to estimate the slope coefficient, r, which corresponds to the average rate at which the dependent variable, $\ln(Y_t)$, changes per period of time (t) under conditions of continuous compounding.

In the context of our discussion, we now refer to equation (8.7) as the **exponential long-run growth model**. (Note that we include the word "exponential" to acknowledge that equation (8.7) was ultimately derived from equation (7.8).) Thus the least-squares estimate of equation (8.7) will determine the constant average per-period growth rate for the economic variable, Y_t.

Our second observation relates to the residuals generated by the least-squares method. Table 8.2 presents the estimates of the dependent variable ($\hat{y}_t$) and the residual ($\hat{e}_t$), for the first ten observations. As we observe, the actual value of personal consumption expenditures on services does not equal its estimated value for any of the years, meaning that $\hat{e}_t$ does not equal 0 for any observation. This demonstrates that the least-squares procedure does not require the "best-fit" line to include any of the actual paired observations in order to estimate the line that most closely describes the change in the dependent variable over time.

Table 8.2 Estimated dependent and residual values for real PCE on services, 1991 Q1–1993 Q3

Year-quarter	$\ln(Y_t)$	t	$\ln(\hat{Y}_t)$	$\hat{e}$
1991-Q2	8.3299	1	7.86669	0.46321
1991-Q3	8.3338	2	7.87455	0.45925
1991-Q4	8.3354	3	7.88242	0.45298
1992-Q1	8.3379	4	7.89028	0.44762
1992-Q2	8.3419	5	7.89815	0.44375
1992-Q3	8.3464	6	7.90602	0.44038
1992-Q4	8.3464	7	7.91388	0.43252
1993-Q1	8.3514	8	7.92175	0.42965
1993-Q2	8.3496	9	7.92961	0.41999
1993-Q3	8.3509	10	7.93748	0.41342

Source: Author's calculations based on data from Chart 8.1.

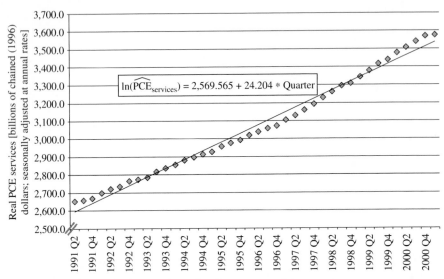

Chart 8.2 Real personal consumption expenditures on services.

Data source: U.S. Bureau of Economic Analysis (2010c), National Economic Accounts, Table 1.2, Real gross domestic product.

The final observation relates to the number of periods in the time series required to obtain reliable least-squares estimates. While the procedure utilizes all observations of the dependent variable in its calculations, it should not be applied to estimate change over a short period of time, which is generally considered to be less than 30 observations. (The reason, which relates to the statistical reliability of the estimates and noise in the data, will be discussed further in Chapter 9.) Because we have 40 observations of the $PCE_{services}$, we can apply the least-squares method to estimate its long-run growth rate between 1991-Q2 and 2001-Q1.

8.3 Growth by constant amounts

We might now ask if it is realistic to describe the long-run change in personal consumption expenditures on services as occurring at a constant rate. What if, instead, $PCE_{services}$ (or any economic variable) is more accurately described as changing by a constant *amount* each period? Fortunately, we can still apply the least-squares method to generate this value, albeit using a different linear relationship between the dependent variable and the independent variable, time.

Let us return to the data for real personal consumption expenditures on services, and now plot the actual values for the variable over the same period of economic expansion. As we observe in Chart 8.2 (p. 222), the variable increases in a linear fashion during the time period, suggesting that it is reasonable to fit a line to the data using the least-squares procedure. Doing so yields a linear trendline estimated by the equation

$$\widehat{PCE}_{services} = 2,569.565 + 24.204t_Q$$

What do the equation's estimated coefficients tell us about long-run change in $PCE_{services}$? Here, again, we note that the estimated intercept term is a statistical artifact, so we will not attach any economic meaning to it. On the other hand, the estimated slope coefficient does have economic meaning, indicating the average per-quarter real dollar change in $PCE_{services}$. Specifically, over the period of 1991-Q2 through 2001-Q1, real $PCE_{services}$ increased, on average, by \$24,204,000,000 per quarter.

We shall refer to the model just estimated as the **linear long-run growth model** for an economic variable, Y_t. In general terms, the linear long-run growth model can be stated as

$$Y_t = a + bt \tag{8.8}$$

where b is the average per-period amount of growth in the variable, Y_t.

We now note two important differences between the exponential and linear growth models. First, the dependent variable in the linear model is measured in the variable's original units, here the real dollar value of personal consumption expenditures on services, rather than the natural log of the variable used in the exponential model. Second, the slope coefficient for this model, b, is the constant real dollar *amount* (rather than rate) by which expenditures on services changed each quarter between 1991-Q2 and 2001-Q1. However, the least-squares procedure used to estimate the slope coefficients in each model is identical.

8.4   Change over time by constant rates or by constant amounts?

While the calculations of both the long-run growth rate and the long-run amount of growth are correct, which type of long-run change better describes the secular trend in real $PCE_{services}$? That is, how do we choose between the exponential long-run growth model and the linear long-run growth model when describing the long-term trend in real $PCE_{services}$ (or any economic variable)?

It is essential that we begin with an important *caveat*: there is *no* definitive test, statistical or economic, that will indicate if an economic variable changes by a constant rate versus a constant amount. Indeed, in both instances above, we have in essence imposed the type of constant growth on the variable by our choice of the underlying relationship between the dependent variable and the independent variable, time.

That said, we have already applied one guiding principle for choosing between the two models: plotting the time-series data for the variable. Displaying how the variable changes over time can suggest the more appropriate description of its long-run secular trend.

Let us now plot real $PCE_{services}$ on the left Y-axis and the natural log of real $PCE_{services}$ on the right Y-axis in Chart 8.3 (p. 224). As we observe, both series increase in a linear fashion over the time period, although neither obviously exhibits a perfect linear relationship between the dependent variable and time. Because the change in each series appears to be constant, it is reasonable to use least squares to estimate a trendline for each $PCE_{services}$ series.

Let us now add a least-squares estimated trendline and underlying equation for each series. As we observe in Chart 8.4 (p. 225), it appears that the least-squares line for the real $\ln(PCE_{services})$ series provides a slightly better fit to the data than does the line for $PCE_{services}$. This would suggest that real $PCE_{services}$ for the time period here may be better described by the exponential long-run growth model. However, because looks can be deceiving, this is not the only factor we want to consider in our selection.

Let us next consider the mathematical process by which an economic variable changes in the long run. Given the compound nature of most economic variables, it is often reasonable to assume that an economic variable changes at a constant rate over time. This manifests itself

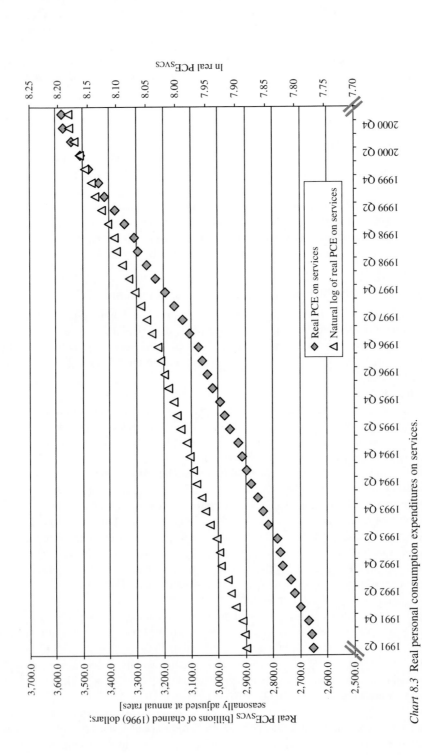

Chart 8.3 Real personal consumption expenditures on services.

Data source: U.S. Bureau of Economic Analysis (2010c), National Economic Accounts, Table 1.2, Real gross domestic product.

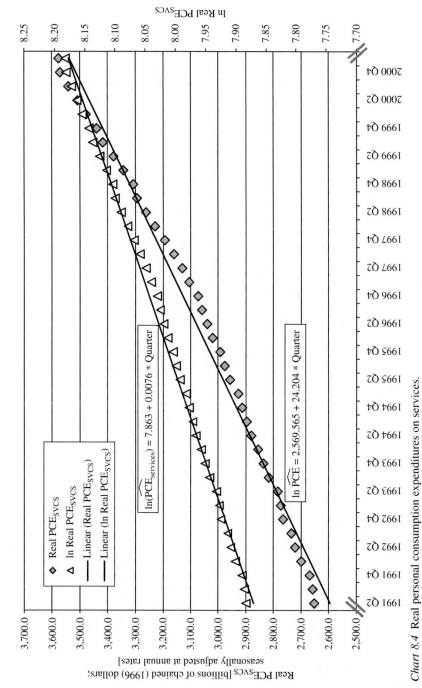

Chart 8.4 Real personal consumption expenditures on services.

Data source: U.S. Bureau of Economic Analysis (2010c), National Economic Accounts, Table 1.2, Real gross domestic product.

in the chart as relatively small changes in the variable at the beginning of the time period, which then increase in size as we move ahead in time. Such a pattern can be attributed to a multiplier-type effect in which, for example, each quarter's expenditures on services build on expenditures from previous quarters. We also typically observe such change in economic variables that are expressed in real or nominal monetary units of measurement.

Not all economic variables will necessarily exhibit this pattern. Take, for example, the number of employees in educational services from the second quarter of 1991 through the first quarter of 2001. As shown in Chart 8.5 (p. 227), employment levels appear to increase by a constant amount, rather than by a constant rate, over the time period. Thus we might now ask if this is consistent with how economists describe long-run change in the labor market for educational services. Here employment levels are largely driven by demand, or the number of available positions, which remained relatively constant over the time period being examined (see, for example, Kaufman and Hotchkiss 2006, pp. 40–41; Lindsay *et al.* 2009, pp. 2–5). This evidence thus reinforces the choice of using the linear long-run growth model to describe long-run change in employment for this sector of the economy.

Mathematical considerations may also guide our choice of long-run growth model. For example, changes in economic variables measured in non-monetary units, such as the number of new homes built, years of educational attainment, and the number of vehicle miles traveled, may be more likely to grow by constant amounts, rather than rates, over time. In addition, long-run trends in rates and indices, such as inflation measured by changes in the CPI-U, may also be more appropriately described by the linear long-run growth model due to the mathematical construction of those variables.

A final consideration relates to which secular trend is more easily understood in the context of the economic questions being analyzed. If both the linear and exponential growth models provide comparable fits to the variable's secular trend, then we would be advised to discuss that trend in more readily understood units. For example, given the large values of macroeconomic variables such as GDP and its components, a long-run growth rate calculated from the exponential growth model will be easier for most of us to comprehend than the trillion dollar changes estimated by the linear growth model. On the other hand, many of us would more readily understand the long-run trend in new homes constructed if it was presented in terms of the average number of homes built each period than if that growth was presented as an average rate.

Note that we have *not* mentioned the R^2 statistic as a guiding principle for selecting between the growth models. As you may recall from your statistics course, R^2 indicates how much of the variation in the dependent variable is explained by the least-squares estimated model. Thus we might be tempted to generate both exponential and linear models, calculate their R^2 statistics, and select the model with the higher R^2 value. This, however, would be inappropriate, since the R^2 statistic is *not* designed to determine the process by which an economic variable changes over time.

Determining which trendline better describes the long-term pattern of an economic variable thus requires that we consider how the variable visually changes over time, as well as any economic information that informs that change and which type of growth is more easily understood. In cases where both trends are statistically and economically plausible and no clear direction is provided by the question at hand, it is wise to report both estimates of long-run growth.

The linear and exponential growth models are not the only functional forms that we can use to describe the long-run growth in a variable. Quadratic and reciprocal functions, among others, may provide a more accurate picture of a variable's secular trend, but we will not

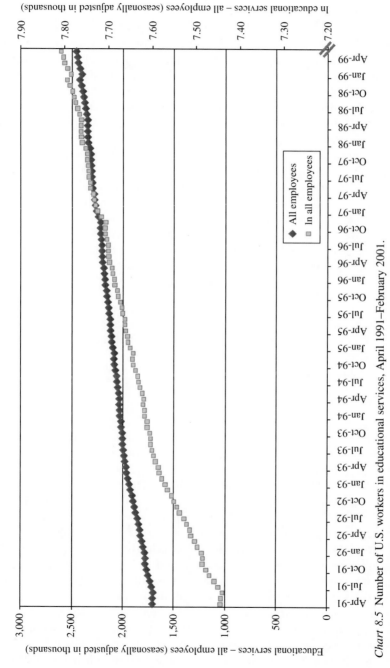

Chart 8.5 Number of U.S. workers in educational services, April 1991–February 2001.

Data source: U.S. Bureau of Economic Analysis (2010c), Current Employment Survey, B-1, Employees on nonfarm payrolls by major industry sector.

explore these functions here. For now, let us consider one additional set of factors that may affect a variable's long-term growth, non-secular factors that disrupt the constant change in an economic variable over time.

8.5   A complete model for describing change

Economic variables do not always exhibit steady change over time. Some economic variables, such as beer sales in summer and personal consumption expenditures in the fourth quarter, show upward "bumps" during the time periods indicated. Other variables, such as expenditures on durable goods, experience declines during recessionary periods, while others, such as unemployment payments, rise above the secular trend during recessions. Still other economic variables, such as defense expenditures during wartime, also deviate from the long-run secular trend as the result of a specific economic episode.

Economists have developed statistical methods for addressing those "bumps in the road." While most are beyond the scope of this text, it is important for us to understand what factors can disrupt a variable's secular trend. Here we will briefly describe each potential source, provide some examples of data series in which each effect appears, and discuss how we might address the presence of these factors in our empirical analyses.

A general model of time-series analysis includes four possible factors. The first, the variable's secular trend, is estimated using one of the methods discussed above and reflects constant change over time. The three remaining potential sources of instability may result from seasonal effects, cyclical effects, or irregular effects. Thus we can state that the long-run change in an economic variable is a function of the four elements acting together:

$$Y = f(\text{trend, seasonal, cyclical, irregular}) \tag{8.9}$$

Two characteristics of equation (8.9) are worth noting. First, we assume that we can measure and estimate each effect, if relevant for the variable Y_t. Second, because we have not specified the functional form of equation (8.9), we have not indicated if the four temporal factors are additive, multiplicative, or otherwise related.

8.6   Seasonal effects

The first category of instability, **seasonal effects**, focuses on regularized differences in a variable over a period of less than one year. The name correctly suggests that some seasonal effects may be caused by weather or environmental conditions, as in the case of beer sales in the summer. As we observe in Chart 8.6 (p. 229), the gallons of beer consumed in the U.S. follow a clear annual pattern: they decline through the fall, reaching a low during the winter months, and then rise through spring, peaking during the summer months. Because beer sales exhibit seasonal effects, neither a constant rate nor a constant amount of growth model would accurately describe the pattern of beer consumption depicted here.

Seasonal effects do not necessarily correspond to natural forces. Some economic variables, such as the personal consumption expenditures driven by holiday shopping during the fourth quarter of each year, reflect long-standing traditions in our society. Others have been determined by cultural factors in business or other institutions.

One example of the latter can be found in the U.S. automobile industry. U.S. automobile manufacturers have historically closed their plants for several weeks in July to retool for new models, which were historically introduced in the fall. As we observe in Chart 8.7 (p. 231), U.S. monthly auto production is typically lowest in July (although we may observe additional dips during recessionary periods). Thus, to describe the secular trend in U.S. auto

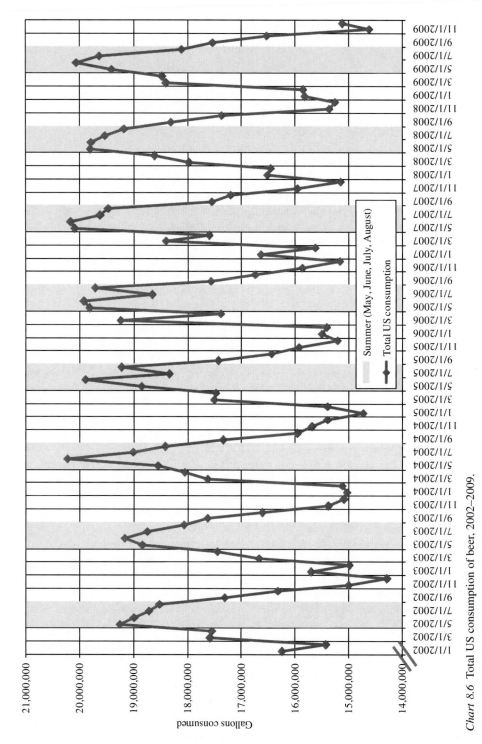

Chart 8.6 Total US consumption of beer, 2002–2009.

Data source: Beer Institute (2010), Brewers Almanac 2010, Total US consumption.

production, we must account for the decline in production each July. If we do not, we will incorrectly attribute production level changes from the July shutdowns to other factors that affect motor vehicle output in the long run.

Let us now turn to the seasonal patterns exhibited by U.S. personal consumption expenditures. As shown in Chart 8.8 (p. 232), not seasonally adjusted (NSA) PCE spikes upward every fourth quarter, driven by sales associated with Halloween in October, Thanksgiving in November, and the numerous religious and secular holidays in December.

We might now ask, statistically speaking, how we handle an economic variable that exhibits seasonal variation. As with all statistical questions, it depends on the focus of our analysis. If, for example, we are interested in explaining the effects of the fourth-quarter jump in personal consumption expenditures, we would certainly want to retain those seasonal effects and use the NSA PCE data. However, if seasonal patterns serve to obscure other patterns of interest, we would like to smooth out those effects from the data.

Seasonal adjustment (SA) essentially spreads the seasonal bumps in a variable across the entire year's set of values. As a result, seasonally adjusted values typically lie above (for upward bumps in the data) or below (for downward bumps, such as air conditioning sales in the winter or furnace sales in the summer) the NSA series. As Chart 8.8 illustrates, the effects of seasonally adjusting quarterly personal consumption expenditures is to smooth out the fourth-quarter "bumps" over all four quarters of each year, and the resulting SA series exceeds the values of the NSA series.

Seasonally adjusted data is published for most economic series, so we rarely need to construct such series ourselves. This is fortunate because seasonally adjusted data allows us to more accurately describe a variable's secular trend. Perhaps more importantly, smoothing out seasonal fluctuations in a data series makes it easier for economists and policy-makers to identify observations that deviate from a variable's secular trend as a result of cyclical changes. Thus seasonally adjusted data is more effective for monitoring the overall health of the economy than data that has not been seasonally adjusted.

8.7 Cyclical effects

Of the three factors that can disrupt an economic variable's secular trend, **cyclical effects** are of greatest concern to economists and policy-makers. Such effects are the result of an economy expanding and contracting over time, and such disruptions, particularly contractions, reduce the economic well-being of many in the economy. Even economic expansions can destabilize economic decision-making.

From a practical perspective, we can identify cyclical variations in an economic variable by plotting its time series. If we observe a distinctly different pattern of change during a particular phase of the business cycle, that variable is said to exhibit cyclical variation.

Chart 8.9 (p. 233) depicts seasonally adjusted real personal consumption expenditures between January 1995 and March 2010. The chart also indicates when the U.S. economy was officially in a recession.[2] As we observe, PCE growth slowed during the 2001 recession and actually declined throughout during the more recent recession.

The above method, combined with our theoretical knowledge of how a particular economic variable changes during the business cycle, helps us to identify if an economic variable is **procyclical** (the variable moves in the same direction as the overall economy), **countercyclical** (the variable moves in an opposite direction from the overall economy), or **acyclical** (the variable does not exhibit a pattern reflective of the overall economy's cyclical changes). If these cyclical movements are important to our analysis, we must account for those effects in any empirical analysis of the variable.

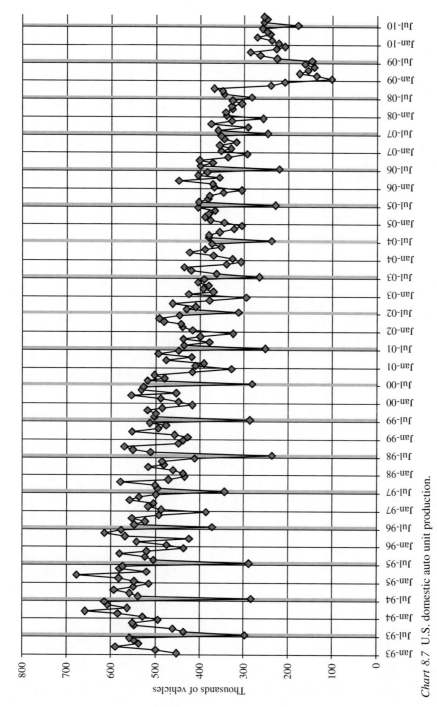

Chart 8.7 U.S. domestic auto unit production.

Data source: U.S. Bureau of Economic Analysis (2010c), National Economic Accounts, Table 7 – Domestic auto nuit production.

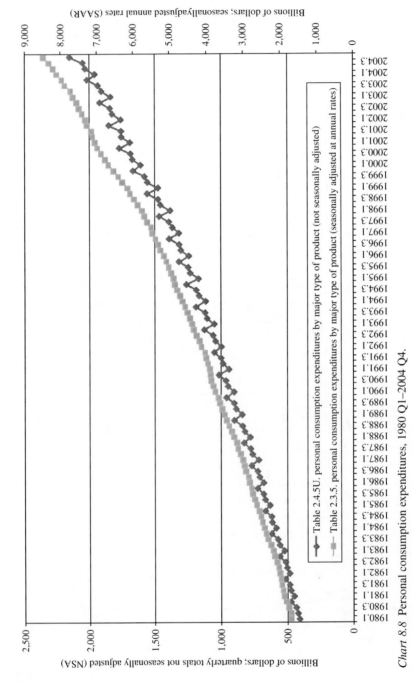

Chart 8.8 Personal consumption expenditures, 1980 Q1–2004 Q4.

Data source: U.S. Bureau of Economic Analysis (2010c), National Economic Accounts, Table 2.4.5U, Personal consumption expenditures by major type of product not seasonally adjusted; Table 2.3.5, Personal consumption expenditures by major type of product seasonally adjusted at annual rates.

Explanatory note: NSA PCE are measuredas quarterly totals, but quarterly estimates of PCE SAAR are expressed at seasonally adjusted annual rates, which assumes that the rate of quarterly activity is maintained for a year. Thus PCE SAAR values are approximately four times as large as the NSA.

Legend (within chart):
— Table 2.4.5U. personal consumption expenditures by major type of product (not seasonally adjusted)
— Table 2.3.5. personal consumption expenditures by major type of product (seasonally adjusted at annual rates)

Axis (top): Billions of dollars; seasonallyadjusted annual rates (SAAR)
Axis (bottom): Billions of dollars; quarterly totals not seasonally adjusted (NSA)

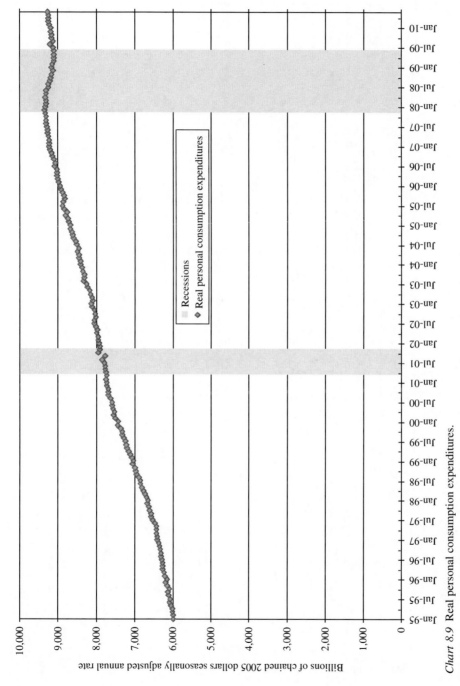

Chart 8.9 Real personal consumption expenditures.

Data source: FRED, Federal Reserve Economic Data, Federal Reserve Bank of St. Louis (2010d), Real Personal Consumption Expenditures, 3 Decimal (PCEC96).

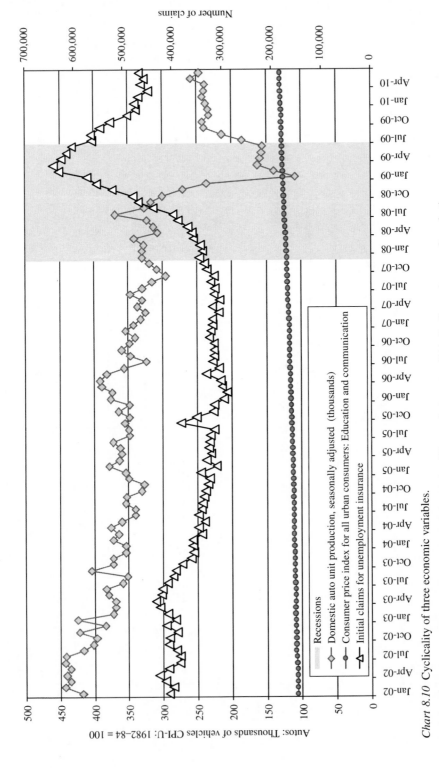

Chart 8.10 Cyclicality of three economic variables.

Data source: U.S. Bureau of Economic Analysis (2010c), National Economic Accounts, Table 7 – Domestic auto nuit production.
FRED, Federal Reserve Economic Data, Federal Reserve Bank of St. Louis (2010e), Initial Claims for Unemployment Insurance Seasonally Adjusted (ICSA).
FRED, Federal Reserve Economic Data, Federal Reserve Bank of St. Louis (2010e), Consumer Price Index for All Urban Consumers: Education and Communication
(CPIEDUSL).

Chart 8.10 (p. 234) presents data for three economic variables, each of which exhibits one of the cyclical patterns defined above. As we observe, U.S. domestic auto production declined significantly during the most recent recessionary period, thereby exhibiting the procyclical tendencies we expect from a durable good. Initial claims for unemployment insurance, not surprisingly, increased during the most recent recession, remaining well above pre-recession levels, thereby exhibiting a countercyclical pattern. We also observe that the consumer prices for education (tuition, books, and supplies) and communication (including telephone services and personal computers) expenditures do not deviate from the secular trend during the recession, suggesting those prices are acyclical.

In those cases for which the secular trend is our primary interest, we may want to consider methods for smoothing out the cyclical effects. One popular method is to use **moving averages**, in which the arithmetic mean values of a time series are "moved" through the observed values to obtain a smooth secular trend, similar to how chain-weighted indices were constructed. We will leave this and other such methods to more advanced texts.

8.8   Irregular effects

We now turn to the final source of instability in time-series analysis, **irregular effects**. Economists divide these variations into two types: **episodic variations** and **residual (random) variations**. Episodic variations are effects due to factors other than seasonal and cyclical variations that can be attributed to a particular event. Episodic variations may occur because of war, weather, disasters, or unusual situations, and they tend to represent an anomaly in the long-run secular trend of a variable.

The episodic nature of some economic variables may leap off the chart of the plotted time series. One well-known example is the jump in the federal government expenditures on national defense between 1942 and 1945. As shown in Chart 8.11 (p. 236), the effects of increased federal government defense consumption expenditures during the United States' involvement in World War II are quite apparent. Thus, in an analysis of the long-run secular trend in defense expenditures, we might apply one of the smoothing techniques cited above to mitigate the effects of this episodic variation, or we might employ a technique known as a dummy variable, which we will discuss in detail in Chapter 18.

A second type of irregular variation is called the residual, or random, variation. Its name comes from how these effects are identified. Technically, they are those irregularities still present after we have accounted for the other types of variation, and they cannot be attributed to known economic or other factors.

Even if we cannot attribute the anomaly to a particular cause, we must acknowledge the presence of these random non-secular effects if they appear in the data. We will, however, leave the process of smoothing out these random effects to those familiar with the sophisticated methods required.

Summary

Long-term changes in an economic variable are of great interest to economists and policymakers. Whether the variable changes by constant rates or constant amounts, the variable's secular trend gives us important information about the behavior of economic phenomena.

The most popular method for determining the long-run growth in an economic variable is the least-squares method, which we can apply to the linear growth model to calculate a variable's constant per-period change and to the exponential growth model to calculate a variable's long-run constant growth rate. To determine which of the two types of growth is more appropriate for describing long-run change in an economic variable, we must consider

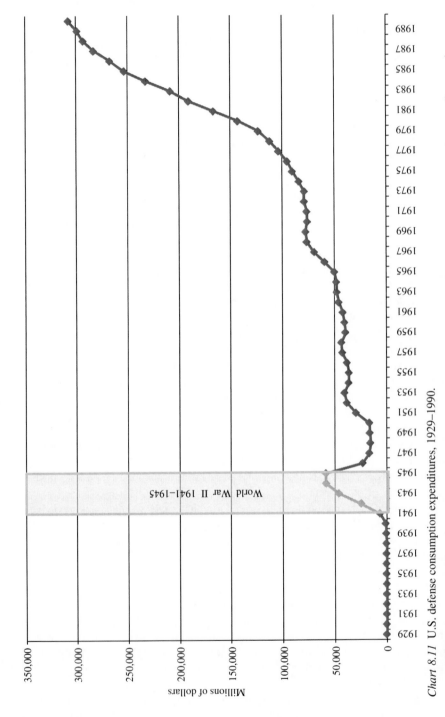

Chart 8.11 U.S. defense consumption expenditures, 1929–1990.

Data source: U.S. Bureau of Economic Analysis (2010c), National Economic Accounts, Table 3.10.5, Government consumption expenditures and general government gross output.

not only the mathematical properties of the variable's time series, but also how economists theorize that change. In addition, we must also account for factors that cause the variable to deviate from its secular trend. Such factors may be attributable to seasonal variations, cyclical movements, episodic events, or sheer randomness in the economy.

Concepts introduced

Acyclical
Countercyclical
Criterion of least squares
Cyclical effects
Dependent variable
Episodic variation
Exponential long-run growth model
Independent variable
Irregular effects
Linear long-run growth model
Long-run secular trend
Moving average
Normal equations
Procyclical
Residual
Residual (random) variation
Seasonal effects

Box 8.1 Using Excel to calculate the least-squares estimates of constant growth

Excel offers two approaches for estimating the long-run growth of an economic variable using least squares. The first approach, which we will simply mention here, applies two built-in functions, LOGEST and LINEST, to the time series to measure constant growth over time. (LOGEST calculates the constant periodic growth rate of the variable, while LINEST calculates the constant periodic change in the variable.)

The second approach fits a trendline to the time series. Since we always begin the analysis of long-run growth by plotting the time series, we will focus our attention on this method.

Determining long-run growth using Excel's Trendlines

As illustrated in the charts in Chapter 8, Excel can add a trendline of either a linear or exponential functional form to the plotted data. It can also display the underlying equation that generates the trendline.

Determining the constant per-period amount of growth

Let us return to real personal consumption expenditures on services (from the U.S. Bureau of Economic Analysis (2010c), National Economic Accounts, Table 1.2, Real gross domestic product) to first determine the constant per-period dollar amount by which real $PCE_{services}$ changes between 1991-Q2 and 2001-Q1.

1. Using an Excel **Line** chart, we plot the variable data but we do *not* include the dates on the *X*-axis. We also remove the line between the data points as follows:

   - Activate any data point, and select **Format Data Series**.
   - Under **Line Color**, select **No Line** and click **Close**.

This yields the following chart:

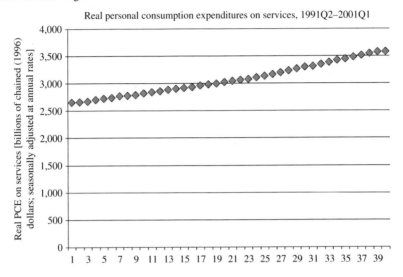

Real personal consumption expenditures on services, 1991Q2–2001Q1

2. Activate a data point, and select **Add Trendline**. Since we want to determine the dollar amount by which PCE on services changes, select the **Linear** trendline (which is the default function), and check the **Display Equation on chart** box:

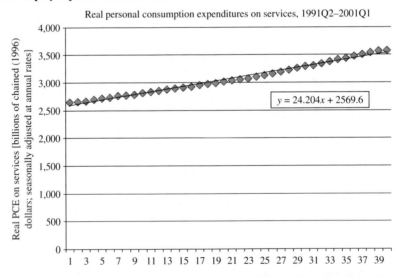

Real personal consumption expenditures on services, 1991Q2–2001Q1

$$y = 24.204x + 2569.6$$

Here we obtain the same equation for the trendline as in Chart 8.2, which indicated that the average per quarter change in real PCE$_{services}$ was $24.204 billion (chained 2005 dollars).
3. The equation generated by Excel retains the general form of a linear equation. However, we can reformat it to make it more meaningful.

- To **increase the decimal places**:
  - ○ Right-click on the equation, and select **Format Trendline Label** from the drop-down menu.
  - ○ Select **Number**, and choose the desired number of decimal places. Here we use four.
  - ○ Click **Close**.

- To **restate the dependent and independent variables** in economically meaningful terms:
  - ○ Right-click on the equation, and select **Edit Text**.
  - ○ Make the desired changes to the equation. Here we change y to $PCE_{services}$ and x to t_Q:

The chart will now look like this:

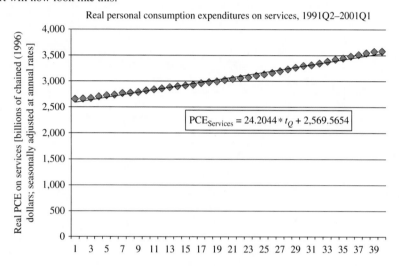

Once we have added the trendline and its equation, we can now insert the actual dates for the time series on the X-axis. (We cannot include the dates in the initial chart because Excel will use those values in its least-squares calculations.) We get the following chart:

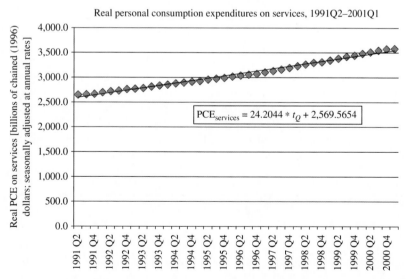

Determining the constant per-period rate of growth

We can also use Excel's **Trendline** feature to determine a variable's long-run rate of growth. Here we follow the same steps, with one exception – we now use an exponential trendline. (We could also fit a linear trendline to the natural log of the dependent variable.)

The exponential trendline corresponds to the exponential long-run growth model developed in Section 8.2, and its underlying equation is derived from equation (7.8):

$$y = a \times e^{rt}$$

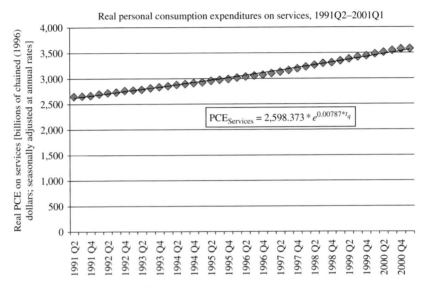

According to the exponential trendline equation, real personal consumption expenditures on services grew at an average per-quarter rate of 0.787 percent between 1991-Q2 and 2001-Q1.

Exercises

1. Not all economic variables exhibit positive growth over time. For example, some types of employment decrease over time due to technological innovations that improve labor productivity or automate processes that then require fewer workers. The agricultural sector in the U.S. economy is a prime example of how both factors have transformed employment. As we see in the table, the number of hired workers in agricultural employment declined from 2.3 million workers in 1948 to 732,000 workers in 2008.

Hired workers in agricultural employment (in thousands), 1948–2007

Year	Hired workers	Year	Hired workers	Year	Hired workers	Year	Hired workers	Year	Hired workers	Year	Hired workers
1948	2,326	1958	1,962	1968	1,233	1978	1,272	1988	1,002	1998	879
1949	2,241	1959	1,944	1969	1,207	1979	1,229	1989	928	1999	913
1950	2,318	1960	1,895	1970	1,224	1980	1,303	1990	919	2000	893
1951	2,189	1961	1,859	1971	1,203	1981	1,256	1991	909	2001	870

Year	Hired workers	Year	Hired workers	Year	Hired workers	Year	Hired workers	Year	Hired workers	Year	Hired workers
1952	2,140	1962	1,844	1972	1,206	1982	1,193	1992	865	2002	870
1953	2,087	1963	1,800	1973	1,245	1983	1,291	1993	857	2003	886
1954	2,111	1964	1,621	1974	1,331	1984	1,161	1994	839	2004	825
1955	2,044	1965	1,499	1975	1,337	1985	1,007	1995	867	2005	780
1956	1,916	1966	1,350	1976	1,372	1986	953	1996	831	2006	752
1957	1,920	1967	1,267	1977	1,313	1987	964	1997	875	2007	747

Data source: *Economic Report of the President* (2009b), Table B-100, Farm input use, selected inputs, 1948–2007. Available from <http://www.gpoaccess.gov/eop/tables09.html#erp8>.

(a) Use the least-squares method to estimate the constant rate of growth in hired agricultural workers between 1948 and 2008.

(b) Use the least-squares method to estimate the constant growth in the number of hired agricultural workers between 1948 and 2008.

(c) Based on these results and the considerations discussed in Section 8.4, discuss which of the two models better describes the long-term trend in agricultural employment for hired workers.

2. Use the extended dataset in the table below to answer the following questions.

Real motor vehicle output (billions of chained (2005) dollars), seasonally adjusted at annual rates

	Final sales of motor vehicles to domestic purchasers	Domestic output of new autos	Sales of imported new autos		Final sales of motor vehicles to domestic purchasers	Domestic output of new autos	Sales of imported new autos
2000-I	482.1	118.8	83.1	2005-II	533.1	104.6	84.6
2000-II	447.3	116.4	79.6	2005-III	553.0	107.8	88.9
2000-III	447.6	113.2	78.5	2005-IV	498.3	112.1	88.9
2000-IV	441.2	106.1	78.6	2006-I	528.9	114.5	93.7
2001-I	456.2	103.6	77.2	2006-II	524.1	107.9	93.6
2001-II	446.5	107.1	78.3	2006-III	529.9	105.7	96.2
2001-III	441.5	105.1	75.7	2006-IV	530.3	105.0	97.7
2001-IV	498.0	106.5	81.9	2007-I	523.3	104.6	99.8
2002-I	468.0	112.3	82.9	2007-II	517.0	106.2	100.9
2002-II	471.0	112.5	84.1	2007-III	515.0	103.2	98.2
2002-III	502.1	110.6	86.8	2007-IV	512.1	105.1	98.2
2002-IV	467.6	103.6	84.0	2008-I	489.9	108.7	94.6
2003-I	463.9	95.6	85.0	2008-II	444.6	102.3	102.8
2003-II	483.2	98.1	82.7	2008-III	407.6	106.7	87.6
2003-III	500.1	96.1	83.1	2008-IV	334.7	83.2	70.8
2003-IV	486.9	98.5	79.9	2009-I	318.4	42.2	69.9
2004-I	491.1	95.4	83.5	2009-II	318.6	50.0	69.6
2004-II	495.1	95.0	86.6	2009-III	345.6	68.2	82.7
2004-III	513.1	96.7	82.7	2009-IV	337.4	76.5	74.9
2004-IV	524.6	101.4	88.1	2010-I	355.6	78.2	79.9
2005-I	514.3	106.7	83.1	2010-II	376.0	80.9	79.5

Data source: U.S. Bureau of Economic Analysis (2010e), National Economic Accounts, Interactive Access to National Income and Product Accounts Tables, Table 7.2.6B, Real motor vehicle output, chained dollars. [Online] Available from <http://www.bea.gov/national/nipaweb/>.

(a) Create a chart with data for "Final sales of motor vehicles to domestic purchasers," "Domestic output of new autos," and "Sales of imported new autos" between the first quarter of 2000 and the second quarter of 2010.

(b) Calculate the constant amount by which each variable grew. Be sure to translate these calculations into meaningful statements about the long-run change in each variable.

(c) Calculate the constant rate by which "Final sales of motor vehicles to domestic purchasers," "Domestic output of new autos," and "Sales of imported new autos" grew between the first quarter of 2000 and the second quarter of 2010. Be sure to translate these calculations into meaningful statements about the long-run change in each variable.

(d) Compare your results from (b) and (c). Discuss which type of long-run change, constant amount or constant rate, better describes the long-run change in each variable.

(e) Examine the time series depicted in part (a) to determine which of the variables exhibit non-secular variations. For those that do, identify which factors are present.

Part III

Statistical inferences about a single variable

The statistical methods discussed thus far have been procedures for describing an economic variable across elements or over a given period of time. These methods may systematically organize large numbers of observations of an economic variable or distill those observations into a single measure of a variable's central tendency or dispersion. Other statistical methods describe how an economic variable changes over time, whether it be through constructing an index number, calculating growth rates, or estimating the secular trend of a time series.

Such descriptive techniques are essential tools of empirical analysis. In applying these methods, we have implicitly assumed that our data can be used to describe the entire population under analysis. In actuality, however, social scientists typically rely on sample data in their statistical analyses because of the prohibitive costs associated with collecting census data for the entire population.

Using sample data in our empirical analyses requires that we develop a new set of statistical methods known as "statistical inference" methods. Such techniques will allow us to make statements about a population based on a representative sample. And because we are now working under conditions of incomplete information, statistical inference methods explicitly acknowledge the possibility that any conclusions based on sample data may not reflect the underlying population.

We will launch our discussion of statistical inference in Part III by describing procedures frequently used by economists to derive statistical inferences about a single economic variable. In Part IV, we will expand those methods to describe relationships between two economic variables, and in Part V, we will introduce the most frequently used statistical inference method in economics for describing relationships between more than two economic variables, multiple regression analysis.

9 Basic concepts in statistical inference

The term **statistical inference** refers to statistical methods that describe a population, which we do not observe, using measures from a sample, which we do observe. This means we will calculate statistics based on a sample and then infer from them the values of the population's parameters. To begin our discussion of statistical inference, let us revisit the concepts of populations and samples introduced in Section 5.2.

9.1 Populations and samples revisited

Suppose we want to know if the average U.S. family's standard of living increased between 1990 and 2010. Assuming that family income is a good indicator of the standard of living, we decide to answer the question empirically. After identifying the appropriate population (here, all families in the U.S.), we next specify an economic variable X (here, income received by a family) that is appropriate for our empirical analysis. We further assume we can answer our question satisfactorily by determining the variable's central tendency as measured by its arithmetic mean.

As we recall from Chapter 4, the arithmetic mean sums all values of X (here, all unique values of family income), weighted by their relative frequency (here, the number of families with each unique income value), and then divides by the number of observations (here, families). If we have data for the entire population, we can then calculate the arithmetic mean using equation (9.1):

$$\mu = \left(\sum_{i=1}^{N} f_i X_i \right) \Big/ N \tag{9.1}$$

where f_i refers to the frequency with which each variable value (X_i) appears in the population, and N represents the number of elements in the population.

Now suppose we cannot afford to conduct a census of the population and must instead rely on a representative sample. As with the population, we observe a value for the variable (here, family income) for each element (here, family) in the sample.

We can now describe the sample using a relative frequency distribution, where f_i is the frequency associated with each unique value of the variable. Thus, to find the mean of the sample, we sum the unique values of the variable (x_i) weighted by their frequencies in the distribution (f_i), and then divide by the number of elements (n) in the sample:

$$\overline{X} = \left(\sum_{i=1}^{n} f_i x_i \right) \Big/ n \tag{9.2}$$

As we observe, we utilize the same procedures to calculate the arithmetic mean for the population (μ) and for the sample ($\overline{X}$). However, we also see differences in the notation used for the two means.

To distinguish between the characteristics of populations and samples, we will use standard notation when referring to the particular set of elements concerned. Let us first consider a population and its relative frequency distribution with respect to the variable, X. As shown in equation (9.1), we denote the number of elements in a population by a capital N and its arithmetic mean by μ (the Greek lower-case letter "mu"). We also recall from Table 5.3 that the population's variance is denoted by σ^2 (the square of the Greek lower-case letter "sigma"). The numerical values of these three characteristics of a population (N, μ, σ^2) are referred to as its **parameters**.

A sample drawn from a population also has an underlying frequency distribution, and, after gathering the available observations, we can estimate the values of its descriptive characteristics. For the n elements in a sample, $\overline{X}$ (X-bar) represents the sample's arithmetic mean, and s^2 represents the variance within the sample. Here we use the general term **statistics** to refer to the values of n, $\overline{X}$, and s^2 (as well as any other characteristics of a sample).

The distinction between the statistical characteristics of populations and samples is not merely one of notation. As we shall learn in Section 9.9, there are important differences between the statistical measures of the frequency distribution for a population and that for a sample. For now, we establish the following: while some notation, such as $\overline{X}$ and s^2, is identical to that used to denote certain descriptive statistics, these measures now have a more technical meaning. That is, in the context of statistical inference, the word "statistics" will refer to those values obtained from a sample from which we make inferences about the values of a population's parameters.

9.2 Sampling procedures

How, then, do we obtain a sample representative of a population? Ideally we want the sample to mirror the population; that is, we would like an **ideal representative sample**. However, as the following example demonstrates, obtaining an ideal representative sample is not possible.

Suppose we want to determine the mean income of families in the United States for a given year. Once we have defined what constitutes a family and family income, we next need to select a group of families whose characteristics reflect those found in the population. But herein lies the dilemma. To capture *all* of the family characteristics that arise in the population, we must first know what the population's characteristics are, information we can only obtain through a census. Since this is why we are taking the sample in the first place, how do we proceed?

Statisticians have fortunately developed methods less costly than a census for drawing samples from the population. Because several alternatives are available, let us now establish two criteria for what constitutes the "best" sample. The first addresses the inevitable error associated with drawing a sample, while the second introduces the concept of **probability** into sampling procedures.

The first criterion derives from the fact that any sample drawn from a population will rarely, if ever, perfectly reflect that population. The resulting difference between the sample statistic and the corresponding population parameter is called the **sampling error**. Since this error is known to exist, we would like to measure that error so we can incorporate it into the inferences we make about the population. This then establishes the first criterion for the best alternative to the unattainable "ideal representative sample": we must be able to measure numerically the errors associated with sample statistics.

A second criterion relates to the likelihood that a given element will be chosen for the sample and derives from the procedures we could use to draw a sample from a specific population. In practice, such sampling procedures fall into two general categories, *purposive* and *random*.

A **purposive sampling procedure** involves the active participation of the researcher, who asks questions of participants. The simplest purposive sampling procedures are those in which the researcher asks questions of whomever walks by, a "person-on-the-street" kind of interview. A slightly more refined approach would take the researcher to different locations across a given area in an effort to expand the diversity of sample participants. In both cases, the researcher determines who will, and who will not, be included in the sample. This, in turn, means that the elements are not equally likely to be chosen because even researchers with the best intentions are unlikely to select all relevant categories without conducting a census. Consequently, sample data collected through purposive sampling procedures will contain unmeasurable errors, thereby violating the first criterion. Such procedures also violate the second criterion, because the elements are not equally likely to be selected for the sample.

The alternative approach is a **random sampling procedure**, which is the type preferred by statisticians and economists. Such procedures remove the researcher, in a sense, from element selection. Instead, a mechanism, such as drawing numbers out of a hat or using a random number generator, is used. This, in turn, ensures that each element has an equal probability of being selected for the sample. It is this characteristic that meets the second criterion for choosing the "best" sample.

Random sampling procedures also allow the researcher to measure the error associated with the sampling process. Because each element is equally likely to be selected, the researcher can determine the probability that a given element will be included in the sample. When combined with statistics from the sample's distribution, the researcher can numerically measure the sampling error, meaning that random sampling procedures also meet the first criterion.

There are several types of random sampling procedures, but economists typically use simple random samples in their analyses. In technical terms, a **simple random sample** is one for which each element in the population has an equal probability of being selected. In addition, all elements are selected independently of one another, meaning that the selection of one element does not depend upon the selection of any other element. Finally, each sample of the same size (n) must have an equal probability of being chosen from the population.

Randomly drawn samples are at the heart of statistical inference procedures. This means that any statement made about a population parameter based on a sample statistic includes a potential but measurable sampling error. Thus, because we cannot make a definitive statement about the population, every statistical inference is really a statement of uncertainty. To measure this uncertainty in quantitative terms, we now turn to probability theory.

9.3 Concepts of probability

Probability theory allows us to make quantitative statements about the likelihood that particular outcomes will obtain. While numerous concepts of probability are available, those used in empirical economics are called **objective probabilities**. These probabilities are based on prior knowledge of possible outcomes from an experiment or on actual observations of the economy. Objective probabilities are also quantifiable, which means we can attach a numerical value to the likelihood associated with a particular experiment.

The first type of objective probability is *a priori* **probability**. Here the researcher knows the likelihood of each possible outcome occurring prior to conducting an experiment, with each outcome equally likely to obtain. The coin-toss experiment is a well-known example of *a priori* probability. In this experiment, only two outcomes – heads or tails – can possibly occur for each toss, and, assuming a fair coin, each outcome is equally likely to result. This then allows us to determine the probability that heads (or tails) will occur by dividing the actual outcome (either heads or tails) by the possible number of outcomes (heads and tails). Thus we know prior to the experiment that we have a 50 percent probability of obtaining heads and a 50 percent probability of obtaining tails for any given coin toss.

The second type of objective probability, **empirical probability**, is based on actual observation of a randomly drawn sample. Here we determine the probability of an outcome only after we have collected and observed the sample data. For example, suppose we draw a sample of 50 members of the labor force, and we observe that six of them are unemployed. This information allows us to determine the empirical probability that one person, selected at random, is unemployed by calculating the ratio of the number of elements that have the specified outcome (six unemployed persons) to the total number of elements from which one is to be selected (50 members of the labor force). That is, we have a probability of 12 percent (six out of 50) that a person in the given group is unemployed.

It is conceptually useful to think of *a priori* and empirical probabilities as observable measures associated with a single element. Thus we might identify the *a priori* probability of a tossed coin showing heads as equal to 0.50, or the empirical probability of an unemployed person being selected from the labor force as equal to 0.12, but, in actuality, we cannot associate a specific probability with a single outcome. In other words, a single coin toss will show either heads or tails, and the person selected from the labor force will be either employed or unemployed. Consequently, probability is a useful operational tool *only* if we think in terms of a large number of observations rather than a single event.

How, then, do we summarize the probabilities of possible outcomes for a large number of observations? We do so using the concept of a probability distribution.

9.4   Probability distributions

A variable associated with a randomly drawn sample is referred to as a **random variable**, and, as such, its values are associated with some probability of being observed. (If the random variable has a limited number of possible values, it is called a **discrete random variable**; if the possible values lie along a continuum, it is called a **continuous random variable**.) Once we know all the possible values that a random variable can assume (that is, once we know all possible outcomes) as well as the probabilities associated with each outcome (that is, the likelihood of each outcome being observed), we can construct a **probability distribution** that lists of all the values and their associated probabilities.

Let us now reconsider the coin-toss experiment. Here, the random variable is defined by the numerical values associated with the two possible outcomes, heads or tails, from a single toss. We have also established that, after a repeated number of tosses, each outcome has a probability of 0.50 (50 percent) of occurring.

We can now define the **discrete probability distribution** associated with an experiment involving two coin tosses. As delineated in Table 9.1, we might obtain heads from both tosses, heads only on the first or the second toss, or no heads, thus yielding three possible outcomes for the two-toss experiment.

Table 9.1 Probability distribution for a two-coin-toss experiment

Possible result	Coin toss		Number of heads (X)	Probability of the outcome (Pr(X))
	1st	*2nd*		
1	H	H	2	$Pr(2) = 1/4 = 0.25$
2	T	H	1 }	$Pr(1) = 1/2 = 0.50$
3	H	T	1 }	
4	T	T	0	$Pr(0) = 1/4 = 0.25$

The results associated with the two-toss experiment exhibit the two characteristics of all probability distributions. First, the probability of a single outcome can be described as a value between 0 and 1, which holds here since the probability of each outcome is either 0.25 or 0.50. Second, the sum of the probabilities of all **mutually exclusive** outcomes will equal 1. Here only one outcome (for example, heads (H) and then tails (T)) can occur, and the likelihood of all four possible outcomes (0.25, 0.50, and 0.25) sum to 1.

The coin-toss experiment is often used to demonstrate the basic concepts of a probability distribution. However, when applying statistical inference theory in empirical economics, we assume economic variables are continuous random variables. (Note that here "continuous" has a different meaning from that in Chapter 3, where it referred to a variable that could theoretically be measured in smaller and smaller increments. Here "continuous" refers to the probabilities attached to the numerical values that define a random variable.) As a result, the probability that a continuous random variable falls within a specified interval is given by the area under the **continuous probability distribution** (sometimes called the **probability density function**) within that interval. Let us now turn to the most commonly used continuous probability distribution in statistical inference, the normal distribution.

9.5 Continuous probability distributions: the normal distribution

The **normal distribution** is the most frequently employed continuous probability distribution in statistical inference because it can be completely described by only two statistics, its mean and its standard deviation. As such, the conventional notation for a normally distributed random variable, X, is

$$X \sim \mathcal{N}(\mu_X, \sigma_X^2)$$

where $\sim$ means "distributed as," $\mathcal{N}$ signifies the normal distribution, and its arguments are the parameters of the distribution. The mean, or expected value of the random variable, is denoted by μ_X, and its variance by σ_X^2.

The first characteristic of the normal distribution is the shape of its curve. As depicted in Chart 9.1 (p. 250), the normal distribution is represented by a bell-shaped curve that has a single peak at the exact center of the distribution corresponding to the distribution's arithmetic mean, median, and mode. It is important to note that in statistical inference, the arithmetic mean is the *expected value* of the population mean and thus may not (and likely will not) correspond to the mean calculated from the sample. Because we are now considering the probability that a random variable will take on a particular value, this statistic is theoretically different from the arithmetic mean discussed in Chapter 4.

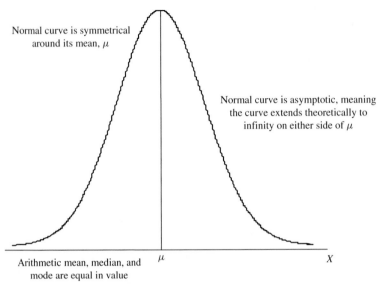

Normal curve is symmetrical around its mean, μ

Normal curve is asymptotic, meaning the curve extends theoretically to infinity on either side of μ

Arithmetic mean, median, and mode are equal in value

μ

X

Chart 9.1 Normal distribution.

The normal distribution is also **symmetric**, which means that half the area under the curve lies to the right of the peak and the other half lies to the left. In the language of probability, this tells us that the likelihood of observing a value of the random variable that is less than the expected value, μ_X, is equal to 0.50, while the probability of observing a value that is greater than μ_X also equals 0.50. This characteristic allows us to focus on either side of the distribution in our statistical inference analyses.

A third characteristic is that the normal distribution is **asymptotic**. This means that, as X moves farther from its expected value, the probability distribution will approach the X-axis but will never actually touch it. In probability terms, this means that the likelihood of observing a value of the random variable increasingly distant from the expected value approaches zero, but will never actually equal zero.

Because the normal distribution is a probability distribution, we know that the total area under the normal curve equals one, but we also observe that most of the curve's area falls close to the distribution's mean. This observation leads us to one additional property, the **Empirical Rule**, which holds for any symmetrical, bell-shaped probability distribution such as the one depicted in Chart 9.2 (p. 251). Specifically, the Empirical Rule says that

- 68.26 percent of the observations will lie within one standard deviation of the mean, μ_X;
- 95.44 percent of the observations will lie within two standard deviations of the mean, μ_X; and
- 99.74 percent of the observations will lie within three standard deviations of the mean, μ_X.

In more concrete language, the Empirical Rule states that over 99 percent of the values of a normally distributed continuous random variable, X, lie within three standard deviations of the distribution's mean.

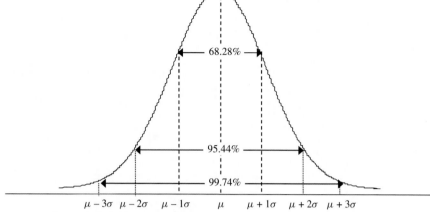

Chart 9.2 Symmetrical frequency distribution.

The question for economists is how many economic variables are actually normally distributed? In general, most economic variables are *not* normally distributed, because they are generally not equally distributed (symmetical) about their means. As we demonstrated in Chapter 5, many economic variables, particularly those measured in dollar terms, are asymmetrical. So we might now ask why economists find the normal distribution to be so important? Before we explore the answer, let us first introduce a special type of normal distribution, the *standard normal distribution*.

9.6 Continuous probability distributions: standard normal distribution

The **standard normal distribution**, denoted by Z, is a normal distribution in which the mean equals zero and the variance equals one:

$$Z \sim \mathcal{N}(0, 1)$$

We can convert any normally distributed random variable into a standard normal distribution by using the **Z-statistic** defined in equation (9.3):

$$Z = \frac{X - \mu_X}{\sigma_X} \tag{9.3}$$

The Z-statistic calculates the distance between the selected value of the normally distributed random variable (X) and its population mean (μ_X), divided by the population standard deviation (σ_X).

As shown in Chart 9.3 (p. 252), the normal and standard normal distributions both exhibit the same symmetrical and asymptotic bell shape about the mean. However, a standard normal distribution necessarily has a mean of 0 (not μ) and a variance of 1 (not σ^2). Thus the standard normal distribution rescales any normally distributed random variable to a random variable with a zero mean and standard deviation equal to one.

The standard normal distribution has numerous economic applications, as the following examples will demonstrate. Suppose we know that the mean household income for

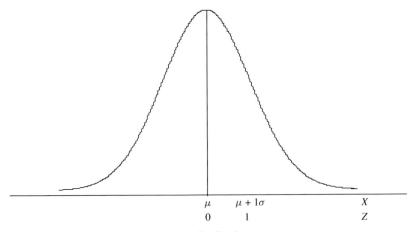

Chart 9.3 Normal and standard normal distributions.

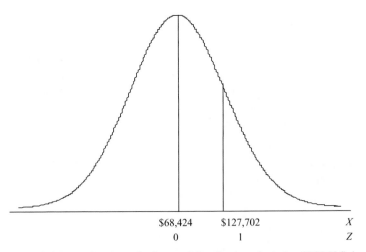

Chart 9.4 Normal and standard normal distributions depicting 2008 U.S. household income.

all U.S. households in 2008 equalled $68,424 and had a standard deviation of $59,278.[1,2] We will also assume (reasonably, as we will discuss in Section 9.9) that income is normally distributed. This yields the normal and standard normal distributions illustrated in Chart 9.4 (p. 252).

Now suppose we want to determine the probability that a randomly selected household in the U.S. had income in 2008 that was greater than or equal to some arbitrary value, say $75,000. In terms of Chart 9.5 (p. 253), this means we want to calculate the area under the normal probability distribution that lies to the right of the point where $X = \$75,000$. However, since it is easier to use the standard normal distribution, we use equation (9.3) to convert the value of X to a Z-statistic:

$$Z_{X=\$75,000} = \frac{\$75,000 - \$68,424}{\$59,278} = 0.1109$$

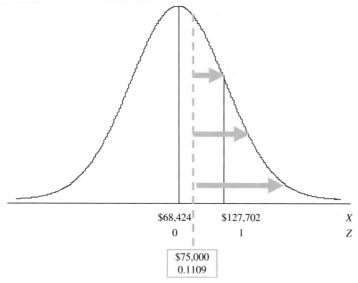

$68,424
0

$127,702
1

X
Z

$75,000
0.1109

Chart 9.5 Area under 2008 U.S. household income distribution for incomes = $75,000.

We now need to determine the area of the standard normal curve that corresponds to $Z \geq 0.1109$. Using Excel (detailed instructions may be found at the end of the chapter), we find that the probability of Z being greater than or equal to 0.1109 is 0.4558. That is,

$$\Pr(X > \$75,000) = \Pr(Z > 0.1109) = 0.4558$$

This result tells us that the probability of a randomly selected U.S. household in 2008 with income greater than or equal to $75,000 is 45.58 percent. An alternative interpretation is that 45.58 percent, or almost one-half of all U.S. households, had income greater than or equal to $75,000 in 2008.

Suppose we now want to determine the proportion of U.S. households that had incomes between $75,000 and $200,000. We have already calculated that 45.58 percent of households had incomes greater than $75,000 in 2008. But how many of these households also had incomes less than $200,000?

To begin, we determine the Z-value that corresponds to $200,000:

$$Z_{X=\$200,000} = \frac{\$200,000 - \$68,424}{\$59,278} = 2.2197$$

Next we determine that the probability Z is less than 2.2197:

$$\Pr(X < \$200,000) = \Pr(Z < 2.2197) = 0.9868$$

This indicates that 98.68 percent of U.S. households had incomes less than $200,000 in 2008. (This also tells us that only 1.32 percent of households had incomes greater than or equal to $200,000 in 2008.)

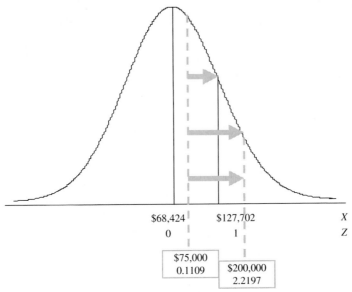

Chart 9.6 Area under 2008 U.S. household income distribution for incomes between $75,000 and $200,000.

We can now determine the proportion of U.S. households with incomes between $75,000 and $200,000:

$$\Pr(\$75,000 < X < \$200,000) = \Pr(0.1109 < Z < 2.2197)$$

Given that the area under the normal distribution equals 1 and that $\Pr(X \geq \$75,000)$ equals 0.4558, we determine that the $\Pr(X < \$75,000)$ equals 0.5442. Thus

$$\Pr(0.1109 < Z < 2.2197) = 0.9868 - 0.5442 = 0.4426$$

According to our calculations, 44.26 percent of all U.S. households in 2008 had incomes between $75,000 and $200,000.

Both of the above examples considered the proportion of households with incomes above the mean, but we can also determine the percentage of households with incomes below the average. For example, suppose we are interested in the proportion of U.S. households in 2008 that had incomes less than the median household income in the poorest state in the U.S., Mississippi. (While the two measures (the mean and the median) of household income are not strictly comparable, we can make statistical inference statements about how they compare.)

After obtaining Mississippi's 2008 median household income, $37,404, from the American Community Survey, 2008 (U.S. Census Bureau 2008a), we want to determine the following:

$$\Pr(X < \$37,404)$$

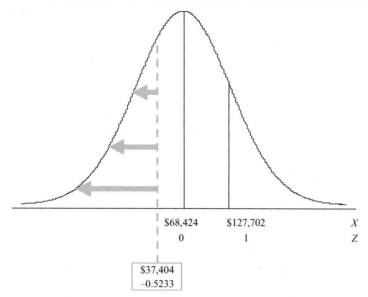

Chart 9.7 Area under 2008 U.S. household income distribution for incomes < $37,404.

As shown in Chart 9.7 (p. 255), the Z-statistic associated with Mississippi's median household income equals

$$Z_{X=\$37,404} = \frac{\$37,404 - \$68,424}{\$59,278} = -0.5233$$

Accordingly, we find

$$\Pr(Z < -0.5233) = 0.3004$$

which indicates that 30.04 percent of U.S. households had average incomes lower than the median household income in Mississippi in 2008.

As the three examples illustrate, the standard normal distribution is a very flexible device for determining the proportion, or percentage, of elements that lie above or below a chosen value. As we shall see in the next two chapters, the standard normal distribution is also useful for statistical estimation and hypothesis testing. Before we turn to those discussions, we have several additional statistical concepts to examine.

9.7 Identifying a normal distribution

We have not yet actually defined what constitutes a normal, or a standard normal, distribution. Thus we might ask if any symmetric, bell-shaped probability distibution is normally distributed. Such determination requires analysis beyond the scope of this text, but the simple answer is "no." That is, for a distribution to be normally distributed with a mean of zero and variance of one, it must conform to the following equation:

$$f = \left(\frac{1}{\sigma\sqrt{2\pi}}\right)e^{-[(x-\mu)^2/2\sigma^2]} \tag{9.4}$$

What this means is that we cannot identify a curve as being normally distributed simply by its shape. Fortunately, economists typically do not need to prove if an economic variable, specifically its mean, is normally distributed. To understand why, we now turn to sampling theory and the conceptually important sampling distribution of means.

9.8   The concept of the sampling distribution of means

Sampling theory is a key component of statistical inference because it studies the relationships between a population and samples drawn from that population. Not only do we use sampling theory to estimate population parameters based on statistics from repeated samples, we also use it to evaluate if our sample results are statistically significant or merely the result of chance variation. When combined with concepts from probability theory, the statements we make about a population from randomly drawn samples are called statistical inferences.

In economics, we are generally most interested in the mean of a population. For each sample that we draw from a population, we can calculate the statistical mean, and once we have drawn repeated samples, we can then arrange those means into what is called a "sampling distribution of means." We now turn to this very important *conceptual* probability distribution to identify several relationships between a statistic ($\bar{X}$) and a parameter (μ).

Let us first establish why sampling distributions are important in statistical inference. Quite simply, their relevance derives from the fact that economists must utilize sample statistics rather than population parameters in their statistical analyses. As previously discussed, when we make inferences about population parameters based on sample statistics, it is likely there will be a difference between the value of the sample statistic and the "true" value of the population parameter. This difference, as we know, is the sampling error.

Another important characteristic of sampling distributions derives from the process inherent in sampling theory. Conceptually, the estimate of a population parameter is based on the result of *repeated* samples being taken from the population. But every time we draw a sample from a population, even when we hold the number of observations (n) in each sample constant, we are unlikely to obtain the same value for the sample's statistic. Thus we need to determine which sample's result to use in our analysis. Fortunately, the concept of the **sampling distribution**, which describes the distribution of all possible sample outcomes for a given sample size (n), provides a standard against which we can compare our sample statistics.

Table 9.2 Specification of all possible samples for $n = 2$ from the population (2, 4, 6)

First element drawn	Second element drawn		
	2	4	6
2	2, 2	2, 4	2, 6
4	4, 2	4, 4	4, 6
6	6, 2	6, 4	6, 6

Table 9.3 Arithmetic means of all possible samples of $n = 2$ from population (2, 4, 6)

First element drawn	Second element drawn		
	2	4	6
2	2	3	4
4	3	4	5
6	4	5	6

Let us now turn to the properties of the sampling distribution of means. While we will illustrate the concept by using a hypothetical numerical example, the sampling distribution of means is *never* empirically specified in detail. As noted above, the sampling distribution of means is a *conceptual distribution* only.

Suppose we are given a population that consists of the numbers 2, 4, and 6, from which we want to draw all samples of size two ($n = 2$). Assuming we allow for the possibility of selecting the same element twice (that is, we **sample with replacement**), we can identify all possible outcomes associated with a random sampling of two of the three elements in the population. As Table 9.2 shows, we can draw nine different samples of two elements from this population.

Now that we have identified all possible samples, we can calculate the mean associated with each. As Table 9.3 shows, the nine possible sample means range in value from 2 to 6, and any one might be the result of drawing a single random sample for $n = 2$.

We can now marry the likelihood of obtaining each possible mean to the means in Table 9.3. We see, for example, that we obtain a sample mean of 3 for two of the nine samples, yielding a relative frequency of 0.222 (or 22.2 percent). Using the nine sample means and the probability associated with each in Table 9.4, we can now describe the probability distribution.

The probability distribution here is the **sampling distribution of means**, which is formally defined as the distribution of the means from all possible samples of a given size (n) drawn from a given population.

As we observe in Table 9.4, the likelihood of obtaining each possible sample mean varies across the different values. This leads us to ask if there is a unique value of the sample mean associated with the greatest likelihood of appearing. As the graphical representation of the sampling distribution of means in Chart 9.8 (p. 258) demonstrates, that is the case, and, in this example, it occurs when the arithmetic mean, $\overline{X}$ equals 4.

Table 9.4 Absolute and relative frequency distribution of means of all possible samples

Sample mean ($\overline{X}$)	Absolute frequency (f)	$f\overline{X}$	Relative frequency
2	1	2	0.111
3	2	6	0.222
4	3	12	0.333
5	2	10	0.222
6	1	6	0.111
Total	9	36	1.000

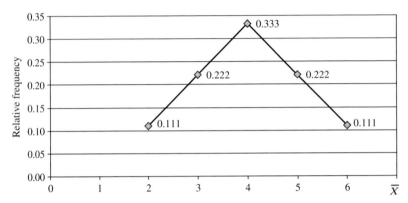

Chart 9.8 Relative frequency polygon of the means from all possible samples where $n = 2$.

This process for generating the sampling distribution of means is very similar to the random sampling process used in the coin-toss experiment. This means that the sample mean, $\overline{X}$, is a random variable since it is the numerical descriptor of the outcome of a random sampling process. It is also true that any random variable whose entire range of possible values and their associated probabilities are known can be described by a probability distribution. Consequently, the sampling distribution of means is not simply a relative frequency distribution, but a probability distribution.

The question now arises as to which probability distribution describes the sampling distribution of means. Here we can draw on statistical theory to make an even more important statement regarding the sampling distribution of means:

> If a random sample is drawn from a normally distributed population with a mean of μ and a variance of σ^2, the sample mean, $\overline{X}$, also follows the normal distribution with the same mean, μ, but a different variance of σ^2/n That is,

$$\overline{X} \sim \mathcal{N}\left(\mu, \frac{\sigma^2}{n}\right)$$

Let us now use our hypothetical population to demonstrate the characteristics of the sampling distribution of means *vis-à-vis* the population's distribution. Here we will define the **estimators** associated with the sampling distribution of means. (In the language of statistics, estimators are the rules (formulas) for calculating the mean and variance of a random variable.)

We first turn to the mean of the sampling distribution of means to examine how it relates to the population mean. As statisticians have shown, the mean of the sampling distribution of means equals the mean of the population. Formulaically, this means that

$$\mu_{\overline{X}} = \mu \tag{9.5}$$

where the symbol $\mu_{\overline{X}}$ represents the **arithmetic mean of the sampling distribution of means**.

Let us now demonstrate this equality using our hypothetical population. First, the arithmetic mean formula from Chapter 4 yields a mean for the hypothetical population (2, 4, 6) of 4, thus

$$\mu = \frac{\sum X_i}{N} = \frac{(2+4+6)}{3} = 4$$

Now let us calculate the weighted average of probability distribution displayed in Table 9.4:

$$\mu_{\overline{X}} = \frac{\sum f_i X_i}{\sum f_i}$$

$$= \frac{(0.111 \times 2) + (0.222 \times 3) + (0.333 \times 4) + (0.222 \times 5) + (0.111 \times 6)}{1.000} = 4.0$$

As we see, the mean of all sample means, or the mean of the sampling distribution of means, also equals 4. This finding is not unique to our example. In fact, the equality of $\mu_{\overline{X}}$ and μ is a general relationship that holds regardless of the shape of the population's distribution and regardless of sample size.

The idea of evaluating a particular statistic in a sampling distribution has a special function in statistical inference methods. Specifically, a statistic of all possible sample values is referred to as the **expected value** of the statistic. Thus we can restate equation (9.5) as equation (9.6):

$$E(\overline{X}) = \mu_{\overline{X}} \tag{9.6}$$

Next we recall from introductory statistics that, when the expected value of a statistic equals the corresponding population parameter, that statistic is said to be an **unbiased estimator**. Thus in the case of the sampling distribution of means, $\overline{X}$ is an unbiased statistic because

$$E(\overline{X}) = \mu_{\overline{X}} = \mu \tag{9.7}$$

This does not, however, imply that the mean of *any* specific sample drawn will necessarily equal the mean of the population, as demonstrated in Chart 9.8. But while there are many possible individual sample means that will not equal μ, it is true that the mean of *all* possible sample means *will* equal μ. Thus the mean of the sampling distribution of means is an unbiased estimator of the population mean.

We can also evaluate the sampling distribution of means in terms of its variance. Let us first calculate the population's variance using the formula introduced in Table 5.3:

$$\sigma^2 = \frac{\sum (X_i - \mu)^2}{N} = \frac{(2-4)^2 + (4-4)^2 + (6-4)^2}{3} = 2.67$$

Next let us calculate the variance of the sampling distribution of means, which we will denote as $\sigma_{\overline{X}}^2$, using the following equation:

$$\sigma_{\overline{X}}^2 = \frac{\sigma^2}{n} \tag{9.8}$$

While this formula describes the variance of the sampling distribution of means, its operational use is limited since we are unlikely to know the value of the population variance, σ^2

Table 9.5 Calculating the variance (s^2) using equation (9.7)

Sample	$\bar{X}$	$\sum(X_i - X)^2$	p_i	$\sum s^2 = p_i(X_i - \bar{X})^2$
2, 2	2	0	0.111	0.000
2, 4	3	2	0.111	0.222
2, 6	4	8	0.111	0.889
4, 2	3	2	0.111	0.222
4, 4	4	0	0.111	0.000
4, 6	5	2	0.111	0.222
6, 2	4	8	0.111	0.889
6, 4	5	2	0.111	0.222
6, 6	6	0	0.111	0.000
		24	1.000	2.667

(or any other population parameter). As a result, we must first estimate the value of σ^2 using equation (9.9):

$$s^2 = \sum f_i(X_i - \overline{X})^2 \tag{9.9}$$

As the calculations in Table 9.5 show, the estimated variance, s^2, equals 2.667. If we substitute this estimated variance into the numerator of equation (9.8), we obtain a variance equal to

$$\sigma_{\bar{X}}^2 = \frac{s^2}{n} = \frac{2.667}{2} = 1.334$$

which does not equal the variance of the population, σ^2. This indicates that the estimator in equation (9.8) does not generate an unbiased estimate of σ^2.

Fortunately, the bias is a systematic type, which means we can correct for it using an adjustment ratio equal to $n/(n-1)$. Thus the unbiased variance for the sampling distribution of means is defined by equation (9.10):

$$\sigma_{\bar{X}}^2 = \left(\frac{n}{n-1}\right)\frac{s^2}{n} = \frac{s^2}{n-1} \tag{9.10}$$

Now let us apply equation (9.10) to our hypothetical population:

$$\sigma_{\bar{X}}^2 = \frac{2.667}{2-1} = 2.667$$

As we see, equation (9.10) yields a value for the variance of the sampling distribution of means that equals the variance for the hypothetical population. This means that equation (9.10) is the unbiased estimator of the variance of the sampling distribution of means.

As we discussed in Chapter 5, economists generally work with a variable's standard deviation rather than its variance. Thus, for analytical purposes, the relevant unbiased estimator of the distribution's variation is the **standard deviation of the sampling distribution of means**:

$$\sigma_{\overline{X}} = \frac{s}{\sqrt{n-1}} \tag{9.11}$$

The standard deviation of the sampling distribution of means has a very special meaning. Referred to as the **standard error of the mean**, this statistic measures the sampling error associated with the sampling distribution of means. It differs from the descriptive standard deviation discussed in Chapter 5, which measured how widely individual observations were dispersed about the mean. Here the standard error of the mean measures how much sample estimates vary from sample to sample. Thus in statistical inference, we will use the standard error of the mean to assess the accuracy of our statistical estimates of a sample's mean.

9.9 Sampling distribution of means and the Central Limit Theorem

Let us now reconsider the probability distribution of $\overline{X}$. We have already established, per equations (9.7) and (9.10), that the expected value of the mean and the variance of the sampling distribution of means equal the population's mean and variance. We also know that, if the underlying population is normally distributed, the sampling distribution of means too will be normally distributed. But we rarely know how the original population is actually distributed. Fortunately, even without such knowledge, the **Central Limit Theorem** tells us that the sampling distribution of means will be normally distributed if our sample size is large enough. More formally, we have the following:

Let $X_1, X_2, \ldots, X_n$ be a random sample drawn from an arbitrary distribution with a finite mean, μ, and variance, σ^2. As the sample size (n) approaches infinity, the sampling distribution of means approaches the normal distribution *regardless* of the shape of the parent population.

We might now ask what value of n is "large enough." A sample where $n \geq 30$ is generally accepted as a sufficient size, so in most cases we can treat the sampling distribution of the means as a normal, and thus a standard normal, distribution.

9.10 Sampling distribution of the Z-statistic

Let us now restate the Z-statistic formula (equation (9.3)) as it applies to the sampling distribution of means:

$$Z = \frac{\overline{X} - \mu_{\overline{X}}}{\sigma_{\overline{X}}} = \frac{\overline{X} - \mu_{\overline{X}}}{\sigma/\sqrt{n}} \tag{9.12}$$

As previously established, this distribution is normal with a mean of zero and variance of one. However, working with the Z-statistic assumes that we know the population variance (σ^2), which is rarely, if ever, the case.

Have we hit another roadblock? Fortunately not, because there is a probability distribution that exhibits the same desirable properties as the normal and standard normal distributions, but also simultaneously uses the unbiased sample variance estimator defined in equation (9.10). This distribution, known as the Student's t-distribution, is more commonly used by economists in statistical inference because we rarely know the value of the population variance.

Summary

Empirical analysis of an economic issue requires that we search for the relevant facts associated with that issue. While such analysis would ideally be based on all observations of the underlying population, practical considerations typically force economists to use a sample drawn from that population. Consequently, we need to examine how we can make inferences about the population from a given sample.

We begin by drawing a sample from the population using random selection procedures so as to remove one source of potential bias in the sample. However, because we still face the possibility of making an incorrect inference about the population from our sample, we want to assess how certain, or confident, we are about our conclusions. This leads us to another key element in statistical inference, probability.

The likelihood of obtaining each value of a random variable is its probability. Once we know the probabilities associated with all possible values of a random variable X, we can construct a probability distribution. While probability distributions can be discrete, continuous probability distributions, especially normal and standard normal distributions, are more often used by economists engaged in statistical inference.

The concept known as the sampling distribution of means provides us with an important theoretical basis for statistical inferences. The statistics from this distribution are unbiased estimators of its parameters, and because of the Central Limit Theorem, we can assume that the sampling distribution of means is normal if the sample size is sufficiently large. However, because we rarely know the population variance, economists most frequently use a related distribution, the t-distribution, which we will now examine in Chapter 10.

Concepts introduced

A priori probability
Arithmetic mean of the sampling distribution of means
Asymptotic
Central limit theorem
Continuous probability distribution
Continuous random variable
Discrete probability distribution
Discrete random variable
Empirical probability
Empirical rule
Estimator
Expected value
Ideal representative sample
Mutually exclusive
Normal distribution
Objective probability
Parameters
Probability
Probability density function
Probability distribution
Purposive sampling procedure
Random sampling procedure
Random variable

Sample with replacement
Sampling distribution
Sampling distribution of means
Sampling error
Sampling theory
Simple random variable
Standard deviation of the sampling distribution of means
Standard error of the mean
Standard normal distribution
Statistical inference
Statistics
Symmetric
Unbiased estimator
Z-statistic

Box 9.1 Using Excel to generate the area under a standard normal probability distribution

The area under the standard normal curve tells us the probability of a numeric value being greater than or less than some specified value. While we could use a standard normal table to determine that probability, we can also use built-in functions in Excel to obtain a more precise measure.

In **Excel 2007**, the NORMSDIST function returns the area associated with the standard normal *cumulative* distribution function. Specifically, it calculates the probability that Z is *less than or equal to* the z test statistic:

$$\text{NORMSDIST}(z) = \Pr(Z \leq z)$$

That is, the NORMSDIST function calculates the gray area underneath the standard normal distribution:

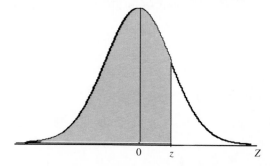

NORMSDIST and the standard normal distribution

Suppose, for example, that z equals 2.0. Entering that value into the NORMSDIST function yields a value equal to 0.97725, which indicates that the probability of Z being less than or equal to 2.0 is 97.725 percent:

$$\Pr(Z \leq 2.0) = 0.97725$$

In cases where we want to know the probability that Z is greater than a particular value, we need only recognize that the area under the standard normal distribution equals one. Thus we can then simply subtract the NORMSDIST value from one:

$$Pr(Z \geq z) = 1 - Pr(Z \leq z) = 1 - \text{NORMSDIST}(z)$$

For example, the probability that $Z \geq 2$ can be calculated using $1 - \text{NORMSDIST}(2)$, which yields a value of 0.02275. Thus the probability that $Z \geq 2$ is 2.275 percent.

The corresponding function in **Excel 2010** has a new name and different construction:

NORM.S.DIST(z, cumulative)

The first argument required, z, is the value for which we want the distribution (and is the same value used in the NORMSDIST function).

The second required argument, **cumulative**, determines the form of the function. Since we want the value associated with normal cumulative distribution function, use the logical argument, **TRUE**.

In the above example, enter the following in Excel 2010:

NORM.S.DIST(2, TRUE)

to obtain the 97.725 percent probability.

Because of the extra precision that can be obtained with Excel, we will utilize the NORMS-DIST/NORM.S.DIST function to calculate the probabilities associated with the standard normal distribution. In recording such results, we will follow convention and report each value to four decimal places.

Exercises

1. Determine the area under the standard normal probability distribution for the following values:

    (a) $z \leq 2.0$;
    (b) $z \geq 0.0$;
    (c) between $z = 0.0$ and $z = 2.0$;
    (d) between $z = 0.8$ and $z = 3.0$.

2. Suppose x is a normally distributed random variable with $\mu = 11$ and $\sigma = 2$. Find the following:

    (a) $Pr(10 \leq x \leq 12)$;
    (b) $Pr(7.8 \leq x \leq 12.6)$;
    (c) $Pr(x \geq 13.24)$.

3. The crop yield for a particular farm in a particular year is typically measured as the amount of the crop produced per acre. Cotton, for example, is measured in pounds per acre. It has been demonstrated that the normal distribution can be used to characterize crop yields

over time (Just and Weninger 1999). Historical data indicate that 2008's cotton yield for a particular Mississippi farmer can be characterized by a normal distribution with mean 853 pounds per acre and standard deviation 142 pounds per acre. The farm in question will be profitable if it produces at least 900 pounds per acre.

(a) Calculate the probability that the farm will lose money next summer.
(b) Calculate the probability that the cotton yield falls within two standard deviations of 853 pounds per acre next summer.

10 Statistical estimation

10.1 Sample surveys as a source of data

Most empirical economic analysis necessarily relies on data obtained through random sample surveys of populations. Such surveys are conducted by numerous research organizations, and they utilize professionally accepted procedures for minimizing sampling and non-sampling errors. One popular survey, sponsored by the U.S. Federal Reserve Board, is the triennial *Survey of Consumer Finances*, which collects detailed information on the finances of U.S. families.

Table 10.1 presents data from the 2007 *Survey of Consumer Finances*. Drawn from a geographically based random sample of 4,422 families in the United States, the table shows percentile distributions of median and mean before-tax family income from 2007. We will use this data to explain the method of statistical estimation.

First, let us reiterate that the data obtained through sample surveys are necessarily imperfect. Thus we must carefully interpret the results obtained from the sample by fully understanding their limitations.[1] One limitation, identified in Chapter 9, is the **sampling error**, which results from incomplete observation of the population. Because this type of error is measurable, we want to incorporate it into the statistical inferences we make. Let us now turn to the procedures for estimating the mean of a population from a sample drawn from that population.

Table 10.1 Before-tax family income and distribution of families, from the 2007 *Survey of Consumer Finances*

	(Thousands of 2007 dollars)		
	Median	Mean	Percentage of families
All families	47.3	84.3	100.0
Percentiles of income			
less than 20.0	12.3	12.3	20.0
20.0–39.9	28.8	28.3	20.0
40.0–59.9	47.3	47.3	20.0
60.0–79.9	75.1	76.6	20.0
80.0–89.9	114	116	10.0
90.0–100.0	206.9	397.7	10.0
Total number of families in survey: 4,422.			

Data source: Bucks *et al.* (2009), A5.

10.2 Interval estimates of the population mean when the variance of the population is known

Suppose we want to determine the arithmetic mean of before-tax income for all U.S. families in 2007. According to the results derived from the 2007 *Survey of Consumer Finances*, the reported mean for the sample is $84,300. However, this mean is derived from a single sample, not all possible samples of 4,422 families that could be drawn from the U.S. population. Thus if we drew another sample of 4,422 families from the population, it is unlikely we would obtain the same value for the mean.

The $84,300 value is called a **point estimate**, which is the single value estimate of a parameter derived from a single sample. Economists generally do not use such estimates to reach conclusions about the population. Not only do different samples (of the same size) yield different estimates of a population's parameters, each point estimate is a descriptive statistic based only on that one sample. As such, we have no procedure for estimating by *how much* it differs from the population parameter. This means that a point estimate is not a statistical inference since the error of the estimate cannot be measured.

Because we cannot make inferences about the population from a point estimate, economists prefer using interval estimates. Each **interval estimate** is a range of values based on a sample's statistics combined with the probability of the interval containing the (unknown) population parameter under conditions of repeated random sampling. While we can never be certain that the population parameter is contained in the estimated interval (statistical inference means we are never 100 percent confident), we can establish the probability such that we have a high level of confidence in the conclusions we reach.

To illustrate why an interval estimate is a statistical inference, let us use the sample mean ($\overline{X} = \$84,300$) and sample size ($n = 4,422$) from Table 10.1. In addition, we will assume we know the population variance (σ^2) and that its value is 265,430,550 dollars squared.

We now combine this information in Chart 10.1 (p. 268). As we observe, every statistical inference is based upon parts of three different, but related, frequency distributions. The first frequency distribution is that for the population, which exists but has not been observed. The population's distribution has a mean μ, which we are trying to estimate, and a variance σ^2, which, for our example, we assume is known. Because the population has not been observed in a census, we cannot specify the horizontal scale for the population in Chart 10.1.

The second distribution is the sample's distribution. As depicted in Chart 10.1, here we can specify the scale of numerical values for the horizontal axis. We can also find the location of the sample mean, $\overline{X}$, and, although the value of s^2 is not necessary for the present problem, we could estimate this value as well.

The third frequency distribution is the conceptual-only sampling distribution of means. While we do not have the resources to draw all possible samples of 4,422 families from the population of 78,425,000 U.S. families in 2007, we do have theoretical information about this distribution. First, we know from Section 9.8 that the expected value of its mean equals the known population mean (that is, $E(\mu_{\overline{X}}) = \mu$). We also know that its variance has an expected value equal to the population variance divided by the sample size (that is, $E(\sigma_{\overline{X}}^2) = \sigma^2/n$), and, because $n = 4,422$, we can assume the sampling distribution is normal per the Central Limit Theorem. In addition, since the sample observed is one member of the set of all possible samples, we know that the observed value of $\overline{X}$ ($84,550) lies somewhere along the horizontal axis. This means we can now rescale the horizontal axis of the sampling distribution of means as Z-values.

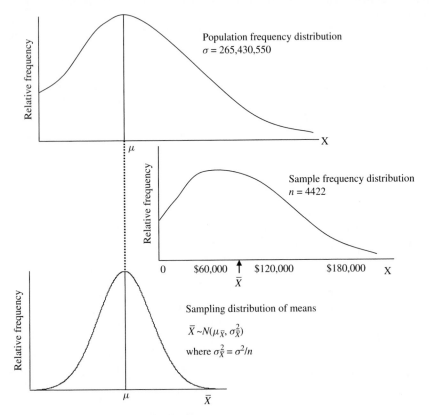

Population frequency distribution
$\sigma = 265,430,550$

Sample frequency distribution
$n = 4422$

0    $60,000 $120,000 $180,000   X
         $\bar{X}$

Sampling distribution of means

$\bar{X} \sim N(\mu_{\bar{X}}, \sigma_{\bar{X}}^2)$

where $\sigma_{\bar{X}}^2 = \sigma^2/n$

Chart 10.1 Three frequency distributions.

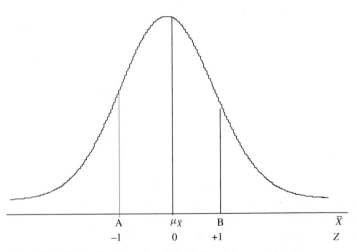

| A | $\mu_{\bar{X}}$ | B | $\bar{X}$ |
| -1 | 0 | +1 | Z |

Chart 10.2 Illustrating the rationale behind an interval estimate.

How do we go about the rescaling process? Let us turn to Chart 10.2 (p. 268), which is the sampling distribution of means from Chart 10.1. Here we have identified two points, A and B, from the $\overline{X}$ scale that have values of $84,055 (A) and $84,545 (B), respectively.

We can determine the corresponding Z-values after calculating the standard deviation of the sampling distribution of means, $\sigma_{\overline{X}}$, using equation (9.8):

$$\sigma_{\overline{X}}^2 = \frac{\sigma^2}{n} = \frac{265,430,550}{4,422} = 60,025$$

such that $\sigma_{\overline{X}} = 245$. We can now apply equation (9.12), renumbered here as equation (10.1):

$$Z = \frac{\overline{X} - \mu_{\overline{X}}}{\sigma_{\overline{X}}} \tag{10.1}$$

Let us first determine the Z-value associated with point A:

$$Z = \frac{A - \mu}{\sigma_{\overline{X}}} = \frac{\$84,055 - \$84,300}{\$245} = -1.00$$

Similarly, the value of Z associated with point B is

$$Z = \frac{B - \mu}{\sigma_{\overline{X}}} = \frac{\$84,545 - \$84,300}{\$245} = +1.00$$

What is the probability that $\overline{X}$ is located at any point between A and B? Using Excel, we determine that

$$\Pr(-1.00 \leq Z \leq +1.00) = 0.8413 - 0.1587 = 0.6827$$

Thus the probability that $\overline{X}$ is located at any point between A and B can be stated as

$$\$84,055 \leq \mu \leq \$84,545 \text{ at a confidence level of } 68.27 \text{ percent}$$

In light of the empirical rule, this is the result we would expect. That is, we know that, if a distribution is normal, 68.27 percent of all observations will lie within one standard deviation of the mean. In this example, we have a 68.27 percent chance[2] that the population mean of before-tax family income fell between $84,055 and $84,545 in 2007.

The above statement is a statistical inference in the sense that we can infer something about the value of the average 2007 family income in the U.S. based on the observed sample mean and for a certain level of confidence. Again, we remind ourselves that the population mean, μ, is a single value, which may or may not actually lie within the interval between $84,055 and $84,545, and that, in the absence of a census, we will never know its true value. However, we shall be able to go further with our conclusions after we discuss confidence levels in more detail.

Table 10.2 Alternative interval estimates and confidence levels for $\overline{X} = \$84,300$ and $\sigma_{\overline{X}} = \$245$

Z	$(\overline{X} - Z\sigma_{\overline{X}})$ ($)	$(\overline{X} + Z\sigma_{\overline{X}})$ ($)	Width of the interval estimate for μ ($)	Confidence level (%)
1.00	84,055.00	84,545.00	490.00	68.26
1.50	83,933.50	84,668.50	735.00	86.64
1.96	83,820.80	84,780.20	960.40	95.00
2.00	83,810.00	84,790.00	980.00	95.44
2.50	83,688.50	84,913.50	1,225.00	98.76
3.00	83,565.00	85,035.00	1,470.00	99.72

10.3 Confidence levels and the precision of an interval estimate

Let us first generalize the interval estimation procedure used with points A and B above. If we let Z take on values other than ± 1, we can determine the associated interval estimates using equation (10.2):

$$\overline{X} - (Z \times \sigma_{\overline{X}}) \leq \mu \leq \overline{X} + (Z \times \sigma_{\overline{X}}) \tag{10.2}$$

where $(Z \times \sigma_{\overline{X}})$ is defined as the **margin of error**. Applying equation (10.2) to the mean 2007 family income for a range of values for Z yields the interval estimates presented in Table 10.2.

Suppose we choose 2.00 as the value for Z such that we obtain an interval estimate for 2007 average family income of

$$(\$84,300 - \$490) \leq \mu \leq (\$84,300 + \$490) \quad \text{or} \quad \$83,810 \leq \mu \leq \$84,790$$

While we know that not all possible means will fall into this range, we can determine what percentage does. According to the empirical rule, 95.44 percent of the observations lie within two standard deviations ($Z = 2$) of the mean, μ From this, we can infer the probability that the mean family income of a random sample of 4,422 families falls between $83,810 and $84,790 is 95.44 percent. That is,

$$\Pr(\$83,810 \leq \mu \leq \$84,790) = 0.9544$$

This equation establishes the **confidence interval** for μ, where the parenthetical term is the **width of the interval estimate** and 0.9544 is the **confidence level**, or probability that the confidence interval includes the unknown population parameter under conditions of repeated random sampling of 4,422 families in the United States.

As indicated in Table 10.2, we can construct additional confidence intervals for μ by varying the confidence level. Suppose, for example, we want to use a 95 percent confidence level. Here we determine that the corresponding Z-statistic (using Excel according to the instructions at the end of the chapter) equals 1.96. This then yields a confidence interval of

$$\Pr(\$83,820 \leq \mu \leq \$84,780) = 0.9500$$

This tells us that, with repeated random sampling, 95 out of 100 samples would yield a mean family income that lies between $83,820 and $84,780.

Table 10.3 Alternative interval estimates and confidence levels for $\overline{X} = \$84,300$ and $\sigma_{\overline{x}} = \$24.50$

Z	$(\overline{X} - Z\sigma_{\overline{x}})$ ($)	$(\overline{X} + Z\sigma_{\overline{x}})$ ($)	Width of the interval estimate for μ ($)	Confidence level (%)
1.00	84,275.50	84,324.50	49.00	68.26
1.50	84,263.25	84,336.75	73.50	86.64
1.96	84,251.98	84,348.02	96.04	95.00
2.00	84,251.00	84,349.00	98.00	95.44
2.50	84,238.75	84,361.25	122.50	98.76
3.00	84,226.50	84,373.50	147.00	99.72

An important observation from Table 10.2 concerns the relationship between the width of the interval estimate and the confidence level. To understand that relationship, let us first note the three factors that contribute to the width of the interval estimate. Two of the factors, the sample mean ($\overline{X}$) and its standard deviation ($\sigma_{\overline{x}}$), are derived from the sample and are thus not subject to change by the researcher. The third factor, the Z-statistic, can be determined by the researcher to correspond to a desired confidence level. Since high levels of confidence are preferred, we typically select high values for Z.

But we recognize that, as we increase the Z-statistic, we also increase the margin-of-error ($Z \times \sigma_{\overline{X}}$) portion of the confidence interval. This, in turn, reduces the **precision** of the interval estimate (defined by how narrowly clustered the estimates are about the mean) because we are moving farther away from the point estimate, $\overline{X}$. Thus we observe a crucial tradeoff:

> the greater the confidence level, the less precise, or reliable, the interval estimate of μ will be.

In the language of economics, we might say that the cost of a relatively precise interval estimate (that is, a narrow width) is a low level of confidence. Thus, with a given sample of size n, we can only attain a higher level of confidence if we give up some precision in our confidence interval estimates.

We can, however, be both confident and precise. To compute the Z-statistic in equation (10.1), we must estimate $\sigma_{\overline{X}}$, which is calculated by dividing the population variance, σ^2, by the sample size, n. While we cannot modify σ^2 once the population has been specified, we may be able to increase the sample size, and, by so doing, increase the precision associated with a given confidence interval.

Let us return to before-tax mean family income in 2007 to illustrate the point. Recall that this statistic was derived from a sample of 4,422 families and that we assumed the population's variance equaled 265,430,550. Now suppose we increase the sample size 100-fold, to 442,200 families. As we observe in Table 10.3, the sample standard deviation now equals $24.50, which reduces the margin of error along with the width of the interval estimate associated with each confidence level.

For example, the width of the interval estimate associated with a 95 percent confidence level now equals $96.04 rather than the $960 range associated with the sample size of 4,422 families. Thus by increasing the sample size, we have gained a higher degree of precision without sacrificing the high level of confidence.

While increasing the sample size resolves the tradeoff between confidence and precision, we are aware that doing so is not a costless proposition. In addition, economists typically work with all available data in their empirical analyses, so enlarging the sample size may not be feasible. Thus, in practice, economists generally choose to use high confidence levels, most typically 95 percent and 99 percent, even though they recognize that doing so may reduce the precision of the confidence intervals.

Up to this point, we have described the confidence level as a measure of how confident we are that the value of the unknown population parameter falls within a given interval estimate. However, we cannot conclude from a single interval estimate, which is based on a point estimate generated from a single random sample, that the probability of this interval containing the true value of μ is, say, 95 percent. In other words, the estimated interval either contains the true population mean or it does not.

To correctly interpret the confidence level, we must think of the single interval estimate as one member of a large set of possible estimates. That is because the sampling distribution is a relative frequency distribution of *all possible values* of the sample mean. Because every possible sample has a sample mean and each sample mean could be the basis of an interval estimate, we recognize that, theoretically, there are as many possible interval estimates as there are possible samples of a given size. Thus the proper interpretation of the confidence level is as follows:

> By observing a single sample mean from a sample of 4,422 families, the value of μ has been estimated as
>
> $$\$83,820 \leq \mu \leq \$84,780 \quad \text{with 95 percent confidence.}$$
>
> If a very large number of sample means obtained from samples of $n = 4,422$ were observed, and similar estimates were constructed, 95 percent of the estimated intervals would contain the true value of μ.

This statement is a statistical inference because the interval estimate is derived from a randomly generated sample mean, and we *infer* that it contains μ. In addition, we acknowledge the uncertainty associated with this inference, expressed by the confidence level we select. Now we have one final important issue to address – what happens if we do not know the population variance?

10.4   The *t*-distribution

Assuming a population's variance is known allows us to estimate the margin of error associated with an interval estimate. However, that information is rarely available, leading us to estimate the population's variance from the sample drawn. In that process, we generate a second source of error, the so-called **estimating error**.

Because we want to measure any error associated with our statistical inference procedures, we now turn to the **Student's *t*-distribution**,[3] which simultaneously accounts for both the sampling error and the estimating error. The *t*-distribution is also very similar to the standard normal distribution. To demonstrate this, let us return to the Z-statistic defined by equation (10.1):

$$Z = \frac{\overline{X} - \mu_{\overline{X}}}{\sigma_{\overline{X}}}$$

where $\sigma_{\overline{X}} = \sigma/\sqrt{n}$. Because we do not know the population's standard deviation (σ), we must estimate it using equation (9.11):

$$\sigma_{\overline{X}} = \frac{s}{\sqrt{(n-1)}}$$

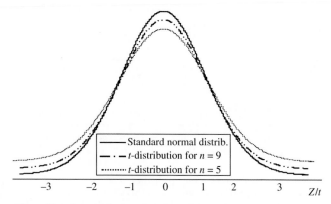

Chart 10.3 Standard normal and two *t*-distributions.

which is the unbiased estimator of the standard deviation. Thus the estimated standard deviation for the sample can be written as

$$\hat{\sigma}_{\overline{X}} = \frac{[s/\sqrt{(n-1)}]}{\sqrt{n}} = \frac{\sigma_{\overline{X}}}{\sqrt{n}} \tag{10.3}$$

which is more commonly known as the **standard error of the mean** for the sample. Note that we have added a hat ($^\wedge$) to distinguish this statistic from the known standard deviation, $\sigma_{\overline{X}}$.

Because we must estimate the standard deviation of the sample, the underlying distribution is now a *t*-distribution, which rescales the $\overline{X}$-statistic using the *t*-statistic:

$$t = \frac{\overline{X} - \mu_{\overline{X}}}{\hat{\sigma}_{\overline{X}}} \tag{10.4}$$

While the numerators of equations (10.1) and (10.4) are identical, the denominators are not. Specifically, in equation (10.4), $(n-1)$ is now a factor in determining the distribution's standard deviation. As a result, we will have a different *t*-distribution for each possible value of *n*. In addition, while the *t*-distribution, like the standard normal distribution, is symmetric about a mean of zero, it is flatter than the standard normal distribution for small values of *n*. As shown in Chart 10.3 (p. 273), this means that more of the *t*-distribution's area lies in its tails.

Let us now define the confidence interval for a given *t*-distribution:

$$\overline{X} - (t_{n-1} \times \hat{\sigma}_{\overline{X}}) \leq \mu \leq \overline{X} + (t_{n-1} \times \hat{\sigma}_{\overline{X}}) \tag{10.5}$$

where $t_{n-1} \times \hat{\sigma}_{\overline{X}}$ is now the margin of error. While similar to equation (10.2), we observe two differences. As we have established, unlike the standard normal distribution, we must calculate the standard error of the *t*-distribution, $\hat{\sigma}_{\overline{X}}$. The second difference is reflected in the subscript on the *t*-statistic, which indicates the degrees of freedom associated with the *t*-distribution.

The concept of **degrees of freedom** refers to the number of mathematical restrictions required to calculate the estimate of one statistic based on the estimate of another statistic.

Table 10.4 The *t*-statistics for selected degrees of freedom and confidence levels

Degrees of freedom (n − 1)	Confidence level		
	0.9000	0.9500	0.9900
10	1.812	2.228	3.169
20	1.725	2.086	2.845
30	1.697	2.042	2.750
∞	1.645	1.960	2.576

Source: All *t*-statistics calculated by author using Excel's TINV function, where the probability value equals 1 − confidence level. See the end of the chapter for detailed directions.

If we place no restrictions on the sample, we would have n degrees of freedom. In the case of the *t*-distribution, however, we must compute $\overline{X}$ to estimate the sample standard deviation, thereby losing one degree of freedom in the process. This means that the *t*-statistic used to estimate the confidence interval for the sample mean will have $(n − 1)$ degrees of freedom.

Let us now examine the *t*-statistics associated with different degrees of freedom and confidence levels that are presented in Table 10.4. We see, for example, that the *t*-statistic associated with 10 degrees of freedom ($n = 11$) and a 95 percent confidence level has a value of 2.228. We can now use these *t*-statistics in equation (10.5) to establish the **confidence limits** of a confidence interval, which are the two values that define the confidence interval's range.

Suppose, for example, we have a sample of 21 elements, and we want to establish the confidence limits for a 90 percent confidence interval. In this case, the *t*-statistic's value would be 1.725. If we increase the sample to 31 observations but retain the 90 percent confidence interval, the value for the *t*-statistic now equals 1.697. As this indicates, the value of the *t*-statistic declines as n increases for a given level of confidence.

We also note that when the sample size approaches (hypothetically) infinity, the *t*-statistic associated with a 95 percent confidence level converges to the Z-statistic's value for that level of confidence. This demonstrates that the limiting values of the *t*-statistic are, in fact, the same values as those for the Z-statistic.

This last result might lead us to conclude that, for sufficiently large values of n, we can use either the *t*-distribution or the standard normal distribution. While statisticians may follow this practice when $n \geq 30$, economists do not. Instead, economists always use the *t-distribution* to calculate confidence intervals, and, as we shall see shortly, to test hypotheses regardless of the sample size. The reason? Because we typically have to estimate the standard deviation. Thus, we will follow economic convention and employ the *t*-distribution throughout the remainder of the text.

10.5 Confidence intervals for the population mean when the variance of the population is not known

Let us now apply the procedures for estimating confidence intervals when the variance must be estimated from the sample. Here we will use the U.S. Consumer Expenditure Survey (CEX), an annual survey conducted by the U.S. Bureau of Labor Statistics that estimates average expenditures by various consumer groups for a variety of budget items.

Table 10.5 Confidence intervals for 2008 consumer expenditures on health care

Confidence level	t_{n-1}	Lower confidence limit ($)	Upper confidence limit ($)	Width of confidence interval ($)
0.900	1.645	2,883.81	3,067.95	184.14
0.950	1.960	2,866.17	3,085.59	219.42
0.990	2.576	2,831.69	3,120.07	288.38

Data source: U.S. Bureau of Labor Statistics (2008a), 2008 Consumer Expenditure Survey, Standard Error Tables, Table 1,300, Age of reference person: annual means, standard errors and coefficient of variation.

According to the 2008 *Consumer Expenditure Survey*, Standard Error Tables, average annual expenditures on health care ($\overline{X}$) for all consumer units was $2,975.88, with a standard error ($\hat{\sigma}_{\overline{X}}$) of $55.97 (U.S. Bureau of Labor Statistics 2008a). We can now estimate the confidence interval of mean expenditures on health care for all U.S. consumer units in 2008 using equation (10.5):

$$\$2,975.88 - (t_{n-1} \times \$55.97) \leq \mu \leq \$2,975.88 + (t_{n-1} \times \$55.97)$$

We now must determine the *t*-statistic, which requires us to identify *n* and to select a confidence level. There are approximately 14,000 consumer units in the integrated sample data collected annually through the CEX Diary and Interview Surveys, so we will use 13,999 (which is tantamount to using ∞) as our sample size. After selecting the commonly used 95 percent confidence level, we obtain a *t*-statistic equal to 1.960. Thus the interval estimate for health-care expenditures in 2008 is

$$\Pr(\$2,866.17 \leq \mu \leq \$3,085.59) = 0.9500$$

This result indicates that, if we were to take repeated random samples from the population, each with 14,000 consumer units, the population mean for annual expenditures on health care for all consumer units would fall between $2,866 and $3,086 for 95 percent of the time.

Now let us examine how the confidence interval changes across different confidence levels. As Table 10.5 illustrates, we first observe the same tradeoff between confidence and precision that we identified for the standard normal distribution. In addition, we can also comment on the likelihood that the population mean will fall within each confidence interval.

According to Table 10.5, there is at least a 90 percent chance that the true mean consumer expenditures on health care in 2008 fell between $2,884 and $3,068, and at least a 99 percent chance that the population mean fell between $2,832 and $3,120 based on the 2008 CEX results.

As we would expect, these confidence intervals will change if we modify the sample size or if we use the health-care expenditures for any subgroup of consumer units. For example, suppose we wanted to construct confidence intervals for health-care expenditures for, say, those between 25 and 34 years of age in 2008 using a 95 percent confidence level. After calculating the sample mean and standard deviation ($1,737 and $67, respectively), we next note the sample size ($n = 2,343$) and then determine the *t*-value (1.9610). This then yields a 95 percent confidence interval of ($1,606, $1,868). In less technical terms, there is a 95 percent probability that the true mean consumer expenditures on health care for those between 25 and 34 years of age fell between $1,606 and $1,868 in 2008.

10.6   Confidence intervals for proportions, percentages, and rates

Economists often work with proportions, percentages, and rates of variables, so it is useful to know how to construct confidence intervals for the three measures. Before turning to the formulas for such confidence intervals, let us first explain how the three measures are related.

A **proportion** (p) is the fraction of a population that has a certain characteristic, such as the number of people in a population who are members of the labor force. Suppose, for example, that we draw a sample (n) of 100 individuals, and we observe that 76 of them are categorized as being in the labor force. Thus the proportion in the labor force is 0.76 as determined by equation (10.6):

$$\text{proportion} \equiv p = \frac{d}{n} \tag{10.6}$$

where d equals the number of elements with the specified characteristic. We can convert this result to a **percentage** by multiplying the proportion by 100:

$$\text{percentage} = p \times 100 \tag{10.7}$$

We could also convert the proportion to a **rate** per a certain number of element (l)

$$\text{rate} = p \times l \tag{10.8}$$

For example, the rate per 1,000 elements is obtained by multiplying the proportion by 1,000.

We now turn to constructing a confidence interval for a proportion. We first note that the following method applies only to proportions. As a result, if our variable is in percentages or is stated as a rate, we must first convert it into a proportion, using equation (10.7) or (10.8), before we apply the confidence interval formula stated below.

We will also assume that we can use the Z-distribution. Technically, the underlying distribution for a proportion is a binomial distribution. However, as n increases, the binomial distribution approaches a normal distribution. Specifically, if $n > 30$, $n \times p > 5$, and $n \times (1 - p) > 5$, we can use the Z-statistic in our calculations. Assuming the above conditions hold, the formula for constructing a confidence interval for a proportion is

$$CI = p \pm \left(Z \times \sqrt{\frac{p \times q}{n}} \right) \tag{10.9}$$

where $q = (1 - p)$ and the radical (square root) term is the **standard error of proportions**.

Suppose we draw a sample of members of the labor force from the March 2009 Current Population Survey (U.S. Bureau of Labor Statistics 2009). Of the 67,014 individuals, we determine that 1,259 were classified as part-time for economic reasons but usually work full-time, yielding a proportion (p) of 0.0188. After determining the values of q (0.9812), the standard error of proportions (0.00052), and the Z-statistic (1.96), we can apply equation (10.7) to obtain the 95 percent confidence interval:

$$95 \text{ percent } CI = 0.0188 \pm (1.96 \times 0.00052) = (0.0178, 0.0198)$$

Accordingly, we find that for every 95 out of 100 randomly drawn samples, the proportion of the labor force that was employed part-time for economic reasons but usually worked full-time fell between 0.0178 (or 1.78 percent) and 0.0198 (or 1.98 percent). As we observe, this is a very narrow width for the confidence interval and is attributable, in part, to the large sample size used.

10.7   Confidence intervals for differences between means and proportions

We conclude our discussion of confidence intervals by examining how to make statistical inference statements about the difference between two means or two proportions. While the procedures are similar to those previously discussed, our focus is now on the difference between a *pair* of sample results.

Suppose, for example, we continue with the March 2009 Current Population Survey (U.S. Bureau of Labor Statistics 2009) of members of the labor force. Sorted by age group, we determine that the average income for the 7,736 individuals in the 50–54 year age group was $58,071.12 (with a standard deviation of $61,224.18), and that the average income for those 5,367 individuals in the 55–59 year age group was $57,131.85 (with a standard deviation of $58,258.61). While the two averages appear to be different from one another, we must account for the errors associated with the sample statistics before we can make such a determination.

Before we turn to the statistical evidence, we can note that the higher average earnings for those in the 50–54 year age group are consistent with the empirical evidence for U.S. incomes. That is, we often observe that after age 54, average incomes tend to decline to below the peak earning years of 45 to 54. While we will leave the discussion of why that obtains to labor economists, our concern is whether the difference between average earnings for the two age groups is statistically different from zero.

One method for answering the question is to construct a confidence interval for the difference. In this case, if we obtain a confidence interval that includes the value of zero, then we cannot conclude, at the specified level of significance, that the average incomes are statistically different from each other.

The confidence interval for the difference between two means is calculated using equation (10.10):

$$CI = (\overline{X}_2 - \overline{X}_1) \pm [Z \times \text{standard error of } (\overline{X}_2 - \overline{X}_1)] \tag{10.10}$$

As we see, we must determine the **standard error of the difference between two means** before we can calculate the confidence interval. Fortunately, the variances of the two means are additive, so we use equation $(10.11)^4$ to obtain the standard error:

$$\text{standard error of } (\overline{X}_2 - \overline{X}_1) = \sqrt{\text{Var}(\overline{X}_1) + \text{Var}(\overline{X}_2)} \tag{10.11}$$

Table 10.6 presents the factors needed to construct the 95 percent confidence interval.

According to equation (10.10), the 95 percent confidence interval for the differences in mean income between the 50–54 year age group and the 55–59 year age group is

$$CI = -\$939.35 \pm (1.96 \times \$1,056.94) \quad \text{or} \quad (-\$3,010.95, \$1,132.24)$$

We observe that the 95 percent confidence interval does, in fact, contain the value of zero. This means that the mean incomes for the two age groups are not statistically different from one another. More formally, we cannot with 95 percent confidence conclude that the mean

Table 10.6 Factors for calculating a 95 percent confidence interval for the difference between mean incomes for two age groups

Difference in mean income for 50–54 years and for 55–59 years	−$939.35
Standard error of difference between two means (equation (10.1))	$1056.94
Variance for 50–54 years	484,602.45
Standard deviation for 50–54 years	$61,224.18
Number of persons 50–54 years	7,736
Standard error for 50–54 years (equation (10.3))	696.13
Variance for 55–59 years	632,513.09
Standard deviation for 55–59 years	$58,258.61
Number of persons 55–59 years	5,367
Standard error for 55–59 years (equation (10.3))	795.31
Z-statistic	1.96

Source: Author's calculations based on March 2009 Current Population Survey sample (U.S. Bureau of Labor Statistics (2009)).

income for those 55–59 years of age was statistically different from the mean income for those in the 50–54 year age group in March 2009.

The final type of confidence interval we will construct is for the difference between two proportions. We begin by stating the two general formulas for such calculations. The first defines the confidence interval for the difference in two proportions, p_1 and p_2, which is

$$CI = (p_1 - p_2) \pm (Z \times \text{standard error of difference in proportions}) \qquad (10.12)$$

while the formula for calculating the **standard error of the difference between two proportions** is

$$\text{standard error of difference in proportions} = \sqrt{\frac{p_1 q_1}{n_1} + \frac{p_2 q_2}{n_2}} \qquad (10.13)$$

Let us now construct a 99 percent confidence interval to examine the difference in the proportion of male and female workers who work, per their typical schedule, part-time for economic reasons. As we observe in Table 10.7, this status pertained to 1,042, or 3.01 percent, of males and 1,312, or 4.05 percent, of females. The question now is whether the difference between the two proportions is statistically different from zero.

Table 10.7 Number of persons by labor force status and sex

	Male	*Female*
Full-time schedules	26,563	21,236
Part-time for economic reasons, usually FT	834	425
Part-time for non-economic reasons, usually PT	2,878	7,262
Part-time for economic reasons, usually PT	1,042	1,312
Unemployed FT	2,855	1,713
Unemployed PT	415	479

Data source: Author's calculations based on March 2009 Current Population Survey sample U.S. Bureau of Labor Statistics (2009).

Let us formally construct the 99 percent confidence interval (the standard error's calculation is left to the reader):

$$CI = -0.0103 \pm (2.58 \times 0.00275) \quad \text{or} \quad (-0.1478, -0.1337)$$

Because zero is not in the 99 percent confidence interval, we can conclude that we are at least 99 percent confident that the difference in the proportion of males and females who worked part-time for economic reasons, per their usual work arrangements, is statistically different from zero.

We can assess these results from another perspective. Our calculations indicate that the margin of error, or the parenthetical value $(Z \times SE)$, is 0.00708. Because this value is considerably smaller than the difference between the proportions (-0.0103), adding it to or subtracting it from the difference will not change the sign of the difference. As a result, we can conclude that we are at least 99 percent confident that the difference in the proportion of males and females who worked part-time for economic reasons, per their usual work arrangements, is statistically different from zero. In other words, the percentage of women who work part-time for economic reasons is statistically different from the percentage of men who work part-time for economic reasons.

Summary

Our discussion of statistical estimation has focused on how a sample mean can be used to estimate the mean of the population from which the sample was drawn. The first estimate that was discussed, a point estimate, is a descriptive statement that μ is estimated to be equal to $\overline{X}$. While this provides valuable information, a point estimate is not a statistical inference because there is no procedure for measuring the uncertainty associated with that estimate.

Economists thus prefer the second estimate type, an interval estimate, when utilizing samples, because we can measure the uncertainty in the form of a probability. Interval estimates of the mean can be calculated using one of two underlying probability distributions, the standard normal distribution or the t-distribution, depending on our knowledge of the population's variance. Following the convention used by economists, we use the t-distribution to calculate interval estimates of the population mean, since the population variance is rarely known. However, we did employ the standard normal distribution when working with proportions.

Calculating an interval estimate requires us to specify a confidence level, which indicates the proportion of correct statements about the value of μ if a very large number of interval estimates were made from different sample means for a given sample size. The size of the confidence level is inversely related to the precision of an interval estimate as measured by the interval estimate's width. Consequently, for any given estimation problem, a gain in precision can be achieved only at the cost of a loss in the level of confidence, and vice versa. Gains in both precision and level of confidence can be achieved, although the cost of doing so is to increase the sample size, which may not always be feasible.

Constructing interval estimates for the mean at specific levels of confidence can provide useful information about the range of values that is likely to include the unknown population parameter. We have also seen how confidence interval estimates can be computed for proportions, and differences between means and between proportions. In all cases, such estimates can be used to conduct tests of hypotheses regarding the designated statistic. This important method of statistical inference in empirical economics is the topic of the next chapter.

Concepts introduced

Confidence interval
Confidence level
Confidence limits
Degrees of freedom
Estimating error
Interval estimate
Margin of error
Percentage
Point estimate
Precision
Proportion
Rate
Sampling error
Standard error of the difference between two means
Standard error of the difference between two proportions
Standard error of the mean
Standard error of proportions
Student's *t*-distribution
Width of the interval estimate

Box 10.1 Using Excel to determine the critical values associated with a probability distribution

The standard normal distribution

In Chapter 9, we used Excel's NORMSDIST (or NORM.S.DIST) function to calculate the probability associated with a specific numerical value for a standard normal continuous probability distribution (z).

Suppose now we know the probability but want to determine the z-value. Here we can use Excel 2007's NORMSINV function (or the NORM.S.INV function in Excel 2010) to determine the numeric value for Z that corresponds to a given probability associated with the standard normal *cumulative* distribution function.

NORMSINV specifically calculates the value of Z associated with the probability (x) that Z is *less than or equal to* the z test statistic:

NORMSINV(x) = Z

That is, the NORMSINV function determines the point along the horizontal axis of the distribution for which the gray area lies to the left of that point.

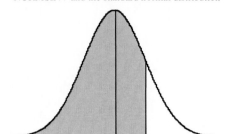

NORMSINV and the standard normal distribution

Suppose, for example, that we want to find the Z-value associated with $x = 0.95$. If we insert 0.95 into the NORMSINV function, we find that z equals 1.6449. In other words,

NORMSINV(0.95) = 1.6449

Because of the extra precision that can be obtained with Excel, we will utilize the NORMSINV function to calculate the z-values associated with the standard normal distribution. In recording such results, we will follow convention and report four decimal places.

The t-distribution

Economists more typically work with the t-distribution. Both Excel 2007 and Excel 2010 have built-in functions that calculate the t-value associated with a specific probability.

The built-in function in Excel 2007 is the TINV function, which requires two arguments:

TINV(x, degrees_freedom) $= t^{\alpha}_{n-1}$

where **x** is the probability associated with the *two-tailed* Student's t-distribution and **degrees_freedom** are the *degrees of freedom* for the sample size.

Suppose we wanted to know the t-value associated with 95 percent probability when $n = 100$. Inserting 0.05 and 99 into TINV yields

TINV(0.05, 99) = 1.9842

Excel's t-distribution functions in Excel 2010

Both the TDIST and TINV functions in Excel 2007 have been replaced in Excel 2010 by two new functions to better reflect the various uses of the t-distribution in statistical estimation and hypothesis testing.

Determining the critical t-values for estimating confidence intervals and conducting hypothesis testing

In Excel 2007, we use the **TINV** function to obtain the critical t-values for estimating the confidence intervals associated with the sample mean and for testing hypotheses about the sample mean. Excel 2010 includes two functions that perform the same tasks.

The first, the **T.INV.2T** function, is used to obtain the critical t-value for estimating confidence intervals and the critical value used in the decision rule for two-tail hypothesis tests of the sample mean.

The **T.INV.2T** function returns the two-tail critical t-value for a particular probability associated with the t distribution:

> **T.INV.2T** (probability, deg_freedom)

where **probability** corresponds to the level of significance ($= 1$ – level of confidence) chosen and **deg_freedom** are the degrees of freedom associated with the t-distribution.

The **T.INV** function in Excel 2010 is used to determine the critical t-value stated in the decision rule for one-tail hypothesis tests.

T.INV returns the left-hand critical t-value for a particular probability:

> **T.INV** (probability, deg_freedom)

where **probability** corresponds to the level of significance ($= 1$ – level of confidence) established and **deg_freedom** are the degrees of freedom associated with the distribution.

To obtain the right-hand critical value, use **–T.INV** (probability, deg_freedom).

Determining the area underneath the t-distribution and the p-value for a test statistic

In Excel 2007, we use the **TDIST** function to obtain the probability associated with a t-distribution and the p-value associated with the test statistic in hypothesis testing.

Excel 2010 includes two functions, **T.DIST.2T** and **T.DIST.RT**, which perform those tasks. The **T.DIST.2T** function determines the area of the t-distribution that lies between the plus and minus critical t-values (the Acceptance Region). We will primarily use this function to determine the p-value associated with the test statistic in two-tail hypothesis tests.

The **T.DIST.2T** function returns the area underneath a two-tail t distribution:

> **T.DIST.2T** (x, deg_freedom)

where x is the test statistic and **deg_freedom** are the degrees of freedom associated with the distribution.

The **T.DIST.RT** function in Excel 2010 is used to determine the area of the t-distribution that lies to the right of the critical value, x (the Rejection Region). We are more likely to use this function to determine the p-value associated with the test statistic for a one-tail hypothesis test.

The **T.DIST.RT** function returns the area (or p-value) associated with the right-tail of the t distribution:

> T.DIST.RT(x, deg_freedom)

where x is the test statistic and **deg_freedom** are the degrees of freedom associated with the distribution.

Applications using the Excel 2010 functions

Suppose we collect data for 100 recently sold homes in a particular city, and we find that the average sales price for the sample was \$225,000. Given a sample standard deviation of \$50,000, the corresponding test statistic equals 0.9950.

Action	Excel 2007 function	Excel 2010 function	Result
Determine the critical *t*-value for constructing a 95 percent confidence interval for the mean sales price	**TINV(0.05,99)**	**T.INV.2T(0.05,99)**	1.9842
Determine the critical *t*-value for testing if the average sales price from the sample is statistically *different from* the state average of \$220,000	**TINV(0.05,99)**	**T.INV.2T(0.05,99)**	1.9842
Determine the critical *t*-value for testing if the average sales price from the sample is statistically *less than* the state average of \$220,000	**–TINV(0.1,99)**	**T.INV(0.05,99)**	−1.6604
Determine the critical *t*-value for testing if the average sales price from the sample is statistically *greater than* the state average of \$220,000	**TINV(0.1,99)**	**–T.INV(0.05,99)**	1.6604
Determine the *p*-value for the two-tail hypothesis test (the average sales price from the sample is statistically *different from* the state average of \$220,000)	**TDIST(0.9950,99,2)**	**T.DIST.2T(0.9950,99)**	0.3222
Determine the *p*-value for the two-tail hypothesis test (the average sales price from the sample is statistically *less than* the state average of \$220,000)	**–TDIST(0.9950,99,1)**	**–T.DIST.RT(0.9950,99)**	−0.1611
Determine the *p*-value for the two-tail hypothesis test (the average sales price from the sample is statistically *greater than* the state average of \$220,000)	**TDIST(0.9950,99,1)**	**T.DIST.RT(0.9950,99)**	0.1611

Differences between TINV and NORMSINV

• **NORMSINV** uses the cumulative probability distribution, but **TINV** does not.

The **NORMSINV** function returns the Z-value associated with the area of the *cumulative* probability distribution, but **TINV** returns the t-value associated with the area in the *two tails* of the non-cumulative probability distribution.

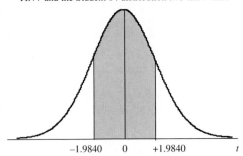

TINV and the Student's *t* distribution two-tail *t*-value

Suppose we want the *t*-value associated with the probability (α) of 0.05 and 100 degrees of freedom. In terms of the chart, this means we want the *t*-value that will define an area equal to 0.025 in *each* end of the distribution. That value, according to Excel's **TINV** function, is 1.9840. (We must manually add the $\pm$ signs to the results.)

To generate the *cumulative* probability distribution for the *t*-statistic, enter α **times 2** ($\alpha \times 2$) for the **x** argument.

- **TINV** uses a different probability value from the **NORMSINV** function.

For the **NORMSINV** function, we used the probability associated with the confidence level. Thus for a 95 percent level of confidence, we entered a value of 0.95.

For **TINV**, the *probability* equals

$$x = 1 - \text{confidence level}(\alpha)$$

To determine the *t*-statistic associated with a 95 percent confidence level, we enter a value for **x** of 0.05.

- When using **TINV**, we must include the degrees of freedom associated with the *t*-distribution.

	NORMSINV	TINV
Function specification	=NORMSINV(z)	=TINV(x, degrees_freedom)
Value of probability	z = confidence level (α)	$x = (1 - \text{confidence level } (\alpha))$
Area defined by statistic	$\Pr(Z \leq z)$	$\Pr(\|x\| > t)$, or
		$\Pr(x < -t)$ and $\Pr(x > t)$
Degrees of freedom	not applicable	$n - 1$

For a 95 percent confidence level and a sample size of 4422 (as from the 2007 *Survey of Consumer Finances*), the two functions would be entered as follows:

NORMSINV(0.95)   TINV(0.05, 4421)

Exercises

1. Suppose you have selected a random sample of $n = 5$ from a normal distribution. Compare the standard normal Z-values with the corresponding t-values used to construct the following confidence intervals:

   (a) 80 percent confidence interval;
   (b) 90 percent confidence interval;
   (c) 95 percent confidence interval;
   (d) 99 percent confidence interval.

   Discuss what you observe about the similarities and differences between the two distributions in words and using a sketch of the two distributions.

2. The mean and standard deviation of a random sample of n measurements are equal to 33.9 and 3.3, respectively.

   (a) Find a 95 percent confidence interval for μ if $n = 100$.
   (b) Find a 95 percent confidence interval for μ if $n = 400$.
   (c) Find the widths of the confidence intervals calculated in parts (a) and (b). Discuss the effect on the width of a confidence interval if we quadruple the sample size while holding the confidence level constant.

3. Consumer expenditures on health care vary noticeably across age groups. According to the 2008 Consumer Expenditure Survey, the U.S. had 11,481 consumer units (CUs) in the 75 years and older age group, whose average annual expenditures (μ) were \$4,413, with a population standard deviation ($\sigma_{\overline{X}}$) of \$16,354.

   (a) Determine the mean ($\overline{X}$) and standard error ($\sigma_{\overline{X}}$) for a sample size of 1,331 consumer units.
   (b) Determine the probability that the mean ($\overline{X}$) of a random sample of 1,331 consumer units fell between \$4,000 and \$5,000. Sketch a chart where the horizontal axis depicts both the $\overline{X}$ and Z scales to illustrate the results.

4. Average annual expenditures on health care for young people are considerably lower than those for the 75 years and older age group. Suppose we draw a random sample of 954 consumer units for the under-25 years age group, and calculate a mean ($\overline{X}$) of \$682 and a sample standard deviation (s) of \$5,350.

   (a) Explain which one of the two probability distributions discussed in Chapter 10 is more appropriate to use for estimating the confidence intervals in parts (b) and (d) below.
   (b) Compute the 95 percent confidence interval for the population mean.
   (c) Fully interpret the results in part (b).
   (d) Demonstrate how the confidence interval changes for

      i. a 90 percent confidence level
      ii. a 99 percent confidence level.

   (e) Discuss what the differences in the results in parts (b) and (d) indicate about the statistical estimation of the population mean.

5. According to the "New Residential Sales in July 2010" press release from the U.S. Census Bureau, the number of new homes sold in the United States fell by 12.4 percent between

June 2010 and July 2010. During the same period, new home sales fell by 13.9 percent for the Northeast portion of the U.S.

(a) For both regions, construct a 90 percent confidence interval for new homes sold. Use the following standard errors: 6.57 percent for the U.S. and 22.37 percent for the Northeast.

(b) Explain why these results indicate that we are at least 90 percent confident that the number of new homes sold declined in the U.S. between June 2010 and July 2010, but we are not 90 percent confident that new home sales in the Northeast fell during the same time period.

6. Use the data in the table below to determine if the difference in average income is statistically different from zero for these specified age groups:

(a) 60–61 years and 62–64 years;
(b) 60–61 years and 65–69 years;
(c) 70–74 years and 75 years and over.

	Mean income ($)	Standard deviation ($)	Number in age group
60–61 years	61,990.99	69,595.94	1,607
62–64 years	60,668.29	47,139.28	1,451
65–69 years	62,259.80	65,957.98	1,147
70–74 years	57,202.96	66,429.38	464
75 years and over	52,175.34	66,429.38	292

7. The following data on members of the U.S. labor force was obtained from the March 2009 Current Population Survey: total number of men in labor force $= 34,587$; total number of women in labor force $= 32,427$; number of men who were unemployed but usually worked full-time $= 415$; and number of women who were unemployed but usually worked full-time $= 479$.

(a) Construct and interpret the 99 percent confidence interval for the difference in the proportion of men and women who were unemployed but usually worked full-time in March 2009.

(b) Explain what the results indicate about differences in the labor force experiences of men and women.

11 Statistical hypothesis testing of a mean

The confidence intervals defined in Chapter 10 are, by their very nature, two-sided constructions. That is, we calculate two values, the confidence limits, along the horizontal axis of the underlying distribution to determine the range of probable values associated with a population parameter, the mean. But economists are generally more interested in the likelihood that a sample result is greater than, or less than, the population parameter for the variable. Thus we would like a "one-sided" method that would permit such an assessment, which we find in the statistical hypothesis testing of an individual mean.

11.1 Testing hypotheses in economics: an analogy to criminal trials

Discussions of hypothesis testing in economics often begin with an analogy to criminal trials in which the guilt or innocence of a defendant is judged. We, too, will begin here in order to provide a brief review of the concepts underlying this important statistical method.

In both criminal trials and statistical hypothesis testing, the procedures used have evolved from social processes, and society has accepted these procedures as being those best suited for guiding the difficult task of decision-making. The elements of this decision-making process in criminal trials are as follows.

- *Truth*, or the facts of the criminal act, which could be known under ideal conditions of observation. As jury interviews and fictional portrayals suggest, "the whole truth" is seldom known.
- A *hypothesis*, or tentative belief concerning truth. In criminal trials in the United States, the hypothesis of "not guilty" is expressed by the well-known principle "innocent until proven guilty."
- *Evidence*, or facts that may be useful in reaching a decision even though some important facts are unknown. Evidence is always incomplete. Furthermore, some evidence is consistent with the hypothesis while other evidence is not.
- A *decision* to accept or reject the hypothesis after evaluating the evidence. In trials, the decision expressed by the verdict "not guilty" means the hypothesis of "not guilty" is accepted, while the verdict of "guilty" means the hypothesis is not accepted.
- A *criterion* for accepting or rejecting the hypothesis. The criterion used in U.S. criminal trials is expressed by the judge's instruction to the jury to return a verdict of "not guilty" if there is any "reasonable doubt" that a "guilty" decision is incorrect.
- The *implications* of the decision as a reflection of truth. Ideally, the logical process of decision-making should lead to an accurate statement of truth. However, the possibility of making incorrect decisions must be taken into account, because the available evidence

is not complete and because the truth cannot be known with certainty. Thus it is possible for a guilty defendant to be found "not guilty" or for an innocent defendant to be found "guilty."

The process of reaching a decision in a trial by jury is essentially a process of choosing between two alternative (and unknown) states of truth. In point of fact, the defendant is either innocent or guilty. The jury decides either to acquit or to punish the defendant. In each decision made by a jury, there are thus two ways to make correct decisions (the defendant is innocent and the jury decides not guilty; or the defendant is guilty and the jury decides guilty) and two ways to make errors (the defendant is innocent but the jury decides guilty; or the defendant is guilty but the jury decides not guilty). Despite the possibility that an error can occur, our society has deemed the process outlined in the steps above to be an acceptable decision-making process in criminal trials. As we shall now see, economists follow similar steps when employing the statistical method of hypothesis testing.

11.2   An overview of hypothesis testing in economics: evaluating truth in advertising

We now turn to the procedures that economists use to test statistical hypotheses. While there are many types of hypotheses we could test, here we describe the procedure for the type of statistical hypotheses we will test in regression analysis, where we ask if the statistical estimate generated from a particular sample is "close" in value to the population parameter.

Answering this question suggests, however, that we know the value of the population mean. As we have discussed, knowing the value of the population mean, μ, requires census data, which is rarely available. Economists must therefore set a hypothetical value for μ. But which hypothetical values of μ are appropriate to use? Here we must move from the realm of statistics to economic theory and policy. This is to say that economists must be guided by economic theory or economic policy in establishing appropriate hypothetical values of μ.

We can illustrate this key point with the following example. Here hypothesis testing will be used to evaluate if an advertising claim accurately reflects the average cost of a company's product.

Suppose a health insurance company uses average expenditures from the 2008 Consumer Expenditure Survey (CEX) to claim in its advertising that its clients spend less on health insurance than the average elderly (65 years and older) American. After receiving complaints from consumers that their insurance rates from this company were, in fact, roughly equal to or higher than the national average, the U.S. Federal Trade Commission (FTC), the federal agency that evaluates truth-in-advertising claims in the health insurance industry, opens an investigation. One component of the investigation is evaluating the truthfulness of the company's claim using hypothesis testing.

Following the protocol for such an evaluation, an FTC analyst begins by establishing the pair of hypotheses to test. Specifically, she wants to test if the average cost of this health insurance company's product for elderly clients is less than $2,844.47, the 2008 national average cost of health insurance for those 65 years and older. This means that the analyst will test the following null hypothesis (H_0) and alternative hypothesis (H_a):

$$H_0 : \mu \geq \$2,844.47 \quad \text{versus} \quad H_a : \mu < \$2,844.47$$

If the statistical results lead the analyst to reject the null hypothesis, she will conclude that the insurance company's advertising claims appear to be accurate.

Why did the analyst establish the hypotheses such that the *alternative* hypothesis reflects the "truth" posited by the health insurance company? The answer relates to the errors that arise in statistical inference. While the analyst may correctly reject, or not reject, the null hypothesis, she also knows that it is possible she might either incorrectly reject the null hypothesis when it is true or incorrectly "accept" (technically, not reject) the null hypothesis when it is, in fact, false. Let us now consider these possibilities in light of the false advertising claim.

First, suppose that the null hypothesis is true, which means that the health insurance company's premiums are actually greater than or equal to the national average, but the statistical evidence leads to the inaccurate conclusion that the null hypothesis be rejected. Here the insurance company will "get away with" their misleading advertising claims because the empirical evidence does not suggest that the company violated the truth-in-advertising law.

Fortunately, the analyst can control the likelihood that this type of error, known as a **Type I error**, will occur. That is, in the second step of the hypothesis testing procedure, the analyst establishes the probability of making a Type I error when she selects the level of significance (α). As we would expect, the analyst wants this level to be as small as possible, so she selects the frequently used level of 5 percent ($\alpha = 0.05$). In addition, by choosing a value for α, the analyst concurrently specifies the confidence level associated with the test, since $(1 - \alpha)$ equals the level of confidence. Thus the analyst has chosen a 95 percent confidence level for the test.

As a rule, economists are typically more concerned about Type I errors than Type II errors, in which a false null hypothesis is not rejected. Here, this means that the analyst wants to minimize the possibility that a misleading advertising claim will not be detected through the statistical analysis. Such an outcome is consistent with the FTC's mission of protecting consumers from false advertising practices.

The analyst next establishes the rule she will use to evaluate the null hypothesis. This first requires her to identify the underlying probability distribution for health insurance expenditures. Because she must estimate the sample variance, she uses the *t*-distribution. She can now state the conditions under which she will reject the null hypothesis, the so-called decision rule. We note that she does this *before* collecting the sample to insure that the statistical results obtained do not influence the decision rule:

Reject H_0 if $t < -t_{n-1}^{0.05}$

According to the decision rule, the analyst will reject the null hypothesis (that the company's average costs for health insurance are greater than or equal to the national average) if the test statistic based on the sample statistics is less than the critical value, $-t_{n-1}^{\alpha}$.

We note that the analyst states the decision rule to reflect the *alternative* hypothesis, again following economic convention. Thus, if the test results allow the analyst to reject the null hypothesis, she will be at least 95 percent confident in that conclusion.

The analyst next draws a random sample of 100 of the company's elderly customers. She can now determine the critical *t*-value, which for a probability of 0.05 and 99 degrees of freedom $(-t_{99}^{0.05})$ equals -1.6604. This means she uses the following decision rule:

Reject H_0 if $t < -1.6604$

Note that here the critical value is negative because the alternative hypothesis states that the sample mean is *less than* the hypothesized value of \$78,500.

The random sample also yields the statistics needed to construct the test statistic, t, previously defined by equation (10.4) and renumbered here as equation (11.1):

$$t = \frac{\overline{X} - \mu_0}{\hat{\sigma}_{\overline{X}}} \tag{11.1}$$

For this specific sample, the average annual expenditures on health insurance for those in the sample ($\overline{X}$) are calculated to be \$2,769.64, while the sample standard deviation (s) equals \$67.30. From the latter statistic, the analyst then determines that the sample standard error ($\hat{\sigma}_{\overline{X}}$) equals

$$\hat{\sigma}_{\overline{X}} = \frac{s}{\sqrt{n}} = \frac{\$67.30}{\sqrt{100}} = \$6.73$$

Identifying \$2,844.47 as the hypothesized value, μ_0, the analyst now applies equation (11.1) to calculate the test statistic:

$$t = \frac{\overline{X} - \mu}{\hat{\sigma}_{\overline{X}}} = \frac{\$2,769.64 - \$2,844.47}{\$6.73} = -11.1189$$

The analyst is ready to evaluate the null hypothesis. Because the test statistic (-11.1189) is less than the critical t-value (-1.6604), the analyst concludes that she can, with at least 95 percent confidence, reject the null hypothesis. In other words, the statistical results from the sample of 100 clients leads the analyst to conclude, with a high degree of confidence, that the average annual expenditures on health insurance by the health insurance company's clients is not greater than or equal to the national average. Since this conclusion suggests

Table 11.1 Steps used by FTC analyst to test hypothesis regarding health insurance company's advertising claims

Step	General procedure	Analyst's actions
1	State the null and alternative hypotheses to be tested	• Establishes the hypothetical value for μ using 2008 CEX results • States the hypotheses as H_0: $\mu \geq \$2,844.47$ versus H_a: $\mu < \$2,844.47$
2	Establish the level of significance (or probability of committing a Type I error)	• Selects $\alpha = 0.05$
3	State the decision rule used to reject the null hypothesis	• Determines that she will reject the null hypothesis if $t < t_{n-1}^{0.05}$
4	Construct the test statistic from sample results	• Draws sample and calculates values for $\overline{X}$ (\$2769.64) and $\hat{\sigma}_{\overline{X}}$ (\$6.73) • Constructs the test statistic, t (-11.1189)
5	Make a decision about the null hypothesis	• Because $-11.1189 < -1.6604$, rejects the null hypothesis • Concludes, with at least 95 percent confidence, that the health insurance company's average costs are less than the national average for the 65 years and older age group

the company's advertising claims were not misleading, the FTC is unlikely, based on these results, to pursue a criminal complaint against the company for false advertising.

As summarized in Table 11.1, this extended example took us through the steps used by economists in hypothesis testing. (We will add a last step in Section 11.10.)

Let us now discuss in greater detail the factors that must be considered in each of the steps just outlined.

11.3 Economic hypothesis testing: stating the hypotheses

The first step in hypothesis testing – establishing the hypotheses to be tested – is the most crucial. In the truth-in-advertising case discussed above, the hypotheses were clearly dictated by the question posed. That is, the FTC analyst evaluated a health insurance company's claim that its elderly (65 years and over) clients pay less on average for health insurance than the national average for the same age group. Using the national average expenditures on health insurance from the 2008 Consumer Expenditure Survey as the hypothesized value, the analyst then determined with a high degree of confidence that average expenditures for the company's clients were, based on the sample statistics, less than the national average.

Economists and policy-makers also use hypothesis testing to evaluate economic policies. Before we turn to an example regarding affordable housing assistance, we want to keep the following point in mind. Even though we will be illustrating how statistical inference is used to evaluate hypotheses about an economic policy, decisions regarding economic policy *cannot* be based solely on the mechanical application of statistical methods. That is, even though empirical evidence from hypothesis tests can inform policy decisions, such results must *not* be the only consideration in considering a particular course of action. Even in the truth-in-advertising case, it is unlikely the FTC would base its decision about the insurance company's claim solely on the results of the hypothesis test performed. Thus, in the practical arena of decision-making, economists and policy-makers consider all pertinent sociological, economic, and political factors, as well as any additional pertinent empirical evidence on top of what is learned through hypothesis testing.

Let us now consider a policy for providing housing assistance to low-income families. Since the 1930s, a central tenet of U.S. economic policy has been to provide aid to families at the bottom end of the income distribution. This aid has taken many forms, from Social Security and Temporary Assistance to Needy Families (TANF) to subsidies for child care, health care, and affordable housing. Whether in the form of direct transfer payments or as cash subsidies for specific goods or services deemed necessary for a decent life, the purpose of these programs has been to provide economic security to low-income families.

Many of these policies set eligibility requirements based on income thresholds. This means that income must be below a specific dollar level to qualify for such programs as TANF, Food Stamps, or affordable housing. To determine eligibility, income is compared to the established threshold, and, if it falls below the threshold, the recipient gains access to the program.

Suppose policy-makers develop an affordable housing program based on Freddie Mac's Home Possible® mortgages program, which provides access to low-interest, no-downpayment, single-family housing loans to residents of a region.[1] The policy is established such that eligibility is determined by how each region's average family income compares to a predetermined threshold level of income. If the region's average family income falls below the threshold, residents of that region are then eligible to apply for these mortgages.[2]

Officials must now determine what threshold value to use for each region, which they will then use as the hypothesized value for μ. This value will be determined in consultation with economists, urban planners, and other experts, who may consider such factors as the income required to purchase a single-family home in the area, the costs of buying and maintaining a home with certain characteristics, and targets such as a desired rate of home-ownership for the area. For our purposes, we will use the income eligibility values set by Freddie Mac in 2007 as the highest value of average family income that would qualify a region's residents for the low-cost mortgages.

We can now pose the hypotheses we wish to test regarding regional income. Following statistical conventions, we establish both a null hypothesis and an alternative hypothesis. The **null hypothesis**, denoted H_0, is defined as the proposition that the true population parameter, μ, has a specified relationship to a particular hypothesized numerical value, μ_0. The **alternative hypothesis**, denoted as H_a, is then the relationship not specified in the null hypothesis, and it can take one of three forms:

- H_a: $\mu \neq$ *hypothesized value*. This form is referred to as a **two-tail hypothesis test**. Such an alternative hypothesis will evaluate if the true mean value, μ, is *either* statistically greater than *or* statistically less than the hypothesized value, μ_0. Two-tail hypothesis tests may be used to evaluate if a sample mean is statistically different from zero, in which case H_a would be specified as $\mu \neq 0$. Two-tail alternative hypotheses are also used when we have a numerical value (μ_0) against which we want to compare the sample mean but we do not know if the sample mean is less than or greater than the hypothesized value. Here H_a would be specified as $\mu \neq \mu_0$.
- H_a: $\mu <$ *hypothesized value*. This form is referred to as a **one-tail hypothesis test**. Here the alternative hypothesis postulates that the true mean value is statistically less than the hypothesized value, μ_0.
- H_a: $\mu >$ *hypothesized value*. This form is also called a **one-tail hypothesis test**. Here the alternative hypothesis suggests that the true mean value is statistically greater than the hypothesized value, μ_0.

As shown in Figure 11.1, each type of alternative hypothesis has a corresponding null hypothesis that states the other possible relationship(s) between μ and μ_0.

Let us now consider the case of Anoka County, Minnesota, for which Freddie Mac established a 2007 income eligibility threshold of $78,500. In the context of the policy, this means we want to test if the county's average family income is less than that threshold value, leading us to test the following pair of hypotheses:

$\mu < \$78,500$ and $\mu \geq \$78,500$

Type of hypothesis test	Null hypothesis	Alternative hypothesis
Two-tail hypothesis test	$\mu = \mu_0$	$\mu \neq \mu_0$
One-tail hypothesis test	$\mu \geq \mu_0$	$\mu < \mu_0$
One-tail hypothesis test	$\mu \leq \mu_0$	$\mu > \mu_0$

Figure 11.1 Types of null and alternative hypothesis tests, where μ_0 is the hypothesized value of the population parameter.

To determine which is the null hypothesis and which is the alternative, we follow economic convention, and establish the alternative hypothesis to reflect the claim of "truth" under consideration. In determining mortgage loan eligibility, the claim of truth is that the region's mean family income is less than that of the qualifying threshold of $78,500 established by officials. This means that economists would first specify the alternative hypothesis as $\mu < \$78,500$, and then the null hypothesis as the opposite case, $\mu \geq \$78,500$. Formally, the hypotheses to be tested in the affordable housing eligibility example are

$$H_0 : \mu \geq \$78,500 \quad \text{versus} \quad H_a : \mu < \$78,500$$

The alternative hypothesis here tells us that we will be conducting a one-tail hypothesis. This determination will not only help us specify the null hypothesis, but also inform the critical value we use in the third step.

Let us make one final point. Even though economists write their hypotheses such that the alternative reflects the claim of truth, the actual process of hypothesis testing assumes that the null hypothesis, H_0, is true. If the sample results fail to contradict H_0 beyond some high degree of confidence, then we must conclude that we cannot reject H_0.

Not rejecting the null hypothesis does *not*, however, mean that we accept H_0 as the truth. It simply means that the statistical evidence does not allow us to rule out H_0 as *possibly* being true. Only when the evidence strongly contradicts H_0 can we reject the null hypothesis and, for a given level of confidence, conclude that the alternative hypothesis can be accepted. In this example, we want to be strongly confident that the mean family income in a region is less than $78,500 before making the low-cost loans available to families in the region.

11.4   Economic hypothesis testing: selecting the level of significance

The second step in hypothesis testing establishes the **level of significance** we want associated with our test. In less technical terms, this step sets the level of confidence we want to have in our decision about the null hypothesis. That is, we want to be very confident that the conclusion reached about the null hypothesis is not simply the result of sampling variability, or a matter of chance based on the sample drawn from the population.

Formally, this step establishes a specific level of significance, α, which economists typically set at either 0.01 or 0.05 (1 percent or 5 percent, respectively). This value corresponds to the probability of making a Type I error, or rejecting a true null hypothesis, which economists want to minimize.

The reason becomes clear when we consider what officials are trying to achieve in the affordable housing case. Because government funds for low-cost mortgages are limited, officials want to restrict loan access to only those regions where average income falls below the threshold. In other words, they do not want an ineligible region to gain access to these funds. As a result, officials want to minimize the likelihood of rejecting the null hypothesis, which would make a region's residents eligible for the loans, when the region's average income is, in fact, greater than or equal to the threshold value.

By minimizing the likelihood of committing a Type I error through choosing the level of significance, we also determine the level of confidence. This results from the relationship between the level of significance, α, and the level of confidence, which equals $(1 - \alpha)$. It also explains why economists most commonly set the level of significance at 0.05 or 0.01, because those values correspond to 95 percent and 99 percent levels of confidence, respectively. Thus by selecting a low level of significance, we both reduce the likelihood of a Type I error and increase the confidence we have in our conclusion regarding the null hypothesis.

For the affordable housing example, officials select 0.05 as the level of significance. This, then, means that officials have established the likelihood of incorrectly rejecting the null hypothesis (that regional income is greater than or equal to $$78,500) at 5 percent and that they will be at least 95 percent confident in the decision made about the null hypothesis.

11.5   Economic hypothesis testing: establishing the decision rule

After specifying the significance level, the next step in hypothesis testing states the **decision rule**, which is the condition under which the null hypothesis will be rejected. The rule utilizes a critical value associated with the underlying distribution and the test statistic based on the sample's results. It also mirrors the relationship postulated in the *alternative* hypothesis.

In the affordable housing case, officials must estimate the sample standard deviation for each region, so they will apply the following decision rule to each region:

$$\text{reject } H_0 \text{ if } t < t_{n-1}^{\alpha}$$

The statistic, t_{n-1}^{α}, is known as the **critical value** for the hypothesis test, and it corresponds to the value of the t-distribution with $n - 1$ degrees of freedom and a level of significance equal to α. It also establishes the acceptance and rejection regions for the null hypothesis, which we will discuss further in Section 11.8.

To determine the critical value, officials must not only select a significance level, but also determine the degrees of freedom associated with the sample. In the case of Anoka County, they draw a sample of 2,079 residents. This yields a critical value for $t_{2,078}^{0.05}$ equal to -1.6456. We now have the following decision rule for Anoka County:

$$\text{reject } H_0 \text{ if } t < -1.6456$$

As with the truth-in-advertising case, the alternative hypothesis indicates which side of the t-distribution will be relevant for our analysis. Since we want to determine if the sample statistic is less than the hypothesized value, the critical value will be negative, with the rejection region lying to the left of the t-distribution's mean of zero.

11.6   Economic hypothesis testing: constructing the test statistic and making a decision about the null hypothesis

The sample of the 2,079 residents also yields the necessary arguments for constructing the test statistic, t. Specifically, the survey generates an average[3] family income in Anoka County of $74,304 and a sample standard error of $2,142.[4] Applying equation (11.1) yields the following test statistic for the Anoka County sample:

$$t = \frac{\$74,304 - \$78,500}{\$2,142} = -1.9587$$

Officials can now evaluate the null hypothesis for Anoka County. Because the test statistic (-1.9587) is less than the critical t-value (-1.6456), officials reject the null hypothesis that the county's average family income is greater than or equal to the threshold income value, being at least 95 percent confident in that decision. They thus conclude that residents of Anoka County are eligible for the low-cost mortgage loans through the Home Possible mortgages program.

The hypothesis test results are confirmed by the 95 percent confidence interval constructed for the Anoka County sample. According to equation (10.5), the 95 percent confidence interval of

$$95 \text{ percent CI} = \$74,304 \pm (-1.6456 \times \$2,142) = (\$70,779, \$77,829)$$

indicates that there is a 95 percent probability that the true average family income in Anoka County in 2007 fell between $70,779 and $77,829. Since the upper limit is less than the income eligibility threshold value of $78,500, officials conclude that they are 95 percent confident that average family income in Anoka County is below the income eligibility threshold, thereby qualifying residents for the low-cost housing mortgages.

11.7 Testing hypotheses versus estimating confidence intervals

While it can be useful to have multiple methods for evaluating a decision based on statistical inference, it is also redundant. Thus we now examine more closely why economists use hypothesis testing over estimating confidence intervals as the preferred method for statistical inference.

Economists prefer hypothesis testing because the theoretical structure of hypothesis tests, unlike confidence interval estimation, allows for a precise description of the errors implicit in the process. That is, when we estimate a confidence interval, we can only determine the probability of accepting a true hypothesis, but we cannot measure the probability of rejecting a true hypothesis thereby committing a Type I error. Thus, in the Anoka County case, all we can conclude from the confidence interval is that, if we were to draw repeated random samples of 2,079 families from the population, the true population mean will fall between $70,779 and $77,829 for 95 percent of the time. However, we cannot assess the likelihood of drawing a sample (of size 2,079 families) that generates a sample mean that falls outside that range.

Hypothesis testing does allow us to estimate this likelihood. When we set the level of significance, α, we establish the probability of erroneously rejecting the null hypothesis based on the sample results. Given that we want to measure the inevitable sampling error associated with statistical inference, the ability to determine the Type I error in hypothesis testing renders hypothesis testing a superior statistical inference procedure to interval estimation.

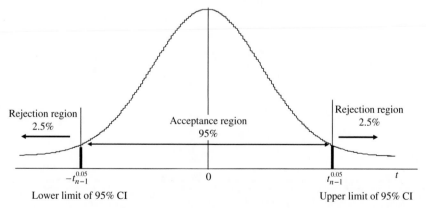

Chart 11.1 Relationship between a two-tail hypothesis test and a 95 percent confidence interval.

There is, however, a strong connection between estimating confidence intervals and two-tail hypothesis tests. As shown in Chart 11.1 (p. 295), a two-tail hypothesis test divides the t-distribution into an acceptance region and rejection regions at each end of the distribution.

The critical values that demarcate the acceptance and rejection regions of the t-distribution are, in fact, equal to the limits of the confidence interval. Thus, when conducting a two-tail hypothesis test, we reach the same conclusion regarding the sample mean as we would by estimating the confidence interval. However, we cannot determine the likelihood of reaching an erroneous conclusion from the information provided by the confidence interval, and we cannot evaluate if a sample mean is statistically greater, or less, than a hypothesized value for the population parameter.

11.8 Evaluating a statistical rule in terms of a Type I error

In the affordable housing example, officials established a 5 percent level of significance. This meant officials were willing to accept that an ineligible region would receive access to the low-cost mortgages, thereby committing a Type I error, 5 percent of the time.

Establishing the probability of a Type I error, and thereby the critical value for the decision rule, also determines the rejection and acceptance regions associated with a true null hypothesis. The **rejection region** corresponds to the set of values of a test statistic that leads to rejecting the null hypothesis, while the **acceptance region** refers to the set of test statistic values for which the null hypothesis cannot be rejected. The demarcation value between the two areas is the "critical value." As Chart 11.2 (p. 296) demonstrates, the rejection region for Anoka County when $\alpha = 0.05$ is the set of all t-values that lie to the left of -1.6456, while the acceptance, or non-rejection, region is the set of all t-values that are greater than or equal to -1.6456.

Now suppose that officials decide that a 5 percent level of significance is too high, since it leads to too many ineligible regions having access to funding. They thus decide to reduce the level of significance to 1 percent (0.01), which changes the decision rule to:

reject H_0 if $t < t_{2,078}^{0.01} = -2.3281$

As we observe in Chart 11.3 (p. 297), lowering the level of significance shrinks the rejection region by moving the critical t-value farther away from the mean.

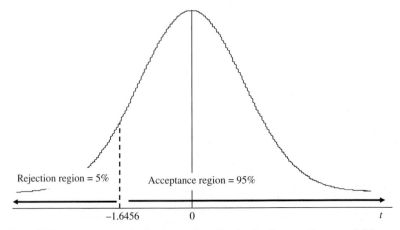

Rejection region = 5%   Acceptance region = 95%

-1.6456 0 t

Chart 11.2 Acceptance and rejection regions for Anoka County when $\alpha = 0.05$.

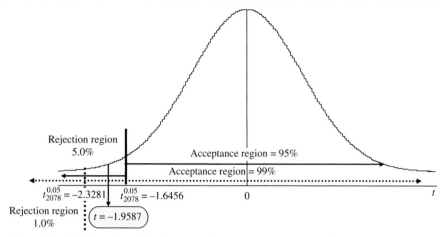

Chart 11.3 Effects of lowering the level of significance for the affordable housing hypothesis test for Anoka County, Minnesota.

Officials now revisit the case of Anoka County. While we will assume for discussion purposes that officials can still use the previously estimated test statistic of -1.9587, doing so is incorrect. The proper procedure is for officials to draw a new sample of 2,079 families from Anoka County and then construct a new test statistic based on the new sample's results. As this indicates, doing so avoids the suggestion that we modified the decision rule after constructing the test statistic. Such practice is unacceptable because it implies that we might have manipulated our results to reach a particular conclusion regarding the null hypothesis, thereby violating the principle of honesty at the heart of responsible statistical analysis.

The officials now can no longer reject the null hypothesis for the established level of significance. In other words, officials cannot reject that average family income in Anoka County is greater than or equal to $78,500, so its residents are no longer eligible for the low-cost mortgages.

As this example demonstrates, reducing the significance level to further reduce the Type I error also makes it more difficult to reject the null hypothesis. Thus, while officials are more confident in their conclusions regarding the null hypothesis, they have also effectively further limited the likelihood that a region will be declared eligible for the low-cost loans.

Given the limited funds available, officials might further reduce the level of significance below 0.01 to further reduce the probability of committing a Type I error. However, as is often the case in economics, doing so is not without costs. Specifically, reducing the probability of making a Type I error *increases* the probability that we might accept, or not reject, a hypothesis that is actually false. That is, it raises the likelihood of committing a Type II error.

11.9 Evaluating a statistical rule in terms of a Type II error

Once officials reduced the level of significance to 0.01, they could no longer reject the null hypothesis for Anoka County. But that does not mean they actually accepted that average income in Anoka County was greater than or equal to $78,500. Indeed, there is a chance they might erroneously not reject the null hypothesis when, in fact, average income was less than $78,500.

Decision	State of nature	
	H_0 is true	H_0 is not true
Reject H_0	Decision is incorrect; Type I error; probability $= \alpha$	Decision is correct
Do not reject H_0	Decision is correct	Decision is incorrect; Type II error; probability $= \beta$

Figure 11.2 Consequences in hypothesis testing.

As shown in Figure 11.2, the cost of not rejecting the null hypothesis when H_0 is false is called a **Type II error**. Denoted by β (Greek lower-case "beta"), the probability of committing a Type II error is analogous to acquitting a guilty defendant in a criminal trial. In the case of mortgage assistance for regions, the Type II error would be committed if families in an eligible region were not granted mortgage assistance.

Determining the value of β is considerably more complex than choosing a value for α. It essentially requires that we know the true value of the population mean, μ, which is then tested against the hypothesized value, μ_0. As we have learned, it is very unlikely we will know the true value of μ. In addition, because most economists consider Type I errors to be of greater concern than Type II errors, we will not discuss the latter in any technical detail.

Several observations about Type II errors are, however, in order. First, if we reduce the likelihood of committing a Type I error (that is, if we reduce the value of α), we simultaneously increase the likelihood of committing a Type II error *as long as the sample size remains the same*. (While this might suggest that $\beta = (1 - \alpha)$, that is not the case.) Thus for a given sample size, we face a tradeoff between the two errors.

The only way to avoid this tradeoff is to increase the sample size, which has the effect of reducing both Types I and II errors. Unfortunately, increasing the sample size is often not feasible in economics, where we rely on samples drawn by others. In addition, even if we could increase the sample size, such action is not costless. Given how economists establish and test their hypotheses, economists focus primarily on reducing Type I errors and largely ignore Type II errors.

11.10 The *p*-value and hypothesis testing

The multi-step hypothesis testing procedure discussed thus far has been at the heart of statistical inference in economics for many years. Recent developments in computing power and statistical software have introduced a relatively new concept in statistical inference, the *p*-value, which provides a second method for evaluating the truth of the null hypothesis. The **p-value** measures the significance of the statistical evidence *assuming* that the null hypothesis is true. More concretely, the *p*-value tells us the likelihood of obtaining a test statistic at least as large (in absolute value) as that calculated from the sample. In other words, it is a measure of how convincing our statistical evidence is regarding our conclusion about the null hypothesis.

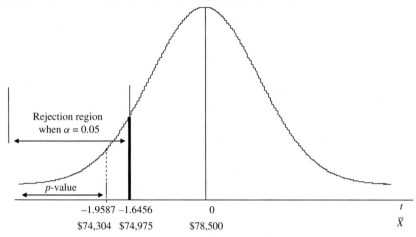

Chart 11.4 The *p*-value for Anoka County when $\alpha = 0.05$.

Let us now turn to Chart 11.4 (p. 299) to further explain how the *p*-value is used in hypothesis testing. Again, we use the results from Anoka County for a 5 percent level of significance, and, as we observe, the corresponding rejection region is demarcated by a critical *t*-value of -1.6456. We have also determined that the test statistic of -1.9587 lies in the rejection region to the left of the critical value, which led us to reject the null hypothesis.

The test statistic of -1.9587 demarcates a second region under the *t*-distribution. While we deduce that this area represents less than 5 percent of the *t*-distribution's area, we can also measure its size precisely by taking the inverse of the *t*-distribution. The resulting area equals the probability of obtaining a test statistic that is at least as extreme as the value calculated from the sample *given that the null hypothesis is true*. This probability, or the *p*-value, is also known as the **observed level of significance**.

For Anoka County, the *p*-value determines the probability of obtaining a test statistic that is less than -1.9587 when the null hypothesis ($\mu \geq \$78,500$) is true. (In terms of the actual sample statistics, this is akin to determining the probability of obtaining a sample mean that is less than the sample average family income of $74,304 for Anoka County.) Thus the *p*-value is

$$\Pr(t > |-1.9587|) = 0.0251$$

This result indicates that the probability of calculating a test statistic at least as large as $|-1.9587|$ is 2.51 percent *assuming* the true value of mean family income is $78,500.

We also observe that the *p*-value associated with Anoka County's average family income ($p = 0.0251$) is smaller than the 5 percent level of significance ($\alpha = 0.05$) we used for the hypothesis test. This relationship provides us with a second way for making a decision about the null hypothesis. Specifically, it is as follows:

- If the *p*-value (p) is smaller than the significance level (α), we can reject H_0.
- If the *p*-value (p) is larger than the significance level (α), we cannot reject H_0.

Thus another interpretation of the *p*-value is that it gives us the smallest level of significance at which a null hypothesis can be rejected given the observed sample statistic.

The *p*-value provides us with additional information about our conclusion regarding the null hypothesis. Because the *p*-value determines the likelihood of obtaining a *t*-value less than our test statistic when the null hypothesis is true, it measures precisely the probability of committing a Type I error. Thus for the Anoka County sample, when the null hypothesis is true, we have a 2.51 percent probability of erroneously rejecting the true null hypothesis.

Given the relationship between the level of significance (α) and the level of confidence $(1 - \alpha)$, we can also interpret the *p*-value in terms of how confident we are in our decision about the null hypothesis. For the Anoka County example, this means we can say we are 97.49 percent confident that average family income in Anoka County in 2007 was less than \$78,500 given the results obtained from the sample. Thus determining the value for $(1 - p) \times 100$ indicates how confident we are in rejecting H_0.

Let us now compare the two methods used in hypothesis testing to examine a new region's eligibility for the Home Possible mortgages, that of St. Louis County, Minnesota. Here, officials have evaluated local housing conditions and determined that the income eligibility threshold will be set at \$55,700. This yields the following pair of hypotheses to test:

$$H_0 : \mu \geq \$55,700 \quad \text{versus} \quad H_a : \mu < \$55,700$$

Rejecting the null hypothesis again means that residents will be eligible to apply for Home Possible mortgages.

The level of significance for the test is established next. Suppose that the officials evaluating St. Louis County are less concerned about committing a Type I error than were those officials evaluating Anoka County, so they set the level of significance equal to 0.10. They then draw a sample of 2,523 residents from the county, and establish the following decision rule:

$$\text{reject } H_0 \text{ if } t < t_{2,522}^{0.10}, \quad \text{where } t_{2,522}^{0.10} = -1.2819$$

Alternatively, officials could use the following decision rule based on the sample's *p*-value:

$$\text{reject } H_0 \text{ if } p < \alpha, \quad \text{where } \alpha = 0.10$$

From the sample drawn, officials estimate average family income to equal \$55,809, with a sample standard error of \$1,803, from which they construct the following test statistic, *t*:

$$t = \frac{(\$55,809 - \$55,700)}{\$1,803} = 0.0605$$

Using the traditional hypothesis test (*t*-statistic) approach, we observe that the test statistic (0.0605) is not less than the critical value (-1.2819). This means that we cannot, with 90 percent confidence, reject the null hypothesis. We must therefore conclude that average family income in St. Louis County is *not* statistically less than the income eligibility threshold set by Freddie Mac.

Let us now determine the *p*-value for St. Louis County using Excel (see the directions at the end of this chapter):

$$p = 0.4759$$

According to the corresponding decision rule, we observe that the sample's *p*-value is greater than the 10 percent (0.10) level of significance. Thus we cannot reject the null hypothesis that average family income in St. Louis County is greater than or equal to the income eligibility threshold. In fact, we are only 52.41 percent confident that we could correctly reject the null hypothesis when the null hypothesis is true, a value considerably lower than the 90 percent level of confidence established for the hypothesis test.

Which procedure do economists employ to test the null hypothesis? It is standard to include both the test statistic and the *p*-value when reporting our statistical results so they can be compared with the critical *t*-statistic and level of significance, respectively. As for interpreting the *p*-value in terms of the level of confidence, it is more common to do so when we can reject the null hypothesis. As the last statement in the previous paragraph suggests, once we determine that we cannot, for the specified level of significance, reject the null hypothesis, there is little to be learned from determining the level of confidence that does obtain.

We conclude with one final point. The *p*-values generated by statistical software packages may be reported as a value of zero (0.0000), although if enough decimal places are added, we will eventually obtain a non-zero value. This is expected due to the asymptotic characteristic of the *t*-distribution. Because the *p*-value will never actually equal zero, we can never be 100 percent confident in our conclusion regarding the null hypothesis, since there is always some probability of sampling error.

Summary

Economists begin the statistical procedure of hypothesis testing with an assumption about a hypothetical value of a population's parameter. After establishing an alternative hypothesis to reflect an expected result for the population parameter, they then evaluate the null (non-alternative) hypothesis using statistics generated from a random sample.

Because statistical inference procedures always include the possibility of reaching an incorrect conclusion about the null hypothesis, economists try to minimize such errors through their choice of the significance level for the test. This choice also establishes the lowest level of significance at which the null hypothesis can be rejected. This measure of reliability, known as the *p*-value, allows economists to assess more precisely the test results *vis-à-vis* the hypothesized value of the population parameter.

Concepts introduced

Acceptance region
Alternative hypothesis
Critical value
Decision rule
Level of significance
Null hypothesis
Observed level of significance
One-tail hypothesis test
p-value
Rejection region
Two-tail hypothesis test
Type I error
Type II error

Box 11.1 Using Excel 2007 to determine the *p*-value in hypothesis testing

As discussed in Box 10.1, Excel 2010 has two built-in functions, T.DIST.2T and T.DIST.RT, that can be used to determine the area associated with the *p*-value. We now present directions for using Excel 2007's built-in function, TDIST, to measure the area associated with the *p*-value.

The TDIST function determines the probability (or area) of the *t*-distribution for a numerical value (*x*) associated with a *t*-test statistic. This function

TDIST(X, deg_freedom, tails)

requires three arguments: **X** is the numerical value of the statistic to be evaluated, **deg_freedom** is the degrees of freedom associated with the sample size, and **tails** indicates the number of distribution tails to return.

In hypothesis testing, these three arguments correspond to the following:

- **X** is the test statistic obtained from the sample results;
- **deg_freedom** equals $(n-1)$, the number of observations minus the degree of freedom lost because we must estimate the sample mean; and
- **tails** is the type of hypothesis test being conducted. In both the FTC and affordable housing examples, we conducted one-tail hypothesis tests, so tails $= 1$; if we had conducted a two-tail hypothesis test, then tails $= 2$.

We can use **TDIST** to determine the *p*-value associated with the FTC hypothesis test:

TDIST$(11.1189, 99, 1) = 2.0179$E19   or $p = 0.000000000000000000020179$

This result confirms the FTC analyst's decision to reject the null hypothesis. That is, she is confident in rejecting the claim that the health insurance company charges their clients as much as, or more than, the national average for health insurance premiums.

Notes

- Excel's TDIST returns the probability that $(X > x)$. Given the symmetry of the *t*-distribution, we can also use its result to determine $\Pr(X < x)$. (See the next note.)
- The test statistic must be entered as a positive value. If the test statistic is negative, use its absolute value as the argument for X.
- The *p*-value generated will be a proportion. To obtain the percentage form, multiply the result by 100.

Exercises

1. Sketch the *t*-distribution for the FTC truth-in-advertising results. Include all relevant horizontal axis values for $\overline{X}$ and t, along with the rejection and acceptance regions for $\alpha = 0.05$. Discuss how this chart supports the conclusion reached by the FTC analyst.
2. A sample of 152 gas stations within the I-494/694 corridor of the Minneapolis–St. Paul, Minnesota, metropolitan area yields an average price for a gallon of regular gasoline

of $2.77, with a standard deviation of $0.05. (Source: Data obtained by author from <http://minnesotagasprices.com/GasPriceSearch.aspx> for the period between 3 p.m., 19 September 2010, and 3 p.m., 20 September 2010.)

(a) Evaluate the following hypothesis at a 95 percent level of confidence: For the week of 20 September 2010, was the price of a gallon of regular gasoline within the I-494/694 corridor of the Minneapolis–St. Paul, Minnesota, metropolitan area greater than the Minnesota state average of $2.766? (Source: U.S. Energy Information Administration (2010), Weekly Retail Gasoline and Diesel Prices.)

(b) Evaluate the following hypothesis at a 95 percent level of confidence: For the week of 20 September 2010, was the price of a gallon of regular gasoline within the I-494/694 corridor of the Minneapolis–St. Paul, Minnesota, metropolitan area greater than the U.S. national average of $2.723?

3. In 1999, Kelley Pace and other researchers corrected housing data from the Boston area used by Harrison and Rubinfeld to study the willingness to pay for clean air as manifested in hedonic housing prices. That dataset was used to generate the following statistics.

	Mean	*Standard deviation*	*Number of observations*
Boston area	$22,533	$9,197	506
Cambridge	$23,650	$11,963	30

(a) Evaluate the hypothesis that the average median value of owner-occupied units in the Boston area was statistically less than the median value of $24,100 for owner-occupied homes in the United States in 1973. Use a 99 percent confidence level.

(b) Evaluate the hypothesis that the average median value of owner-occupied units in the town of Cambridge was statistically less than the median value of $24,100 for owner-occupied homes in the United States in 1973. Use a 99 percent confidence level.

4. A sample of 15,620 observations of women in the labor force from the March 1993 Current Population Survey generated the following descriptive statistics.

	Mean (in weeks)	*Standard deviation*	*Number of observations*
Average weeks of work by all wives in the labor force	36.45	10.11	15,620
Average weeks of work by wives who *did* receive health insurance through husband's job	34.63	10.72	7,337
Average weeks of work by wives who *did not* receive health insurance through husband's job	38.07	9.25	8,283

(a) Using a 95 percent confidence interval, test the following hypothesis: Did the wives who *did* receive health insurance through their husband's job work fewer hours per week than wives who *did not* receive health insurance through their husband's job?

(b) Provide an economic explanation for your findings in part (a).

5. Each year, the U.S. Census Bureau publishes family income data for various characteristics, including geographical region.

Mean family income, 2008

U.S. National	$79,634
Northeast	$88,722
Midwest	$76,432
South	$73,640
West	$85,596

Data source: U.S. Census Bureau (2010c), Income, Table F-6, Regions – Families (all races) by median and mean income.

Suppose we want to compare average family income in specific states with average family income for the U.S. as a whole and by the region in which the state is located. Collecting data on family income from the 2008 American Community Survey, we obtain the following statistics.

	Region	State mean family income ($)	Number of families in sample	Sample standard deviation ($)
Pennsylvania	Northeast	84,911.11	14,498	75,584.48
Minnesota	Midwest	77,127.76	25,468	72,465.63
North Carolina	South	77,680.66	9,991	66,859.27
Oregon	West	$79,095.76	33,908	$71,135.45

Data source: U.S. Census Bureau (2008a), American Community Survey, 2008.

(a) Test the hypothesis that each state's average family income was different from the U.S. national average in 2008. Use a 95 percent confidence level.

(b) Test the hypothesis that each state's average family income was different from its regional average in 2008. Use a 90 percent confidence level.

Part IV

Relationships between two variables

Description and statistical inferences

Our discussion thus far has focused on describing the statistical characteristics of economic variables across elements (Part I) and across time (Part II). We have also examined (Part III) the statistical inference procedures for estimating confidence intervals and testing hypotheses about the mean of an economic variable based on a sample drawn from a given population.

In Part IV, we turn to statistical methods used by economists for assessing the quantitative relationships between two economic variables. These methods are particularly popular because economic theory typically posits how two variables are related to one another.

The process of translating economic theory into quantitative statements is a key element in the statistical methods that follow. However, economists do not begin their analysis by specifying the exact quantitative terms that describe such theoretical statements. Instead, they begin by hypothesizing the directional relationship anticipated between two variables based on applicable economic theory. Once such a theoretical relationship is established, economists collect appropriate data and then apply appropriate statistical methods for describing the hypothesized relationship. They conclude their analysis by evaluating the strength of the statistical relationship between the two variables based on their theoretically informed hypotheses. However, while our statistical results might support the theorized relationship between the variables, such methods *can never* be used to *prove* that the underlying economic theory is correct.

12 Correlation analysis

12.1 Statistical relationships between two variables

Economic analysis often focuses on understanding the relationship between two economic variables. As beginning students in economics, we learn, for example, how economists try to explain such two-variable relationships as those between product prices and the quantities buyers demand, the cost of factor inputs and output production levels, and changes in GDP and unemployment rates. In all these **bivariate relationships** (two-variable relationships), economic theory offers at least one explanation of how the two variables are related. Thus, in the first case mentioned, economists hypothesize that, if the price of a product increases, an individual consumer, *ceteris paribus*, will likely reduce how much of the product they purchase, thereby reflecting the so-called "law" of demand. In mathematical terms, this theory leads economists to postulate a negative, or inverse, relationship between prices and quantities demanded.

But merely asserting such a relationship is insufficient. Consequently, economists employ statistical methods to evaluate if this hypothesized theoretical relationship is consistent with the empirical evidence. As a first step, we want to determine if a pair of economic variables are, in fact, statistically related to one another using correlation analysis.

Let us begin our discussion of correlation analysis by examining a relationship that has interested economists since Adam Smith's time: the relationship between human capital and a nation's economic health. According to Smith and subsequent human capital theorists, the greater the average level of education attained by a nation's adult population, the more skilled and trainable that population will be. A more highly skilled workforce, in turn, means more productive workers who can generate higher levels of output. Mathematically, this leads economists to hypothesize a positive relationship between educational attainment and national product.

Postulating that levels of human capital and national output in both developed and developing nations are related, we collect data for 140 countries in 2000 from the World Bank. (See Box 12.1 for information about the World Bank's Data Catalog.) Here we will measure human capital by the average years of schooling completed by adults (aged 15 years and over), and national output by GDP per capita, purchasing power parity (PPP)[1] (reported in constant 2005 international dollars). We then construct a scatter plot to visually inspect the relationship.

As Chart 12.1 (p. 308) suggests, it appears that educational attainment and GDP per capita are positively related across the 140 nations in our sample. However, because we recall from Chapter 8 that "eyeballing" the data does not constitute a legitimate method for assessing a statistical relationship between two variables, we now turn to the formal statistical method

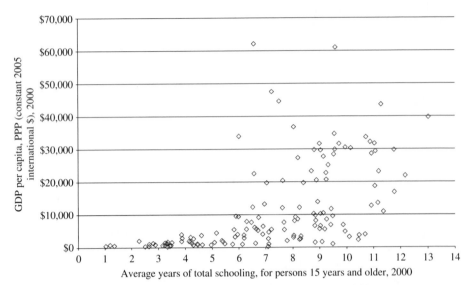

Chart 12.1 Average years of schooling completed and GDP per capita, 2000.

Data sources: GDP per capita: World Bank (2010b), World Development Indicators. Average years of schooling: World Bank (2010a), EdStats Database.

of correlation analysis. As we develop this method, we must remember that we can *never* prove that a causal relationship exists using statistical inference methods.

12.2 Correlation analysis: descriptive statistics

The method of **correlation analysis** allows us to assess both the direction and the strength of a *linear* relationship between two variables. To use this method, we must first introduce two new statistics, the covariance and the correlation coefficient. The **covariance** measures if the relationship between two variables is direct (positive) or indirect (negative), while the **correlation coefficient** measures the relative strength of that relationship.

Calculating an unbiased estimate of the covariance requires one of two formulas, the choice of which depends on whether we are working with a sample (equation (12.1)) or a population (equation (12.2)):

$$s_{XY} = \frac{\sum (X_i - \overline{X}) \times (Y_i - \overline{Y})}{n - 1} \tag{12.1}$$

$$\sigma_{XY} = \frac{\sum (X_i - \mu_X) \times (Y_i - \mu_X)}{N} \tag{12.2}$$

As we observe, both covariance formulas pair each value of X with the corresponding value of Y.

Let us now calculate the covariance between adult educational attainment and GDP per capita using the World Bank sample. We first determine that the average years of schooling completed by adults $(\overline{X})$ in the 140-nation sample was 7.28 years and that average GDP per capita, PPP $(\overline{Y})$ was 11,625.39 (constant 2005 international dollars). Applying equation (12.1) yields a covariance for our sample of 18,961. The positive value of the covariance indicates a positive (or direct) relationship between educational attainment and GDP per capita, which *appears* to support the relationship postulated above. However, we must complete further testing to determine if this is a statistically significant result.

In reporting the above covariance, we did not specify the units of measurement. Given the units for education and GDP, we determine that the covariance is measured in units of "years of schooling completed by adults" times "constant 2005 international dollars." This makes little sense, economic or otherwise. In fact, even where the variables are measured in more compatible units, it is difficult to attach economic meaning to the covariance statistic. Thus economists primarily calculate the covariance of a pair of variables solely for the purpose of calculating another statistic, the correlation coefficient.

The correlation coefficient – sometimes referred to as the **simple correlation coefficient**, indicating a relationship between a *pair* of variables – utilizes the covariance between X and Y as well as the standard deviations of both X and Y to measure the relative strength of the relationship between the two variables. As with the covariance, we have one correlation coefficient formula for samples (equation (12.3)) and one for populations (equation (12.4)):

$$r_{XY} = \frac{s_{XY}}{s_X s_Y} \tag{12.3}$$

$$\rho_{XY} = \frac{\sigma_{XY}}{\sigma_X \sigma_Y} \tag{12.4}$$

where the Greek letter rho (ρ) is used to denote the population correlation coefficient. Unlike the covariance, the correlation coefficient is a *pure number* (that is, it has no units of measurement), which facilitates interpretation of the statistic.

Let us now compute the sample correlation coefficient for our education–national product example. After calculating the sample standard deviations (using equation (5.6)) for both educational attainment ($s_X = 2.68$ years) and GDP per capita, PPP ($s_Y = 13{,}192$ in constant 2005 international dollars), we obtain a sample correlation coefficient equal to 0.5409.

What does this tell us about the relationship between educational attainment and per-capita GDP across 140 nations in 2000? First, its positive value reinforces our findings from the covariance that the two variables are positively related. We also recognize that the sign of the covariance will determine the sign of the correlation coefficient, since s_X and s_Y are both positive values. In addition, the correlation coefficient's value of 0.5409 indicates a moderate positive relationship between the two variables per the guidelines presented in Figure 12.1.

Concluding that 0.5409 indicates a moderate correlation derives from the range of the correlation coefficient's possible values. It can be shown, for example, that, if all data points lie on an upward-sloping straight line, r_{XY} (and ρ_{XY}) will have a value of $+1.0$, and if all points lie on a downward-sloping straight line, r_{XY} (and ρ_{XY}) will equal -1.0. Further,

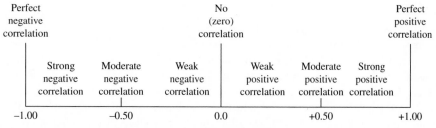

Figure 12.1 Range and strength of the correlation coefficient.

in the case of no linear relationship between X and Y, it can be shown that r_{XY} (and ρ_{XY}) will equal zero. This gives us a range for r_{XY} (and ρ_{XY}) of $+1.0$ to -1.0. Thus, because 0.5409 is closer in value to $+0.5$ than it is to $+1.0$ in the example here, we conclude that there is a moderate positive relationship between educational attainment and GDP per capita.

While Figure 12.1 does provide information about the strength of the relationship between two variables, terms such as "strong" and "weak" are not statistically precise. Consequently, we introduce one additional statistic, the coefficient of determination, which is both more precise and easier to interpret.

The **coefficient of determination**, or r^2, is calculated by squaring the correlation coefficient r_{XY} (or ρ_{XY} when using population data). This statistic measures the proportion of the variation in Y that is accounted for by the variation in X. In our example, r^2 equals 0.2926, which indicates that 29.26 percent of the variation in per-capita GDP was accounted for by the variation in years of education completed by adults in the 140 sample countries. Before we evaluate if this represents a statistically significant relationship between the two variables, let us bear in mind that, while our language might suggest causality, we *cannot* verify a causal relationship between the two variables using correlation analysis.

12.3   Testing the significance of the correlation coefficient

We have used the three statistics from correlation analysis to establish that GDP per capita and years of schooling completed by adults are positively and moderately related to each other. However, it is possible that our sample results are the product of chance and do not reflect the true relationship between the two variables found for the population.

To test if the correlation results are statistically reliable, let us now evaluate our results using the hypothesis test procedure. We begin by evaluating if the correlation coefficient associated with the paired variables is, in fact, statistically different from zero. (We will test the statistical significance of the positive relationship in the next section.) If our statistical evidence supports this claim, we can conclude, for the specified level of significance, that the two variables are correlated with one another.

We again follow the conventions used by economists in establishing the pair of hypotheses. This means we write the alternative hypothesis to reflect the relationship postulated by economic theory. Here we establish H_a to reflect that the level of education and national output are expected to be statistically related to one another, or, technically, that the sample correlation coefficient is statistically different from zero:

- H_0: $\rho = 0$, i.e. there is no linear relationship between GDP per capita and educational attainment.
- H_a: $\rho \neq 0$, i.e. there is a linear relationship between GDP per capita and educational attainment.

As we see in Figure 12.2, this means that we will be conducting a two-tail hypothesis test.

We next set the significance level (α) for the test. As in Chapter 11, we want to set α at a low value to minimize the likelihood of committing a Type I error. That is, we want to be very confident we can correctly reject the statement that the correlation coefficient is zero. Here, this leads us to establish a level of significance equal to 1 percent ($\alpha = 0.01$).

The next step establishes the decision rule for rejecting the null hypothesis. We first recognize that we will use the t-distribution for our test since the correlation coefficient is constructed from the sample covariance and sample standard deviations of both variables.

- To test if two variables are statistically correlated with one another, conduct a two-tail hypothesis using the following hypotheses:

$$H_0: \rho = 0 \quad \text{and} \quad H_a: \rho \neq 0$$

- To test the direction of the correlation between two variables, conduct a one-tail test in which the alternative hypothesis reflects the postulated relationship. To test if the correlation is statistically positive, the hypotheses are specified as:

$$H_0: \rho \leq 0 \quad \text{and} \quad H_a: \rho > 0$$

- To test if the correlation is statistically negative, the hypotheses are specified as:

$$H_0: \rho \geq 0 \quad \text{and} \quad H_a: \rho < 0$$

Figure 12.2 Establishing hypotheses about the correlation coefficient.

Next we determine the critical t-value against which we will compare our test statistic. As we observe in Figure 12.3, we note a difference from the critical t-value used to test the sample mean.

In order to compute the correlation coefficient, we must estimate the sample means ($\overline{X}$ and $\overline{Y}$) for both variables, which means that we lose two degrees of freedom in the process. Thus the critical t-value for testing if the correlation coefficient is statistically different from zero is t_{n-2}^{α}, and the decision rule becomes

reject H_0 if $|t| > t_{n-2}^{\alpha}$

- The most frequently used levels of significance in correlation analysis are 1 percent and 5 percent. These levels reflect the high degree of confidence desired when rejecting the null hypothesis.
- The decision rule may reflect whether any correlation exists between the two variables or if that correlation exhibits a particular direction:
  o the decision rule used to assess if the correlation between two variables is statistically significant is

$$\text{reject } H_0 \text{ if } |t| > t_{n-2}^{\alpha}$$

  o the decision rule used to assess if the correlation between two variables is statistically positive is

$$\text{reject } H_0 \text{ if } t > t_{n-2}^{\alpha}$$

  o the decision rule used to assess if the correlation between two variables is statistically negative is

$$\text{reject } H_0 \text{ if } t < t_{n-2}^{\alpha}$$

Figure 12.3 Setting the significance level and establishing the decision rule.

Note that we take the absolute value of the test statistic, since the alternative hypothesis locates the rejection region in both tails of the t-distribution.

For our particular example, the critical t-value, $t_{138}^{0.01}$, equals 2.6119. Thus the decision rule for evaluating the null hypothesis regarding a relationship between education and national output is

reject H_0 if $|t| > 2.6119$

We are now ready to construct the test statistic for the hypothesis test. For tests of the correlation coefficient, the relevant test statistic utilizes both the correlation coefficient and the coefficient of determination:

$$t_{n-2} = \frac{r \times \sqrt{n-2}}{\sqrt{1-r^2}} \qquad (12.5)$$

Let us examine equation (12.5) more closely in light of the correlation coefficient's range. We have already established that, if X and Y are not linearly correlated with one another, $r = 0$, which in turn yields a test statistic equal to zero ($t_{n-2} = 0$). Since zero is smaller than any conceivable critical t-value we might use,[2] we will never reject the null hypothesis when the correlation coefficient is zero. That result leads us to conclude that the two variables are not linearly correlated with each other.

Now suppose the two variables are perfectly correlated with each other, or that r equals either $+1$ or -1. Equation (12.5) now yields a value of ∞, which is larger than any critical t-value we might obtain. In such a case, we will reject the null hypothesis of no linear correlation between the two variables with virtually 100 percent confidence, leading us to conclude that the statistical evidence very strongly supports a relationship between the two variables.

As we might expect, it is rare that we would obtain a correlation coefficient for a pair of variables equal to either zero or one. Thus our decision about the null hypothesis will depend on both the test statistic (r_{XY}) and the critical t-value (t_{n-2}^{α}), as shown in Figure 12.4.

Let us now calculate the test statistic for the educational attainment–GDP per capita example. Given that the correlation coefficient equals 0.5409 and the coefficient of determination equals 0.2926, we obtain a test statistic equal to

$$t = \frac{0.5409 \times \sqrt{138}}{\sqrt{1 - 0.2926}} = 7.5545$$

We can now make a decision about the null hypothesis:

since $|t| = 7.5545$ is greater than the critical $t_{138}^{0.01}$ value of 2.6119, we can, with at least 99 percent confidence, reject H_0.

In economic terms, this means that we are at least 99 percent confident that the relationship between GDP per capita and adult educational attainment is not due to chance.

We may now ask how confident we are in the above conclusion. This question leads us to calculate the p-value associated with the correlation test results.

As we recall, the p-value is the probability of obtaining a test statistic that is at least as extreme as that obtained from the sample *when the null hypothesis is true*. For the correlation

- The equation for determining the test statistic for correlation analysis is

$$t = \frac{r \times \sqrt{n-2}}{\sqrt{1-r^2}}$$

  where r is the sample correlation coefficient and r^2 is the coefficient of determination.
- Possible decisions about the null hypothesis:
  - For a two-tail test,
    - if $|t| > t_{n-2}^\alpha$, reject the null hypothesis and conclude, with at least $1 - \alpha$ percent confidence, that the two variables are statistically related;
    - if $|t| \leq t_{n-2}^\alpha$, cannot reject the null hypothesis.
  - For a one-tail hypothesis test of a positive correlation between the two variables,
    - if $t > t_{n-2}^\alpha$, reject the null hypothesis and conclude, with at least $1 - \alpha$ percent confidence, that there is a statistically significant positive correlation between X and Y;
    - if $t \leq t_{n-2}^\alpha$, cannot reject the null hypothesis.
  - For a one-tail hypothesis test of a negative correlation between the two variables,
    - if $t < t_{n-2}^\alpha$, reject the null hypothesis and conclude, with at least $1 - \alpha$ percent confidence, that there is a statistically significant negative correlation between X and Y;
    - if $t \geq t_{n-2}^\alpha$, cannot reject the null hypothesis.

Figure 12.4 The correlation coefficient test statistic and decisions regarding the null hypothesis.

coefficient, we want to calculate the probability of obtaining a t-value that is greater (in absolute value) than the test statistic's value when the two variables are *not* statistically correlated with each other. Mathematically, as shown in Figure 12.5, this means we want to determine

$$\Pr(|t| > t \mid H_0 : \rho = 0)$$

Let us now calculate the p-value for the education–GDP example:

$$\Pr(|t| > 7.5545 \mid H_0 : \rho = 0) = 5.2404\text{E-}12 = 5.2404 \times 10^{-12}$$

This result leads us to conclude that the likelihood of obtaining a correlation coefficient as large as that generated by the sample, when there is no correlation between education and national product, is very small. Given this very small value of p, we can say, with great confidence, that our conclusion about the correlation between GDP per capita and adult educational attainment is supported by the statistical evidence. However, we also recognize that we cannot actually be 100 percent confident about our conclusion owing to the nature of statistical inference procedures.

- The *p*-value for a two-tail hypothesis test, $\Pr(|t| >$ test statistic $|H_0: \rho = 0)$, determines the probability of obtaining a correlation test statistic at least as large in absolute value as that obtained from the sample when the null hypothesis is true.
- The *p*-value for a one-tail hypothesis test of a positive correlation between two variables, $\Pr(t >$ test statistic $|H_0: \rho = 0)$, determines the probability of obtaining a correlation test statistic at least as large as that obtained from the sample when the null hypothesis is true.
- The *p*-value for a one-tail hypothesis test of a negative correlation between two variables, $\Pr(t <$ test statistic $|H_0: \rho = 0)$, determines the probability of obtaining a correlation test statistic at least as small as that obtained from the sample when the null hypothesis is true.

Figure 12.5 Determining the correlation coefficient's *p*-value.

12.4   Testing the sign on the correlation coefficient

While we have established a statistically significant relationship between education and national product based on the World Bank sample and data from 2000, economic reasoning led us to postulate a positive relationship between the two variables. In cases where a directional relationship is suggested, economists typically perform a one-tail hypothesis test.

Before we articulate the steps involved with a one-tail hypothesis test, we must stress the following point. It is *not appropriate* for us to use the statistical results obtained from the two-tail hypothesis test to test a new pair of hypotheses. To test the newly stated hypotheses, we must draw a *new sample* and construct a new test statistic based on that sample's statistics. This leads us to draw a new sample (of 142 countries) from the World Bank for the year 2005.[3]

Let us now test if there is a positive relationship between the average years of schooling completed by adults and GDP per capita PPP (constant 2005 international dollars). Since we have previously discussed each step, we will report each here only briefly.

- *Step 1.* $H_0: \rho \leq 0$ and $H_a: \rho > 0$ (GDP and education are positively related).
- *Step 2.* Set $\alpha = 0.01$.
- *Step 3.* Reject H_0 if $t > t_{142-2}^{0.01}$, where $t_{142-2}^{0.01} = 2.3533$.
- *Step 4.* Calculate the test statistic, where $r_{XY} = 0.5664$ and $r^2 = 0.3208$:

$$t = \frac{0.5664 \times \sqrt{(142-2)}}{\sqrt{(1-0.3208)}} = 8.1312$$

- *Step 5.* Reject H_0 since $t = 8.1312 > t_{142-2}^{0.01} = 2.3533$.
- *Step 6.* The *p*-value $= 1.0205\text{E-}13$ indicates that we are very confident that there is a statistically significant positive relationship between GDP per capita and educational attainment for the 142 countries in the 2005 sample.

The statistical results from the two hypothesis tests provide empirical support for the theoretical relationship we postulated between GDP per capita and educational attainment. But we cannot use these results to conclude that there is a causal relationship between the two variables. That is, these results do *not* demonstrate that a higher level of education *causes* a higher level of GDP per capita. Consequently, we *cannot* use correlation analysis, or any method of statistical inference, to *prove* that a theorized causal relationship between two variables exists.

Summary

Correlation analysis moves us into the realm of statistical inference for pairs of economic variables. It specifically allows us to evaluate the statistical strength and direction of a linear relationship between the two. However, because economists typically postulate a theoretical relationship between the two variables, the information obtained from correlation analysis is limited. Consequently, we use correlation analysis primarily as an introduction to the statistical inference method most frequently used by economists. That method, linear regression analysis, is the focus of the remainder of the text.

Concepts introduced

Bivariate relationships
Coefficient of determination
Correlation analysis
Correlation coefficient
Covariance
Simple correlation coefficient

Box 12.1 Where can I find...? The World Bank's Data Catalog

The World Bank collects economic and social data for countries around the world, with particular emphasis on low- and middle-income countries. It has recently made much of this data available free of charge through its Open Data initiative.

According to its website (<http://data.worldbank.org/data-catalog>):

> The World Bank's Open Data initiative is intended to provide all users with access to World Bank data. The data catalog is a listing of available World Bank datasets, include databases, pre-formatted tables and reports. Each of the listings includes a description of the dataset and a direct link to that set. Where possible, the databases are linked directly to a selection screen to allow users to select the countries, indicators, and years they would like to search. Those search results can be exported in different formats. Users can also choose to download the entire database directly from the catalog.

Many datasets are accessible through an easy-to-use query tool called the World Databank. Users can select from over 2000 variables for the countries or aggregates of interest for a wide range of years. In addition to formatting options, users can downloaded the extracted data in several formats, including Excel.

Databanks available from the world bank

The following databanks were available at the World Bank as of 18 December 2010. The descriptions that follow were obtained from the websites listed. Note that this is only a partial listing of available databases. For more databases, go to <http://data.worldbank.org/data-catalog>.

World Development Indicators (WDI)

<http://data.worldbank.org/data-catalog/world-development-indicators>
    This provides a comprehensive selection of economic, social, and environmental indicators, drawing on data from the World Bank and more than 30 partner agencies. The database covers more than 900 indicators for 213 economies, with data back to 1960. These indicators are updated three times a year in April, September, and December.

Global Development Finance (GDF)

<http://data.worldbank.org/data-catalog/global-development-finance>
    This focuses on financial flows, trends in external debt, and other major financial indicators for developing countries. It includes over 200 time-series indicators from 1970 to 2009 for 128 countries. This series is updated annually in January.

Africa Development Indicators (ADI)

<http://data.worldbank.org/data-catalog/africa-development-indicators>
    This provides the most detailed collection of data on Africa, containing over 1,600 indicators, covering 53 African countries and spanning the period 1961 to 2009. Data include social, economic, financial, natural resources, infrastructure, governance, partnership, and environmental indicators. This series is updated annually in March.

Education Statistics (EdStats)

<http://data.worldbank.org/data-catalog/ed-stats>
    This provides data from 1960 through 2008 for 210 countries on over 1,000 internationally comparable indicators for access, progression, completion, learning outcomes, literacy, teachers, expenditure, and background indicators. The indicators cover the educational cycle from pre-primary to tertiary education, and the databank also includes education projections up to 2050. This series is updated quarterly in January, April, July, and October.

Gender Statistics (GenderStats)

<http://data.worldbank.org/data-catalog/gender-statistics>
    This provides data from 1960 through 2008 on indicators on key gender topics for 213 countries. Themes included are demographics, education, health, labor force, and political participation. This series is updated twice a year in April and September.

Global Economic Monitor (GEM)

<http://data.worldbank.org/data-catalog/global-economic-monitor>
    This provides daily updates of global economic developments, with coverage of 196 high-income and developing countries. Daily data updates are provided for exchange rates, equity markets, interest rates, stripped bond spreads, and emerging market bond indices. Monthly data coverage (updated daily and populated upon availability) is provided for consumer prices, high-tech market indicators, industrial production, and merchandise trade. This series is updated daily.

Health Nutrition and Population (HNPStats) statistics

<http://data.worldbank.org/data-catalog/health-nutrition-and-population-statistics>

This provides key health, nutrition and population statistics gathered from a variety of international sources for 213 countries. Themes include population dynamics, nutrition, reproductive health, health financing, medical resources and usage, immunization, infectious diseases, HIV/AIDS, DALY (disability-adjusted life year), population projections and lending. HNPStats also includes health, nutrition, and population statistics by wealth quintiles. This series is updated twice a year in April and September.

Millennium Development Goals (MDG)

<http://data.worldbank.org/data-catalog/millennium-development-indicators>

This includes relevant indicators for 213 countries drawn from the World Development Indicators, reorganized according to the goals and targets of the Millennium Development Goals (MDGs). The MDGs focus the efforts of the world community on achieving significant, measurable improvements in people's lives by the year 2015: they establish targets and yardsticks for measuring development results. This series is updated twice a year in April and September.

International Comparison Program (ICP)

<http://data.worldbank.org/data-catalog/international-comparison-program>

This provides a collection of comparative price data and detailed expenditure values of countries' gross domestic product (GDP) and purchasing power parity (PPP) estimates of the world's economies for 191 countries. Started in 2005, this series was last updated in February 2008.

GEM Commodities

<http://data.worldbank.org/data-catalog/commodity-price-data>

This provides data on monthly commodities prices and indices from 1960 to present, updated on the third working day of each month, as presented in the Commodity Price Data (a.k.a. Pink Sheet).

Joint External Debt Hub (JEDH)

<http://data.worldbank.org/data-catalog/joint-external-data-hub>

This provides external debt data and selected foreign assets from international creditor/market sources for 218 countries. Started in 1990, this series is updated quarterly in January, April, July, and October.

Quarterly External Debt Statistics (QEDS/SDDS)

<http://data.worldbank.org/data-catalog/quarterly-external-debt-statistics-ssds>

This provides detailed external debt data that are published individually by countries that subscribe to the IMF's Special Data Dissemination Standard (SDDS) as well as countries participating in the General Data Dissemination System (GDDS) that are in a position to produce the external debt data prescribed by the SDDS. Starting in 1998, this series is updated quarterly in January, April, July, and October and is currently available for 73 countries.

Quarterly Public Sector Debt (QPSD)

<http://data.worldbank.org/node/518>

This database, jointly developed by the World Bank and the International Monetary Fund, brings together detailed public sector debt data of selected developing/emerging market countries (Czech Republic is an advanced economy; the rest of the advanced economies will be invited to participate in this initiative in 2011). The QPSD database includes country and cross-country tables, and enables users to query and extract data, by country, group of countries, and specific public debt components. The data represent the following sectors on an as-available basis: general government; otherwise central government; otherwise budgetary central government; non-financial public corporations; financial public corporations; and a table presenting the total public sector debt.

Box 12.2 Using Excel to perform correlation analysis

Excel can quickly calculate all three statistical measures used in correlation analysis as well as the critical *t*- and *p*-values used in testing hypothesis about the correlation coefficient.

Calculating the covariance

Excel's formula for the covariance, **COVAR**, uses a hybrid of equations (12.1) and (12.2):

$$\text{COVAR}(X, Y) = \frac{\sum (X - \overline{X}) \times (Y - \overline{Y})}{n}$$

where $\overline{X}$ and $\overline{Y}$ are the sample means of the two variables and n is the sample size.

We can convert Excel's covariance into the sample covariance from equation (12.1) by multiplying Excel's result by $n/(n-1)$. This may be required when using very small samples ($n < 30$). In such cases, the difference in the two formulas could lead to different estimates of the sample covariance.

Calculating the correlation coefficient

Excel's correlation coefficient formula, **CORREL**, calculates the sample correlation coefficient (r_{XY}) from equation (12.3):

CORREL(array 1, array 2)

where array 1 is the set of values for either X or Y, and array 2 refers to the other set of values.

We could also use the **Correlation Tool** in the Data ToolPak. Here the data for the two variables must be in adjacent columns.

Calculating the coefficient of determination

Excel also has a statistical function for the coefficient of determination (r^2), **RSQ**:

RSQ(known_*y*'s, known_*x*'s).

This function squares the sample correlation coefficient generated by CORREL.

Calculating the critical t-value

In Excel 2007, the **TINV** function is used to calculate the critical t-statistic for the correlation coefficient.

- The arguments needed here for **TINV** are the probability associated with the t-distribution (α established in step 2) and the degrees of freedom ($n - 2$):

    $$\text{TINV}(\alpha, n - 2)$$

    o To obtain the critical t-value for a *two-tail* hypothesis test, set **Probability** $= \alpha$.
    o To obtain the critical t-value for a *one-tail* hypothesis test, set **Probability** $= \alpha * 2$.

In Excel 2010, the **T.INV.2T** function is used to determine the critical t-value for both the two-tail and one-tail hypothesis tests. It uses the same arguments as those for the TINV function in Excel 2007.

Calculating the p-value

TDIST function can be used to calculate the correlation coefficient's p-value.

- The **TDIST** function uses the test statistic value, t; the degrees of freedom, $n - 2$; and the number of distribution tails:

    $$\text{TDIST}(t, n - 2, \text{tails})$$

    o To test if two variables are correlated with each other (H_0: $\rho = 0$ and H_a: $\rho \neq 0$), set **Tails** $= 2$.
    o To test if the correlation is positive (H_a: $\rho > 0$), set **Tails** $= 1$.
    o To test if the correlation is negative (H_a: $\rho < 0$), set **Tails** $= 1$.

In Excel 2010, the choice of built-in function depends on the hypothesis being tested.

- Use the **T.DIST.2T** function to determine the p-value associated with a two-tail hypothesis test:

    $$\text{T.DIST.2T}(t, n - 2)$$

    where t is the test statistic value and $n - 2$ is the degrees of freedom.
- Use the **T.DIST.RT** function to determine the p-value associated with a one-tail test:

    $$\text{T.DIST.RT}(t, n - 2)$$

    where t is the test statistic value and $n - 2$ is the degrees of freedom.

Exercises

1. Use the datasets below to calculate and interpret the correlation coefficient for the following pairs of variables.

(a) Annual CPI-U for coffee (100 percent, ground roast, all sizes, per pound) and annual retail pounds of coffee consumed per capita between 1967 and 2007.

Year	CPI-U for coffee	Per-capita coffee availability – retail (pounds)	Year	CPI-U for coffee	Per-capita coffee availability – retail (pounds)
1967	26.6	11.14524	1988	115.0	7.31467
1968	26.5	11.19602	1989	120.4	7.52990
1969	26.9	10.59619	1990	117.5	7.74313
1970	31.7	10.41316	1991	115.3	7.75125
1971	32.6	9.88813	1992	110.7	7.51333
1972	32.1	10.30358	1993	109.8	6.81392
1973	35.7	10.01227	1994	140.4	6.14004
1974	42.5	9.61740	1995	163.1	5.99376
1975	46.4	9.19701	1996	149.2	6.64658
1976	63.8	9.41918	1997	168.0	6.96663
1977	112.9	6.95488	1998	163.4	7.13460
1978	107.2	7.92930	1999	154.8	7.51365
1979	101.8	8.57716	2000	154.0	7.83523
1980	111.6	7.67540	2001	146.7	7.22710
1981	96.2	7.47060	2002	142.6	7.06508
1982	98.5	7.36837	2003	144.9	7.24997
1983	98.8	7.47301	2004	145.3	7.36611
1984	102.7	7.60344	2005	161.2	7.24150
1985	105.5	7.78835	2006	165.3	7.28982
1986	132.7	7.80720	2007	175.647	7.33471
1987	116.2	7.59699	2008	188.027	7.23830

Data source: CPI-U for coffee: U.S. Bureau of Labor Statistics (2010b), Consumer Price Index, All Urban Consumers (Current Series).

Per capita coffee availability: Economic Research Service, U.S. Department of Agriculture (2010a), Food Availability (Per Capita) Data System.

(b) The 2004 annual income (dollars) and years worked in nursing since receiving first RN license for full-time, non-union, registered nurses (RNs) with advanced degrees who are classified as staff or general duty nurses and working in a hospital.

Nurse	Current annual income from principal nursing position and all other nursing positions (dollars)	Years worked in nursing since first RN license	Nurse	Current annual income from principal nursing position and all other nursing positions (dollars)	Years worked in nursing since first RN license	Nurse	Current annual income from principal nursing position and all other nursing positions (dollars)	Years worked in nursing since first RN license
1	72,000	7	43	60,000	30	85	60,000	2
2	96,000	25	44	48,000	22	86	52,416	12
3	70,000	27	45	75,000	17	87	52,000	7
4	60,000	21	46	74,000	35	88	65,000	17
5	63,000	28	47	50,000	24	89	50,000	20
6	58,000	23	48	73,350	32	90	70,000	29
7	68,000	33	49	67,000	24	91	50,000	7
8	63,000	21	50	44,000	21	92	71,250	8
9	76,000	36	51	52,000	32	93	60,000	10
10	63,000	22	52	52,000	37	94	49,000	11
11	100,000	28	53	39,890	36	95	60,000	17
12	75,000	16	54	60,000	25	96	60,000	25
13	80,000	25	55	75,600	36	97	65,000	27
14	62,000	18	56	60,000	6	98	65,000	11
15	76,000	23	57	50,000	24	99	78,000	2
16	57,000	20	58	53,060	28	100	53,500	22
17	50,000	18	59	35,000	6	101	57,000	17
18	84,000	19	60	44,100	1	102	69,000	18
19	37,000	11	61	68,000	21	103	64,500	22
20	58,000	4	62	58,000	14	104	74,000	33
21	76,250	24	63	53,000	24	105	61,000	25
22	58,872	30	64	80,000	21	106	60,000	28
23	52,000	11	65	67,000	16	107	44,000	6

Nurse	Current annual income from principal nursing position and all other nursing positions (dollars)	Years worked in nursing since first RN license	Nurse	Current annual income from principal nursing position and all other nursing positions (dollars)	Years worked in nursing since first RN license	Nurse	Current annual income from principal nursing position and all other nursing positions (dollars)	Years worked in nursing since first RN license
24	60,000	19	66	60,000	17	108	65,421	23
25	63,000	23	67	10,000	14	109	53,000	39
26	47,500	14	68	48,000	11	110	63,300	13
27	45,000	23	69	69,000	28	111	68,000	27
28	65,000	14	70	60,000	17	112	85,000	7
29	45,000	31	71	47,850	22	113	83,000	40
30	20,000	7	72	94,000	30	114	73,000	23
31	65,000	7	73	95,000	27	115	61,409	23
32	69,000	16	74	36,000	10	116	47,900	26
33	61,000	19	75	58,000	26	117	53,000	20
34	51,000	10	76	80,000	12	118	93,000	17
35	40,000	20	77	54,000	27	119	81,000	8
36	43,000	30	78	50,000	3	120	75,000	22
37	63,000	20	79	35,000	15	121	52,000	8
38	68,000	22	80	57,000	13	122	69,000	34
39	74,000	33	81	60,000	4	123	60,000	16
40	48,672	19	82	48,000	7	124	71,120	27
41	39,000	6	83	75,000	15	125	65,003	10
42	62,000	5	84	55, 000	14			

Data source: U.S. Department of Health and Human Services, Health Resources and Services Administration (2010), 2004 National Sample Survey of Registered Nurses.

(c) Annual number of new houses sold and annual new home mortgage yields between 1991 and 2007.

Year, Month	HSN1F	MORTG	Year, Month	HSN1F	MORTG	Year, Month	HSN1F	MORTG
1991-06-01	516,000	9.62	1994-09-01	677,000	8.64	1997-12-01	793,000	7.10
1991-07-01	511,000	9.58	1994-10-01	715,000	8.93	1998-01-01	872,000	6.99
1991-08-01	526,000	9.24	1994-11-01	646,000	9.17	1998-02-01	866,000	7.04
1991-09-01	487,000	9.01	1994-12-01	629,000	9.20	1998-03-01	836,000	7.13
1991-10-01	524,000	8.86	1995-01-01	626,000	9.15	1998-04-01	866,000	7.14
1991-11-01	575,000	8.71	1995-02-01	559,000	8.83	1998-05-01	887,000	7.14
1991-12-01	558,000	8.50	1995-03-01	616,000	8.46	1998-06-01	923,000	7.00
1992-01-01	676,000	8.43	1995-04-01	621,000	8.32	1998-07-01	876,000	6.95
1992-02-01	639,000	8.76	1995-05-01	674,000	7.96	1998-08-01	846,000	6.92
1992-03-01	554,000	8.94	1995-06-01	725,000	7.57	1998-09-01	864,000	6.72
1992-04-01	546,000	8.85	1995-07-01	765,000	7.61	1998-10-01	893,000	6.71
1992-05-01	554,000	8.67	1995-08-01	701,000	7.86	1998-11-01	995,000	6.87
1992-06-01	596,000	8.51	1995-09-01	678,000	7.64	1998-12-01	949,000	6.72
1992-07-01	627,000	8.13	1995-10-01	696,000	7.48	1999-01-01	875,000	6.79
1992-08-01	636,000	7.98	1995-11-01	664,000	7.38	1999-02-01	848,000	6.81
1992-09-01	650,000	7.92	1995-12-01	709,000	7.20	1999-03-01	863,000	7.04
1992-10-01	621,000	8.09	1996-01-01	714,000	7.03	1999-04-01	918,000	6.92
1992-11-01	614,000	8.31	1996-02-01	769,000	7.08	1999-05-01	888,000	7.15
1992-12-01	650,000	8.22	1996-03-01	721,000	7.62	1999-06-01	923,000	7.55
1993-01-01	596,000	8.02	1996-04-01	736,000	7.93	1999-07-01	900,000	7.63
1993-02-01	604,000	7.68	1996-05-01	746,000	8.07	1999-08-01	893,000	7.94
1993-03-01	602,000	7.50	1996-06-01	721,000	8.32	1999-09-01	826,000	7.82
1993-04-01	701,000	7.47	1996-07-01	770,000	8.25	1999-10-01	872,000	7.85
1993-05-01	626,000	7.47	1996-08-01	826,000	8.00	1999-11-01	863,000	7.74
1993-06-01	653,000	7.42	1996-09-01	770,000	8.23	1999-12-01	873,000	7.91
1993-07-01	655,000	7.21	1996-10-01	720,000	7.92	2000-01-01	873,000	8.21
1993-08-01	645,000	7.11	1996-11-01	771,000	7.62	2000-02-01	856,000	8.33
1993-09-01	726,000	6.92	1996-12-01	805,000	7.60	2000-03-01	900,000	8.24
1993-10-01	704,000	6.83	1997-01-01	830,000	7.82	2000-04-01	841,000	8.15
1993-11-01	769,000	7.16	1997-02-01	801,000	7.65	2000-05-01	857,000	8.52
1993-12-01	812,000	7.17	1997-03-01	831,000	7.90	2000-06-01	793,000	8.29
1994-01-01	619,000	7.06	1997-04-01	744,000	8.14	2000-07-01	887,000	8.15
1994-02-01	686,000	7.15	1997-05-01	760,000	7.94	2000-08-01	848,000	8.03
1994-03-01	747,000	7.68	1997-06-01	793,000	7.69	2000-09-01	912,000	7.91
1994-04-01	692,000	8.32	1997-07-01	805,000	7.50	2000-10-01	933,000	7.80
1994-05-01	691,000	8.60	1997-08-01	815,000	7.48	2000-11-01	880,000	7.75
1994-06-01	621,000	8.40	1997-09-01	840,000	7.43	2000-12-01	983,000	7.38
1994-07-01	628,000	8.61	1997-10-01	800,000	7.29	2001-01-01	936,000	7.03
1994-08-01	656,000	8.51	1997-11-01	864,000	7.21	2001-02-01	963,000	7.05

Data source: FRED, Federal Reserve Economic Data, Federal Reserve Bank of St. Louis (2010b, c), New One Family Houses Sold (HSN1F); 30-Year Conventional Mortgage Rate (MORTG).

Year	Qtr	Real hourly compensation [index, 1992 = 100]	Output per person, [index, 1992 = 100]	Year	Qtr	Real hourly compensation [index, 1992 = 100]	Output per person [index, 1992 = 100]	Year	Qtr	Real hourly compensation [index, 1992 = 100]	Output per person [index, 1992 = 100]
1987	Q1	98.398	85.553	1994	Q1	101.286	109.829	2001	Q1	110.897	155.724
1987	Q2	97.788	86.777	1994	Q2	100.82	111.819	2001	Q2	109.876	156.191
1987	Q3	98.234	87.686	1994	Q3	100.393	112.663	2001	Q3	109.404	155.993
1987	Q4	97.568	89.65	1994	Q4	100.648	114.726	2001	Q4	111.204	156.98
1988	Q1	97.775	90.167	1995	Q1	98.866	116.927	2002	Q1	113.981	161.571
1988	Q2	96.976	90.991	1995	Q2	99.404	117.044	2002	Q2	115.755	167.306
1988	Q3	96.911	90.935	1995	Q3	99.986	118.485	2002	Q3	116.067	170.361
1988	Q4	97.399	92.067	1995	Q4	100.01	120.083	2002	Q4	116.133	173.388
1989	Q1	96.726	92.478	1996	Q1	98.745	120.215	2003	Q1	119.687	175.781
1989	Q2	94.898	91.763	1996	Q2	98.102	123.326	2003	Q2	121.656	177.93
1989	Q3	95.309	91.259	1996	Q3	98.241	126.324	2003	Q3	122.812	182.25
1989	Q4	96.534	90.813	1996	Q4	97.735	127.597	2003	Q4	124.169	183.999
1990	Q1	94.608	91.869	1997	Q1	97.222	130.305	2004	Q1	116.96	184.944
1990	Q2	95.542	92.648	1997	Q2	98.161	132.395	2004	Q2	118.249	184.423
1990	Q3	95.185	93.53	1997	Q3	98.467	135.15	2004	Q3	120.686	184.546
1990	Q4	94.969	91.762	1997	Q4	99.617	138.982	2004	Q4	120.737	185.826
1991	Q1	96.093	90.752	1998	Q1	101.583	140.137	2005	Q1	120.255	187.434
1991	Q2	97.238	92.011	1998	Q2	102.677	140.076	2005	Q2	120.104	190.003
1991	Q3	97.86	94.86	1998	Q3	103.431	143.057	2005	Q3	120.561	194.016
1991	Q4	98.048	95.966	1998	Q4	103.364	146.996	2005	Q4	118.119	200.076
1992	Q1	99.256	96.763	1999	Q1	104.19	150.259	2006	Q1	120.809	202.071
1992	Q2	99.995	99.727	1999	Q2	104.503	152.88	2006	Q2	117.926	206.615
1992	Q3	100.901	101.254	1999	Q3	104.9	154.439	2006	Q3	117.848	210.843
1992	Q4	99.872	102.27	1999	Q4	107.228	157	2006	Q4	122.965	210.429
1993	Q1	99.218	104.304	2000	Q1	112.219	159.091	2007	Q1	124.808	210.499
1993	Q2	99.074	105.134	2000	Q2	110.229	159.897	2007	Q2	123.306	216.136
1993	Q3	99.418	105.279	2000	Q3	112.397	159.642	2007	Q3	122.487	220.675
1993	Q4	99.915	108.047	2000	Q4	111.279	157.37	2007	Q4	122.528	221.343

Data source: U.S. Bureau of Labor Economics (2010), Major Sector Productivity and Costs Index.

(d) Index of "Real hourly compensation" in durable goods manufacturing and the index of labor productivity as measured by "Output per person" in durable goods manufacturing between 1987-Q1 and 2007-Q4.

2. Test if each correlation coefficient calculated in question 1 is statistically significant with respect to the relationship postulated by economic theory.

13 Simple linear regression analysis

Descriptive measures

13.1 Introduction to simple linear regression analysis

Linear regression analysis is perhaps the most widely used statistical method in economics. This empirical technique draws on the same basic ideas behind correlation analysis as well as the least-squares approach that we developed in Chapter 8 for measuring long-run economic growth. To learn what differentiates regression analysis from these previous methods, let us return to the educational attainment–GDP per capita example from Chapter 12.

We begin with Chart 13.1 (p. 327), which reproduces the scatterplot depicted in Chart 12.1 but also includes one additional component – a straight line (or linear trendline in Excel) fitted to the paired data using the least-squares method described in Chapter 8. This line is known as the **simple linear regression line**,[1] and here it has been estimated by the linear equation

$$\widehat{\text{GDP}} = -7{,}783.67 + 2{,}665.99 \times \text{EDUC}$$

The trendline added to Chart 13.1 may appear to be the only feature that differentiates Charts 12.1 and 13.1, but it is not. That is, in Chart 12.1, we could plot GDP per capita on either the Y- or the X-axis to illustrate the correlation between the two variables. In Chart 13.1, however, we deliberately plot GDP per capita on the Y-axis and educational attainment on the X-axis. This illustrates economists' belief that higher levels of adult educational attainment contribute to higher levels of GDP per capita. In technical terms, this means we have specified GDP per capita as the **dependent variable (Y)** in the estimated equation and educational attainment as the **independent variable (X)**, sometimes referred to as the **explanatory variable**.

Let us now interpret the values in the simple regression equation based on the 2000 sample data for 140 countries. A literal interpretation of the estimated intercept term indicates that a country whose adult population has zero years of education is predicted (by the equation) to have a *negative* GDP per capita of $-\$7{,}783.67$ real international dollars (PPP). As we shall further discuss in Section 13.3, this result is neither economically realistic nor statistically reliable.

The estimated slope value indicates how much higher GDP per capita is, on average, for each additional year of education attained by adults ages 15 years and over. Specifically, the regression results indicate that, for every additional year of education attained by adults, GDP per capita PPP is estimated to be, on average, $2,665.99 real international dollars higher.

Once we verify the statistical significance of these results, we might be tempted to conclude that we have proven our hypothesized relationship, but such a conclusion would be incorrect. In fact, we can *never* determine from the statistical results that higher levels of

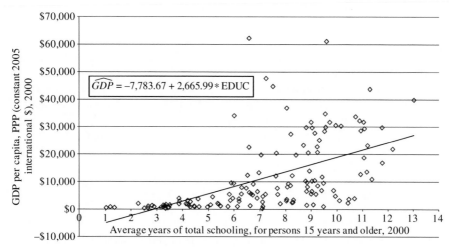

Chart 13.1 Simple linear regression model for GDP per capita as a function of educational attainment.

Data source: GDP per capita: World Bank (2010b), World Development Indicators. Average years of schooling: World Bank (2010a), EdState Database.

education *increase* GDP per capita. This is to say that we can *never* verify a *causal* relationship between the dependent and independent variables using regression analysis (or any other type of statistical inference).

13.2 The algebra of linear relationships for regression analysis

Regression analysis can be applied to bivariate relationships for a given time period (cross-section analysis), over time (time-series analysis), or for the same group of elements over time (panel, or longitudinal, data analysis).[2] For each type of analysis, the linear relationship to be statistically estimated is based on the *theoretical relationship* between the two variables. That is, we use economic theory to guide our choice of dependent and independent variables. As we saw above, for example, economic theory led us to specify GDP per capita as the dependent variable and adult educational attainment as the independent variable.

Let us now generalize the simple linear regression equation:

$$Y = a + bX \tag{13.1}$$

where Y denotes the dependent variable and X is the independent variable. Because equation (13.1) represents a theoretical statement of the relationship between the dependent variable and the independent variable, economists refer to equation (13.1) as the **theoretical (simple linear) regression model**.

Our goal is to estimate values for a, the Y-axis intercept, and b, the slope of the line, using the paired values of Y and X, which we will do using the ordinary least-squares (OLS) method described in Chapter 8. As we discussed in Section 8.2, the line estimated by the OLS procedure will not intersect all pairs of observations unless there is a perfect linear relationship between Y and X. To acknowledge this possibility, we now restate the theoretical regression model to account for the **residual (e)**, or **error term**, between the actual data points and those estimated by the OLS line:

$$Y = a + bX + e \tag{13.2}$$

The presence of the residual signifies that equation (13.2) is the **statistical (simple linear) regression model**, which we will estimate using OLS.

We also recall that the **ordinary least-squares (OLS) method** generates a linear function that *best* fits the data, which is achieved by minimizing the sum of squared residuals, $\sum e^2$. Thus we can describe the OLS procedure as follows:

$$\min \sum e^2 \quad \text{where } e = Y - \widehat{Y} \text{ and } \widehat{Y} = \hat{a} + (\hat{b} \times X)$$

Note that Y represents the observed values of the dependent variable, while $\widehat{Y}$ (Y-hat) represents the estimated values of the dependent variable that lie on the regression line.

While deriving the formulas for $\hat{a}$ and $\hat{b}$ is beyond the scope of this text,[3] it can be shown that the following equations minimize the sum of squared residuals through the OLS procedure:

$$\hat{a} = \overline{Y} - \hat{b}\overline{X} \tag{13.3}$$

$$\hat{b} = \frac{\sum (X - \overline{X})Y}{\sum (X - \overline{X})^2} \tag{13.4}$$

Equation (13.4) is typically rewritten as the more computationally friendly equation (13.5):

$$\hat{b} = \frac{\sum XY - \overline{Y}\sum X}{\sum X^2 - \overline{X}\sum X} \tag{13.5}$$

Because $\hat{b}$ is needed to estimate $\hat{a}$, it is standard practice to solve equation (13.5) first, and then use $\hat{b}$ to solve equation (13.3).

We can now state the **estimated (simple linear) regression model**:

$$\widehat{Y} = \hat{a} + \hat{b}X \tag{13.6}$$

The $\widehat{Y}$ of the estimated model is often referred to as the **predicted value of Y** or the **fitted value of Y**. We also use hats on a and b to indicate that $\hat{a}$ and $\hat{b}$ are the OLS estimates of the true population intercept and slope. This is analogous to our work in Part III, in which we used sample statistics to estimate population parameters. It also indicates that the OLS method of estimating the intercept and slope coefficients is another type of statistical inference.

13.3   Simple linear regression analysis: education and GDP

Let us now formally apply the OLS method to the educational attainment–GDP per capita relationship. We begin the analysis by stating the *theoretical (simple linear) regression model*:

$$\text{GDP} = a + (b \times \text{EDUC})$$

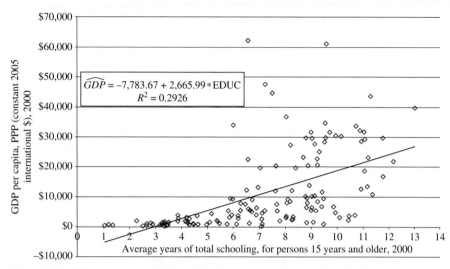

Chart 13.2 GDP per capita-educational attainment regression model and R^2.

Data source: GDP per capita: World Bank (2010b), World Development Indicators. Average years of schooling: World Bank (2010a), EdState Database.

where GDP = per-capita gross domestic product, purchasing power parity (PPP), in constant 2005 international dollars, and EDUC = average years of schooling for adults 15 years and over.

While the theoretical model represents the relationship indicated by economic theory, it is not the model we will estimate using the drawn sample. Because there is a probability that the sample will not accurately reflect the population, the *statistical (simple linear) regression model* includes an error term, e:

$$GDP = a + (b \times EDUC) + e$$

After collecting data appropriate for analyzing the theoretical relationship, we apply the OLS method to estimate the regression equation for our model. As we see in Chart 13.2 (p. 329), the *estimated (simple linear) regression model* is

$$\widehat{GDP} = -7,783.67 + (2,665.99 \times EDUC)$$

Again, the estimated slope coefficient of 2,665.99 indicates that, for each additional year of schooling completed by adults, per-capita GDP will, on average, be 2,665.99 international dollars higher. The positive sign on the estimate suggests that the marginal social return to education is a positive value.

Let us now consider the intercept estimate in more detail. According to the results, a nation in which the adult population has completed, on average, zero years of schooling will have a corresponding per-capita GDP of (*negative*) −7,783.67 international dollars. Such a result makes neither economic nor empirical sense. First, no country in the sample has a completely uneducated adult population, nor does any country have negative GDP per capita, so the result is not consistent with the economic reality facing any of the countries in the sample. In addition, the intercept term of the simple regression model picks up any statistical misspecifications associated with the model, including the effects of omitted variables. Since it is unlikely that differences in adult educational attainment is the only factor to affect the variation of GDP per capita, the intercept term has little statistical credibility.

We observe another statistic associated with the regression model, the R^2. Analogous to the coefficient of determination in correlation analysis, R^2 is the **regression coefficient of determination**. It measures how much of the dependent variable's variation is accounted for by the variation in the independent variable. Thus this statistic allows us to determine how well the OLS-estimated line fits the data.

In our example, $R^2 = 0.2926$. This tells us that 29.26 percent of the variation in GDP per capita for the 140 countries in the sample is accounted for by the variation in adult educational attainment. (Since the maximum value of R^2 is 1.0, it also indicates that 70.74 percent of the variation in GDP per capita is not "explained" by the model.) Let us now examine the statistical concepts underlying this often-cited statistic.

13.4   The algebra of variations in linear regression relationships

The goal of the ordinary least-squares method is to minimize the sum of the squared residuals associated with the regression model. These residuals result from the fact that, when we estimate the linear relationship between the dependent and independent variables, that line will not perfectly fit all paired variables. Thus the regression model will systematically explain only part of the variation in the dependent variable, while the remaining part, measured by the residual, will not be explained. We can therefore decompose each observation of the dependent variable (Y_i) into two parts, the explained $(\widehat{Y}_i)$ and the unexplained $(\hat{e}_i)$:

$$Y_i = \widehat{Y}_i + \hat{e}_i \tag{13.7}$$

where again $\widehat{Y}_i = \hat{a} + \hat{b}X_i$ and $\hat{e}_i = Y_i - \widehat{Y}_i$.

We can now identify each component for three countries – Australia, Paraguay, and Yemen – from the 2000 World Bank sample as presented in Chart 13.3 (p. 331). Here the actual value of the dependent variable (Y_i), indicated by a diamond, may lie above (in the case of Australia and Yemen) or below (Paraguay) the estimated regression value $(\widehat{Y}_i)$ for GDP, indicated by a circle. We also see that none of the actual values of GDP lie on the estimated regression line, which means an estimated residual will be associated with each country. We further note that the residual may be positive (Australia and Yemen) or negative (Paraguay).

Let us now add the means of the dependent $(\overline{Y})$ and independent $(\overline{X})$ variables to the scatterplot. As we observe in Chart 13.4 (p. 332), the means of the paired dependent and independent variables intersect at a point on the estimated regression line, which illustrates one characteristic of the fitted regression line (when the equation contains an intercept term).

We will now use these means to decompose not the individual observations, but the **variation** of the dependent variable, which is defined by equation (13.8):

$$\text{variation} = \sum_{i=1}^{n} (Y_i - \overline{Y})^2 \tag{13.8}$$

(We note that this is identical to equation (5.4) in Figure 5.1.)

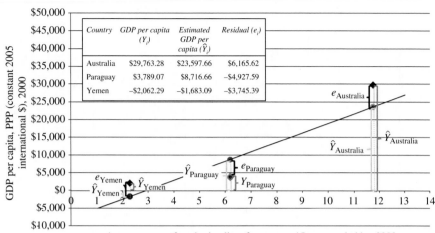

Chart 13.3 Relationship between the actual and estimated values of GDP per capita for Australia, Paraguay, and Yemen.

Data source: GDP per capita: World Bank (2010b), World Development Indicators. Average years of schooling: World Bank (2010a), EdState Database.

We now refer to the parenthetical term $(Y_i - \overline{Y})$ as the **total deviation** (T_i) of each observed value for Y from its mean:

$$T_i = Y_i - \overline{Y} \tag{13.9}$$

If we return to equation (13.7), subtract the sample mean from both sides of the equation, and substitute the residual expression in terms of the dependent variable, we obtain equation (13.10):

$$(Y_i - \overline{Y}) = (\widehat{Y}_i - \overline{Y}) + (Y_i - \widehat{Y}_i) \tag{13.10}$$

Let us now discuss the decomposition of the total deviation (T_i) in terms of the two components on the right-hand side of equation (13.10). The first component, $(\widehat{Y}_i - \overline{Y})$, represents what is explained by the regression equation and is thus denoted as the **explained deviation (R_i)** or **regression deviation**:

$$R_i = \widehat{Y}_i - \overline{Y} \tag{13.11}$$

The second component, $(Y_i - \widehat{Y})$, reflects what is not explained by the regression equation. Denoted as the **residual deviation** (E_i), it is defined as the difference between the observed value of Y (i.e. Y_i) and the OLS-estimated value of the dependent variable (i.e. $\widehat{Y}_i$):

$$E_i = Y_i - \widehat{Y}_i \tag{13.12}$$

This last deviation is sometimes referred to as the **error deviation** or the **unexplained deviation** of Y from its estimated value $(\widehat{Y}_i)$. It tells us the extent to which an observed value of

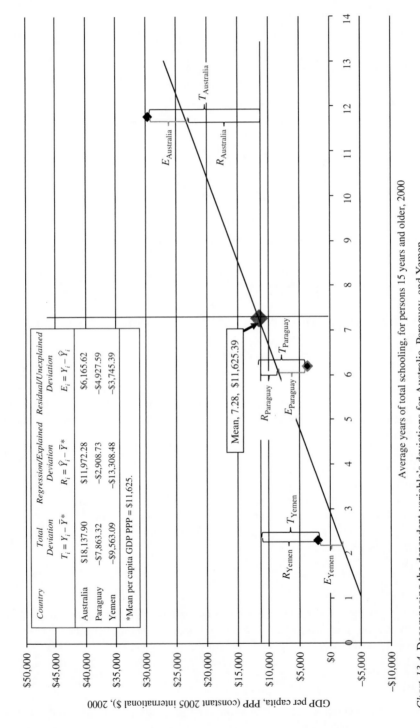

Country	Total Deviation $T_i = Y_i - \bar{Y}_i$	Regression/Explained Deviation $R_i = \hat{Y}_i - \bar{Y}^*$	Residual/Unexplained Deviation $E_i = Y_i - \hat{Y}_i$
Australia	$18,137.90	$11,972.28	$6,165.62
Paraguay	–$7,863.32	–$2,908.73	–$4,927.59
Yemen	–$9,563.09	–$13,308.48	–$3,745.39

*Mean per capita GDP PPP = $11,625.

Mean, 7.28, $11,625.39

GDP per capita, PPP (constant 2005 international $), 2000

Average years of total schooling, for persons 15 years and older, 2000

Chart 13.4 Decomposing the dependent variable's deviations for Australia, Paraguay, and Yemen.

Data source: GDP per capita: World Bank (2010b), World Development Indicators. Average years of schooling: World Bank (2010a), EdState Database.

Y is not equal to the estimated regression value, $\widehat{Y}_i$, or, alternatively, the amount by which the regression equation has "erred" in explaining the deviation of Y from $\overline{Y}$.

We have thus identified three types of deviations for each observation of the dependent variable in our regression model, which are related as follows:

$$\text{total deviation } (T_i) \equiv \text{regression deviation } (R_i) + \text{residual deviation } (E_i)$$

If we now extend these definitions of deviations directly to the definition of variations, or sums of squares, we find that, for *all* observed values of (Y_i, X_i), we have

total variation (sum of squares): $\qquad \text{TSS} = \sum (Y_i - \overline{Y})^2$ \hfill (13.13)

regression variation (sum of squares): $\quad \text{RSS} = \sum (\widehat{Y}_i - \overline{Y})^2$ \hfill (13.14)

residual variation (sum of squares): $\quad \text{ESS} = \sum (Y_i - \widehat{Y}_i)^2 = \sum e_i^2$ \hfill (13.15)

We will now use these concepts to consider the R^2 statistic in detail.

13.5 The coefficient of determination in regression analysis

Let us begin by defining the **regression coefficient of determination** (R^2). Mathematically, R^2 is the ratio of the regression variation of the dependent variable divided by the total variation of the dependent variable:

$$R^2 = \frac{\text{RSS}}{\text{TSS}} \tag{13.16}$$

In less technical language, the R^2 statistic measures the proportion of total variation of the dependent variable (TSS) that is explained by the regression relationship between Y and X (RSS). We can also define the R^2 statistic using the **error sum of squares (ESS)**:

$$R^2 = \frac{\text{TSS} - \text{ESS}}{\text{TSS}} = 1 - \frac{\text{ESS}}{\text{TSS}} \tag{13.17}$$

What constitutes a "good" R^2 statistic? Given that the goal of estimating a regression equation is to explain as much of the variation in the dependent variable as possible, a "good" outcome is one for which the regression model explains a "large" portion of the variation in the dependent variable. But what constitutes "large"? The possible range of values for the regression coefficient of determination will help us decide.

Suppose we obtain the (unlikely) result of perfectly describing the relationship between X and Y with our estimated regression function, meaning that all observed values of Y (and X) lie along the regression line. In such a case, each value of Y (i.e. Y_i) would equal the corresponding $\widehat{Y}_i$, such that the regression sum of squares (RSS) will equal the total sum of squares (TSS). In this scenario, the value of the regression coefficient of determination, R^2, would be one, indicating that the simple regression model perfectly explains the variation in the dependent variable.

Now suppose that the variation in the independent variable explains nothing about the variation in the dependent variable. In this case, the RSS equals zero, making R^2 equal to zero as well. This essentially tells us that our regression model is useless for explaining the dependent variable's variation. In other words, the total variation in Y (i.e. TSS) is completely attributable to the random error sum of squares (ESS).

These two extreme cases provide us with the minimum and maximum values of R^2:

$$0 \leq R^2 \leq 1$$

At this point, it might be tempting to conclude that the closer R^2 is to a value of one, the better the regression model. Such a conclusion would be incomplete and incorrect, as we will now demonstrate for the GDP–education example.

As calculated earlier, the R^2 for the GDP–education model equals 0.2926, which we interpreted as the variation in per-capita GDP that is accounted for by the accompanying variation in educational attainment. Specifically, the variation in adult educational attainment for the 140 nations in our sample explains 29.26 percent of the variation in GDP per capita.

Is 29.26 percent a "good" result? The short answer is that we cannot determine this from the single statistic. Our assessment must consider other, comparable regression model results. That is, we must compare our R^2 statistic with other cross-country linear regression models of the GDP–education relationship to make such a determination. We could, for example, estimate the same statistical regression model using the World Bank data from 2005, which yields the following results:

$$\widehat{GDP}_{2005} = -10,557.94 + (3,064.27 \times EDUC)$$

$$R^2_{2005} = 0.3208$$

We observe that the GDP–education model estimated using the 2005 data explains more of the variation in GDP per capita, 32.08 percent, than did the model using data from 2000. While this result *might* suggest that the model is better for understanding the variation in GDP in 2005 than in 2000, we have several other results to consider. In addition, the model still accounts for only about one-third of the variation in per-capita GDP.

Owing to the unmeasurable variations across elements, cross-section regression models often yield much lower R^2 values than those associated with time-series models. This does not mean that time-series models are superior to cross-section ones. In fact, the high R^2 values associated with time-series models are often attributable to statistical problems in the model.

As alluded to above, we also do not want to assess the overall "goodness" of any regression model by looking *only* at its R^2. Indeed, economists are more typically concerned with the statistical significance of the regression model's slope coefficient ($\hat{b}$), which we develop in the next chapter. Thus, focusing on the R^2 statistic in isolation reflects incomplete statistical practices. The moral of our story? Do not fall victim to judging regression results based solely on the R^2 value obtained.

13.6   Simple linear regression analysis: infant mortality rates and skilled health personnel at birth

We now consider another economic relationship of global importance, that between infant mortality rates and the percentage of births attended by skilled health personnel. Infant mortality rates, which we will measure as the number of infant deaths per 1,000 live births, are viewed as a key indicator of a country's ability to develop economically, and are frequently cited as a marker of the quality of life in a nation. The economic rationale is that fewer infant deaths are positively correlated with a healthier population, which in turn makes for a more productive workforce in both the short and long runs. Consequently, many countries would be very interested in learning what factors might reduce infant mortality rates.

Based on casual observation and reasoned thought, one factor that may reduce the number of infant deaths is the quality of health care available during labor and delivery. If a birth is attended by skilled medical personnel (such as doctors, nurses, and midwives) who have the training and experience to deal with complications, their presence will likely increase the infant's chance of survival, thereby lowering the infant mortality rate, *ceteris paribus*. (Here, *ceteris paribus* means we assume that all other factors that might influence the infant mortality rate – such as the mother's health, her access to sanitation, clean water, and adequate nutrition, as well as her access to prenatal care – are held constant both theoretically and statistically.) But before a country commits to a large and sustained investment in the number of skilled natal personnel, most officials would want further evidence that such a relationship holds. One source for such support may be the systematic empirical evidence that a simple linear regression analysis can provide.

To evaluate the relationship between infant mortality rates (denoted IMR) and the percentage of births attended by skilled medical personnel (denoted ATTEND), let us now proceed through the steps for conducting a simple regression analysis. We begin by establishing the theoretical relationship between the dependent and independent variables:

$$IMR = a + (b \times ATTEND)$$

This initial step embodies appropriate statistical practices in economics, where a regression analysis must begin by stating a theoretical relationship between the dependent and independent variables. For the current example, this means economists began by postulating a relationship between infant mortality rates and the presence of health personnel at births. It also illustrates the most important feature of effective regression analysis: a theoretical relationship *must* be established *before* conducting an empirical analysis of that relationship.

The second step specifies the statistical simple linear regression model to which we will apply the ordinary least-squares method:

$$IMR = a + (b \times ATTEND) + e$$

The presence of the error term not only signifies that linear regression analysis is a statistical inference procedure, but also indicates that we are unlikely to fully describe the statistical relationship between the dependent and independent variables.

After collecting appropriate data series for the dependent and independent variables, we can plot the paired variables in a scatter diagram and then fit an OLS-estimated line to that data. As we observe in Chart 13.5, we estimate the following regression equation using data from the United Nations' World Health Organization for 152 countries:

$$\widehat{IMR} = 120.65 - (1.0519 \times ATTEND)$$

According to the intercept coefficient, a country in which no skilled medical personnel attend child births is estimated to have an infant mortality rate of 120.65 infant deaths per 1000 live births. While it is unlikely that ATTEND will ever be equal to zero, we do observe that the percentage of births attended by medical personnel is below 20 percent for five

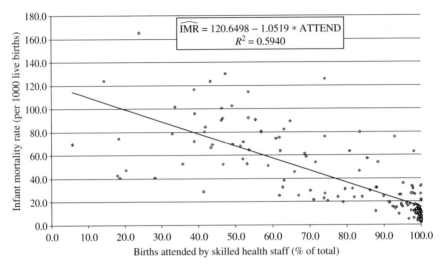

Chart 13.5 Infant mortality rates and births attended by skilled medical personnel.

Data source: World Health Organization (2010b).

countries – Ethiopia (5.7 percent), Chad (14.4 percent), Bangladesh (18.0 percent), Timor-Leste (18.4 percent), and Nepal (18.7 percent) – so the intercept term *may* be empirically reliable.

Before we accept such results, however, we must acknowledge that, while important, the presence of skilled natal personnel is not likely to be the only factor to affect infant mortality rates. Thus the simple regression model is most likely incomplete for explaining the variation in infant mortality rates across the 152 countries. Statistically this means that the intercept term will pick up the effects of omitted independent variables, along with other model mis-specifications. Consequently, we will, as in the previous example, focus our attention on this model's slope coefficient.

The regression model's estimated slope coefficient of −1.0519 indicates that a one percentage point increase in the number of births attended by skilled medical personnel corresponds to a reduction in the infant mortality rate of 1.0519 infant deaths per 1,000 live births. While this result appears to support the theorized relationship between infant mortality and natal care, we have yet to evaluate if the result is statistically significant. (We shall do so shortly in Chapter 14.) Even if the result is determined to be statistically significant, we cannot use the regression results to conclude that skilled natal care does reduce infant mortality.

Chart 13.5 (p. 336) also includes the regression coefficient of determination. Here the R^2 value of 0.5940 tells us that 59.40 percent of the variability of infant mortality rates is accounted for by the variation in the proportion of births attended by skilled medical personnel for the 152 countries surveyed. Again, we cannot assess from this single statistic if this is a "good" fit for models estimating this particular relationship.

We might now wonder if we can compare the R^2 values obtained for the two theoretical regression models presented in Chapter 13 to determine if one model is "better" than the other. The quick answer is "no." The reason? Regression coefficients of determination are not comparable across different theoretical models. Indeed, the only time we could even consider comparing R^2 values is when the linear regression models are evaluating the same theoretical relationship and when they use comparable dependent and independent variables.

Summary

Simple linear regression analysis allows economists to measure the relationship between two economic variables. This method is generally preferred over correlation analysis because regression models are written to reflect a *theoretically* hypothesized relationship between a pair of dependent and independent variables.

Simple linear regression models provide statistical evidence of how well the estimated equation describes the relationship between Y and X. Using the regression coefficient of determination R^2, we can measure what percentage of the variation in the dependent variable is explained by the variation in the independent variable.

Simple linear regression models also estimate the numerical relationship between the two variables in the form of the intercept coefficient, $\hat{a}$, and the slope coefficient, $\hat{b}$ However, because those results may not be statistically significant, we now turn to procedures for evaluating if this is, in fact, the case.

Concepts introduced

Dependent variable (Y)
Error deviation (E_i)
Error sum of squares (ESS)
Error term (e)
Estimated (simple linear) regression model
Explained deviation (R_i)
Explanatory variable (X)
Fitted value of Y ($\widehat{Y}$)
Independent variable (X)
Ordinary least-squares (OLS) method
Predicted value of Y ($\widehat{Y}$)
Regression coefficient of determination (R^2)
Regression deviation (R_i)
Regression variation (sum of squares) (RSS)
Residual (e)
Residual deviation (E_i)
Residual variation (sum of squares) (ESS)
Simple linear regression line
Statistical (simple linear) regression model
Theoretical (simple linear) regression model
Total deviation (T_i)
Total variation (sum of squares) (TSS)
Unexplained deviation (E_i)
Variation

Box 13.1 Using Excel to estimate the simple linear regression model from a scatterplot

We can visually depict the relationship between a dependent variable and an independent variable using Excel's XY (Scatter) chart type.

After plotting the values on the appropriate axes, we can *add a linear trendline* to the data, which Excel estimates using the ordinary least-squares method. (See "Using Excel to calculate the least-squares estimates of constant growth" in Chapter 8 for detailed directions.)

When adding the trendline, we can tell Excel to provide both the OLS-estimated equation and the R^2 statistic by checking the boxes, **Display Equation on chart** and **Display R-squared value on chart**, respectively.

Exercises

1. For each of the following pairs of variables, identify the dependent variable, and specify the statistical regression model.

    (a) Annual CPI-U for coffee (100 percent, ground roast, all sizes, per pound) and annual retail pounds of coffee consumed per capita between 1967 and 2007. (Make sure you can explain why we chose to use the coffee CPI in the demand for coffee model rather than the average dollar price of coffee.)

    (b) Annual income (dollars) in 2004 and years worked in nursing since receiving first RN license for full-time, non-union, registered nurses (RNs) with advanced degrees who are classified as staff or general duty nurses and working in a hospital.

    (c) Annual number of new houses sold and annual new home mortgage yields between 1991 and 2007.

2. Use the data at the end of Chapter 12 to estimate the simple regression model for each of the scenarios in question 1. (Add a linear trendline with its equation and R^2 value to the scatterplot in Excel.) Interpret the slope coefficient for each model.

3. Interpret the R^2 value associated with each model in question 2.

4. Marginal productivity theory holds that a worker's compensation will equal the additional (marginal) output that worker adds to the production process. Use the data (at the end of Chapter 12) for the index of "Real hourly compensation" in durable goods manufacturing and the index of labor productivity as measured by "Output per person" in durable goods manufacturing to specify and then estimate a simple regression model for worker compensation in U.S. durable goods manufacturing between 1987-Q1 and 2007-Q4.

14 Simple regression analysis
Statistical inference

14.1 The need for statistical inference in regression analysis

Both simple linear regression models discussed in Chapter 13 used sample data to estimate the intercept and slope coefficients of the regression models. As a result, we can only make probabilistic statements about a population based on the sample drawn. Thus we now turn to statistical inference theory as it applies to simple linear regression analysis to determine the statistical significance of the estimated parameters.

We will begin by demonstrating the statistical inference procedures used to evaluate the GDP–education model's slope coefficient, $\hat{b}$. While we obtained regression results that appeared to be consistent with our theoretical preconceptions, we must now determine if the consistency between the regression results and the underlying economic theory is statistically sound or simply a matter of chance.

14.2 Testing hypotheses about the GDP–education regression model's slope coefficient

In simple linear regression analysis, the primary focus of statistical inference is testing if the model's slope coefficient is statistically significant. Because economic theory often postulates *how* the dependent and independent variables are related, whether it be positively or negatively, we will typically conduct a one-tail hypothesis test of statistical significance. Let us now demonstrate the process of testing hypotheses about a model's slope coefficient using the GDP–education model estimated in Chapter 13.

We begin by establishing the null and alternative hypotheses to be tested. Unlike the previous hypothesis tests of sample means and correlation coefficients, the hypotheses for regression analysis reflect the underlying relationship posited by economic theory. This reinforces the practice described in Chapter 13, where we began the regression analysis by articulating the theoretically predicted relationship between the dependent variable, Y, and an independent variable, X.

In the GDP–education model, we postulated that the more education attained by a nation's adults, the more general and job-specific skills the labor force will likely have, making the workforce the more productive, thereby raising GDP per capita. Mathematically, this means we expect a positive relationship between education and GDP. This leads to the following pair of hypotheses, where b_{EDUC} represents the population slope coefficient:

$$H_0 : b_{\text{EDUC}} \leq 0 \quad \text{versus} \quad H_a : b_{\text{EDUC}} > 0$$

Here we continue to follow economic convention and state the alternative hypothesis to reflect the underlying economic theory. Remember that the goal of regression analysis is to make inferences from a sample about a relationship between two economic variables in the population. Such inferences often have important real-world implications. In the case here, for example, if the statistical evidence supports, with a high degree of probability, the hypothesized theoretical relationship, it may lead officials in some nations to increase investment in education to improve the economic well-being of its citizens in the form of higher GDP per capita. But before undertaking such a costly investment, officials want to be as confident as possible that higher educational attainment does, in fact, correspond to higher levels of GDP per capita.

After establishing the hypotheses to test, we next set the level of significance (α) for which we want to evaluate the null hypothesis. As previously noted, economists most frequently use 5 percent (0.05) and 1 percent (0.01) levels of significance. Given the policy implications associated with the GDP–education model, we will set the level of significance at 0.01 to minimize the probability of a Type I error. That is, we want to minimize the possibility of rejecting a true hypothesis since that might lead officials to direct resources to educational programs that do not correspond to the desired effect on GDP per capita.

Let us now state the decision rule that we will use to assess the null hypothesis. Again, we establish this rule prior to estimating the regression model to insure our conclusions are not influenced by the statistical results.

Because it states the conditions under which we will *reject* the null hypothesis, the decision rule for the slope coefficient will reflect the underlying economic theory of the model. For the GDP–education model, we expect a positive slope coefficient, so we write our decision rule as:

$$\text{reject } H_0 \text{ if } t_b > t_{n-2}^{0.01}$$

In simple linear regression analysis, the critical t-value has two arguments. The superscript indicates the level of significance specified above, while the subscript indicates the appropriate degrees of freedom. Like correlation analysis, the number of degrees of freedom associated with the simple linear regression model is $n - 2$, reflecting that we must first determine the means of Y and X before we can estimate the slope coefficient, $\hat{b}$.

The critical t-value for the 140-country GDP–education model where $\alpha = 0.01$ and $(n - 2) = 138$ is 2.3537, which yields the following decision rule:

$$\text{reject } H_0 \text{ if } t_b > 2.3537$$

(The critical value was obtained following the directions for "Calculating the critical t-value" in "Using Excel to perform correlation analysis" at the end of Chapter 12.)

The next step is to run the regression for the GDP–education relationship. As we found in Chapter 13, the estimated simple linear regression model is

$$\widehat{\text{GDP}} = -7{,}783.67 + 2{,}665.99 \times \text{EDUC}$$

where the estimated slope coefficient indicates that, in 2000, for every additional year of education attained by a country's adult population, GDP per capita (PPP) was, on average, 2,665.99 (constant 2005 international dollars) higher. We now want to evaluate if this result is statistically greater than zero.

Evaluating the coefficient's statistical significance requires us to calculate the **test statistic for the slope coefficient** using equation (14.1):

$$t_b = \frac{\hat{b} - b}{se_{\hat{b}}} \tag{14.1}$$

Let us now consider the values required here. The estimated slope coefficient, $\hat{b}$, is the first term in the numerator. As we have seen, it has a value of 2,665.99 (constant 2005 international dollars). Next in the numerator, the value for b is determined by the null and alternative hypotheses. Recall that we are testing if the estimated slope coefficient's relationship is statistically greater than *zero*. This then tells us that b will equal 0. Indeed, when testing the statistical significance of a regression coefficient, the hypothesized value will almost always be equal to zero.

We now need one additional piece of information in equation (14.1), the estimated standard error of the slope coefficient, $se_{\hat{b}}$. While the formula for calculating this statistic is presented as equation (14.6) in the following section, we typically let our statistical software generate this value for us. For the GDP–education model, $se_{\hat{b}} = 352.90$ (constant 2005 international dollars), so that the test statistic equals 7.5545:

$$t_b = \frac{2665.99 - 0}{352.90} = 7.5545$$

We are now ready to make a decision about the null hypothesis. Inserting our test statistic into the decision rule:

reject H_0 if $t_b = 7.5545 > t_{138}^{0.01} = 2.3537$

we find that the test statistic, t_b, is greater than the critical t-value, $t_{138}^{0.01}$, so we can reject the null hypothesis for a 1 percent level of significance. This indicates that we are at least 99 percent confident that the positive relationship between adult educational attainment and GDP per capita is supported by the statistical evidence and is not simply a matter of chance.

Note that we have been very careful in the language used to describe the hypothesized relationship between education and GDP. At no time have we claimed that the statistical results *prove* that higher levels of education lead to higher levels of GDP per capita. The most we can conclude is that the linear regression results from this one sample and this very basic (and incomplete) regression model support the postulated relationship between education and GDP. Thus we cannot use the regression results to conclude that higher levels of adult education *cause* GDP per capita to be higher. In other words, we *cannot prove causality* using the statistical inference method of regression analysis.

We now turn to the final step in hypothesis testing, determining the p-value associated with the slope coefficient. Recall that p-values allow us to assess the probability of obtaining the test statistic's value when the null hypothesis is true. In simple linear regression analysis, the p-value is

$$\Pr(|t| > t_{\hat{b}} \mid H_0 : b = 0) \tag{14.2}$$

where $t_{\hat{b}}$ is the test statistic obtained from the regression results.

For our example, the p-value equals the probability of obtaining a test statistic of 7.5545 from a sample of 140 countries when the statistical relationship between GDP and adult educational attainment is either negative or zero. Given the observed difference between the test statistic and critical t-value already determined for the GDP–education model, we expect to obtain a very small value for p.

According to our calculations, the p-value for the GDP–education estimated regression model is

$$\Pr|t| > 7.5545 \mid H_0 : b = 0) = 2.6202\text{E-}12 = 2.6202 \times 10^{-12}$$

As expected, the p-value is quite small. In addition, it is less than the 1 percent level of significance established earlier. Thus because

$$\alpha = 0.01 > p = 2.6202\text{E-}12$$

this result confirms our decision to reject the null hypothesis that $b_{\text{EDUC}} \leq 0$. If we now follow the interpretation of the p-value described in Section 12.3, we can state precisely how confident we are in rejecting the null hypothesis. Specifically, if we apply equation (14.3):

$$\text{level of confidence in decision about } H_0 = (1 - p) \times 100 \qquad (14.3)$$

then we can say that we are

$$(1 - 0.0000000000026) \times 100 = 99.99999999974 \text{ percent}$$

confident there is a statistically positive relationship between adult educational attainment and GDP per capita based on the regression results from the 140-country sample based on data from 2000.

14.3 Sampling distributions of the linear regression's slope and intercept coefficients

Whenever we use statistical inference methods, we must identify the sampling distributions associated with the estimated parameters. Let us now examine the properties of the sampling distributions associated with the simple linear regression model's parameters.

We have thus far specified three simple linear regression models – theoretical, statistical, and estimated – to describe the linear relationship between the dependent and independent variables. All three were based on an expected relationship between the dependent and independent variables in the *unobserved* population. More formally, we can express a tentative belief that there exists in a population a relationship between X and Y that takes the form of

$$Y^{\text{P}} = \beta_0 + \beta_1 X + u \qquad (14.4)$$

In this mathematical model,[1] Y^{P} denotes the dependent variable associated with the population; β_0 is the population parameter corresponding to the sample statistic a; β_1 is the parameter corresponding to the sample statistic b; and u is the residual. As the regression equation for a population, equation (14.4) indicates the values of Y that we could compute from the (unknown) values of β_0 and β_1 if we know the values of X. The error term, u, is included in equation (14.4) because it is likely that, even for the population, Y^{P} will not be perfectly described by the single independent variable, X.

Because we rarely have access to population data, we must estimate values for the intercept and slope coefficients from sample data. Thus we use this statistical model:

$$Y = a + bX + e$$

to estimate a and b, from which we will make inferences about β_0 and β_1, respectively. Now let us discuss the distribution that underlies the sample estimate of the equation's slope. (The following properties also hold for the intercept estimate.)

We begin with the distribution of the dependent variable, Y. As demonstrated in most econometrics textbooks (see, for example, Wooldridge 2009, pp. 117–123), the dependent variable's underlying distribution is normal with a mean that is linear with respect to X and a constant variance. If we also assume that the independent variable X is a random variable, the sampling distribution underlying the slope coefficient, b, will also be a *normal distribution*. This result yields the first desirable statistical property associated with the OLS coefficients – the sampling distribution of b, like all normally distributed variables, can be completely described by its mean and variance.

We next want to consider the statistical properties associated with the estimators we use to calculate the regression coefficients, a and b (Figure 14.1). The term **estimator** refers to the rule – more accurately here, the equation – used to obtain a numerical value for the population parameter from a sample. As we discussed in Chapter 9, one desirable statistical property relates to the expected value of the estimator. In simple linear regression analysis, this means we want the slope coefficient estimated using the OLS method to generate an **unbiased estimator** of the population parameter β such that

$$E(b) = \beta$$

This property indicates that the expected value of the regression's slope coefficient is the value of the population parameter that measures the linear relationship between the dependent and independent variables. Since we must make inferences about a population from a sample, this is a desirable property for the sample statistic. Fortunately, it can be shown that the OLS estimator for b produces an unbiased estimate of the population parameter β.

Another desirable statistical property relates to the variance, and thus standard deviation, of b. Specifically, we not only want the sampling distribution to be centered on the true population parameter, we also want its variance around the mean to be as small as possible (Figure 14.1). Fortunately, the estimator used by the OLS method to generate the variance produces the desired result. That is, the OLS-estimated variance for b will be the smallest possible variance among all linear unbiased estimators. (Statisticians often refer to this variance as the **best**.) In addition, the OLS estimator for b is a **consistent estimator**, which means that, as the sample size increases, the variance of b gets smaller, and the sample statistic approaches the true value of the population parameter.

These properties associated with the regression slope coefficient mean we could theoretically utilize the standard normal distribution to test hypotheses about the coefficient values. However, because we do not typically observe the population standard deviation, we must calculate the standard error of the estimator. As a result, we use the t-distribution to describe the sampling distribution of b (and a). Specifically, the t-statistic associated with the estimated slope coefficient is defined as

$$t_b = \frac{b - \beta}{\text{se}_b} \qquad (14.5)$$

1. The sampling distribution of the population parameters of the simple linear regression model is normal and can be completely described by its mean and variance.
2. The estimators of the population parameters of the simple linear regression model are unbiased.
3. The estimators of the population parameters of the simple linear regression model are best and, as $n \to \infty$, consistent.

Figure 14.1 Desirable statistical properties of the OLS population parameters.

where

$$se_b = \frac{\text{SER}}{\sqrt{(n-1) \times \sum (X - \bar{X})^2}} \tag{14.6}$$

The numerator of equation (14.6) is known as the **standard error of the regression (SER)** and can be used to assess the quality of a regression relationship. More accurately called the **standard error of the predicted Y-value for each X in the regression** and denoted by SER or S_{YX}, this statistic is defined as

$$\text{SER} = S_{YX} = \sqrt{\frac{\sum e_i^2}{(n-2)}} = \sqrt{\frac{\sum (Y_i - \hat{Y})^2}{(n-2)}} \tag{14.7}$$

Here we note a similarity between equation (14.7) and the deviations of Y discussed in Chapter 13. However, the SER represents the deviation of Y from the regression line ($\widehat{Y}$) rather than from the mean. As this suggests, the smaller the SER, the better the regression line fits the data. (However, we cannot look at a single SER to determine a model's goodness of fit.)

What have we learned here? We have established that the sampling distribution for b (and a) mirrors the distribution for the true population parameter, β_1 (and β_0). We have also indicated that the ordinary least-squares method generates estimators for β_0 and β_1 that are unbiased, best, and consistent. Finally, we have identified the t-distribution as the one we use to conduct hypothesis tests on the linear regression model's coefficients.

Having established the connections between the population parameters and the sampling distribution's coefficients, we now return to a more detailed discussion of each step associated with testing hypotheses about the regression model's coefficients. While we will continue to focus on the regression model's slope coefficient, the same considerations apply to testing hypotheses about the intercept coefficient when that is warranted.

14.4   Establishing the null and alternative hypotheses for the slope coefficient

Let us now discuss the possible hypotheses we might test for a simple linear regression model. Since the regression models used by economists reflect a theoretically hypothesized relationship between two economic variables, we often posit either a positive or a negative relationship between the dependent and independent variables. For example, suppose we wanted to estimate the demand for a product, say, coffee. Here we would cite the so-called "law" of product demand and posit a negative relationship between the quantity of coffee demanded by buyers and the price of coffee. In statistical hypothesis testing, this means we would establish the following pair of hypotheses:

$$H_0 : b_{\text{price}} \geq 0 \quad \text{versus} \quad H_a : b_{\text{price}} < 0$$

Because we have specified a sign on the regression's slope coefficient, we will conduct a **one-tail hypothesis test of the slope coefficient**.

Not all economic relationships, however, are so clearly delineated. It may be the case, for example, that economists use the statistical evidence to categorize the relationship between the dependent and independent variables. In the case of the demand for coffee, for example, we may not know *a priori* (literally, before the fact) if an increase in consumer income will increase (indicating coffee is a normal good) or decrease (where coffee would be classified as an inferior good) the demand for coffee. Here, then, economists would test if the relationship between coffee prices and the quantity demanded was statistically different from zero. That is, they would conduct a **two-tail hypothesis test of the slope coefficient** to evaluate its statistical significance.

Figure 14.2 summarizes three hypotheses of the relationship between the dependent and independent variables that we might test for the simple linear regression model.

You may wonder why we conduct a two-tail hypothesis test of the regression slope coefficient when we could assess the statistical significance of the relationship between Y and X using correlation analysis. While we do reach the same outcome from correlation analysis as we do in simple linear regression analysis, correlation analysis cannot be applied to regression models with more than one independent variable. Since virtually all economic theory suggests that a dependent variable is affected by multiple factors (that is, more than one independent variable), we follow economic convention, and use a two-tail hypothesis test on the regression model's slope coefficient rather than testing if the correlation coefficient is statistically different from zero.

Let us now turn to another example in which a two-tail hypothesis test of the regression slope coefficient would be appropriate. Suppose we want to explain the variation in the number of hours worked by husbands and wives in married-couple families. Here we might use a labor supply model in which the hours worked by individual workers are a function of the wages received:

hours worked $= a + (b_{\text{wage}} \times \text{wage})$

	Null hypothesis	Alternative hypothesis
One-tail test		
Positive relationship between Y and X	$H_0: b \leq 0$	$H_a: b > 0$
Negative relationship between Y and X	$H_0: b \geq 0$	$H_a: b < 0$
Two-tail test		
Statistically significant, but theoretically uncertain, relationship between Y and X	$H_0: b = 0$	$H_a: b \neq 0$

Figure 14.2 Possible hypothesis tests for the simple regression slope coefficient.

According to the labor–leisure theory from neoclassical economics, workers face two choices when allocating their time (outside of sleeping and taking care of basic necessities). Workers may engage in leisure activities that directly confer utility, or they may work to generate income for purchasing utility-conferring goods and services. At the margin, if workers value leisure more highly than work, they will choose not to work an extra hour; if they value work more highly, they will work the extra hour.

Analogous to the income and substitution effects in neoclassical consumer demand theory, the decision to work additional hours is affected by two conflicting possibilities when wages increase. On one hand, the *income effect* predicts that an increase in a worker's wage will increase the amount of income at her/his disposal, which then leads the worker to reduce the total number of hours worked to devote more time to leisure. On the other hand, the *substitution effect* predicts that, as the reward (wage) for each hour worked increases, leisure becomes relatively more expensive, which leads the worker to substitute more work hours for leisure time. Thus, when a worker's wage increases, the income effect leads to fewer hours worked, but the substitution effect increases the hours worked.

Because labor–leisure theory does not tell us which effect will be larger (the outcome depends on individual worker preferences and wages received), we cannot postulate the sign for the relationship between hours worked and wages. Thus, when empirically estimating a labor supply model based on labor–leisure theory, we do not test for either a positive or negative relationship between the two variables. Instead, we test if the dependent and independent variables are statistically related to one another. That is, the null and alternative hypotheses for the labor supply model would be

$$H_0 : b_{\text{wage}} = 0 \quad \text{and} \quad H_a : b_{\text{wage}} \neq 0$$

Interestingly, labor economists have discovered a difference between the sexes when estimating labor supply models for married couples. As the following (partial) regression results demonstrate,[2] the slope coefficient for wives' hours of work (WHRS) is positive, whereas the slope estimate for husbands' hours of work (HHRS) is negative:[3]

$$\widehat{\text{WHRS}} = 658.49 + 113.67 \times \text{WWAGES} + \cdots$$

and

$$\widehat{\text{HHRS}} = 2505.33 - 33.45 \times \text{HWAGES} + \cdots$$

The estimated regression model for wives indicates that, for every \$1 increase in their wages, their hours of work increased by 113.67 hours per year, indicating that the substitution effect outweighs the income effect for wives in the sample. For husbands in the sample, a \$1 increase in their wages corresponded to 33.45 hour decrease in hours worked per year, indicating that the income effect for husbands outweighs the substitution effect. Only by empirically estimating the wage effect for wives and husbands can we resolve which effect is larger for each of the two groups. Thus we establish two-tail hypotheses for the wage coefficient to reflect both potential outcomes.

This extended example not only illustrates how economists test hypotheses when a clearly defined directional relationship does not obtain, but also raises a very important caution. It would be incorrect statistical practice to use the coefficient results just obtained to respecify our hypotheses to reflect the empirical results. That is, we must not modify the previously established hypotheses so the alternative hypothesis for the slope coefficients reflects a directional relationship (H_a: $b > 0$ for wives; and H_a: $b < 0$ for husbands). If we do want to test these alternative hypotheses, we must start our empirical analysis over by establishing new pairs of hypotheses to reflect these expected relationships, and then draw a new sample to obtain new sample results. Not doing this violates the random sampling assumption of the OLS method and represents inappropriate statistical practice.

14.5 Levels of significance and decision rules

Determining the level of significance (α) at which to evaluate the null hypothesis corresponds to the likelihood of committing a Type I error, or incorrectly rejecting a true null hypothesis. As demonstrated in the GDP–education example, economists prefer to minimize Type I errors, even though doing so increases the likelihood of Type II errors (incorrectly "accepting" a false null hypothesis) as long as the sample size remains constant.

This practice follows directly from how economists establish their hypotheses about the simple regression model's slope coefficient. Because the alternative hypothesis reflects the relationship between the dependent and independent variables posited by economic theory, economists want to be very confident in rejecting the null hypothesis. Since the level of confidence associated with the decision about H_0 equals $(1 - \alpha)$, economists maximize that level when they minimize the probability of incorrectly rejecting a true null hypothesis.

Once the level of significance is established, we next state the conditions under which we will reject the null hypothesis regarding the regression model's slope coefficient. In simple linear regression analysis, the decision rule indicates how the slope coefficient's test statistic must compare to the critical value associated with the underlying t-distribution in order to reject H_0.

Figure 14.3 presents the decision rules associated with the three most commonly tested hypotheses about the slope coefficient. As we observe, the decision rule reflects the sign found in the alternative hypothesis. In the case of a one-tail test of a positive slope coefficient ($b > 0$), for example, we see that the test statistic for the coefficient (t) must be greater than the critical value for t (i.e. t_{n-2}^{α}) in order to reject the null hypothesis that $b \leq 0$.

For a one-tail test of a negative slope coefficient ($b < 0$), we see that, to reject the null hypothesis, the test statistic for the coefficient (t) must be less than $-t_{n-2}^{\alpha}$. This not only indicates that our focus is on the left-hand side of the coefficient's distribution, it also demonstrates that the slope coefficient must have the theorized directional sign to reject the null hypothesis. For example, if we obtain a positive value for $\hat{b}$ when we hypothesize a negative relationship (H_a: $b < 0$), we necessarily cannot reject the null hypothesis. As will become more apparent in Part V, the so-called "correct" sign on the regression's slope coefficient is the first consideration when evaluating the null hypothesis on a regression model's coefficients.

The sign obtained on the slope coefficient, and therefore that of the test statistic, is not, however, a consideration when conducting a two-tail hypothesis test. In this case, we want to

---

One-tail test

$\quad H_a$: $b > 0$ Reject H_0 if $t > t_{n-2}^{\alpha}$; cannot reject H_0 if $t \leq t_{n-2}^{\alpha}$

$\quad H_a$: $b < 0$ Reject H_0 if $t < -t_{n-2}^{\alpha}$; cannot reject H_0 if $t \geq -t_{n-2}^{\alpha}$

Two-tail test

$\quad H_a$: $b \neq 0$ Reject H_0 if $|t| > t_{n-2}^{\alpha}$; cannot reject H_0 if $|t| \leq t_{n-2}^{\alpha}$

Figure 14.3 Decision rules for testing hypotheses about the simple linear regression model's slope coefficient.

know if the test statistic falls in the rejection regions found in either end of the t-distribution. Thus the decision rule for a two-tail test uses the absolute value of the test statistic and compares it to the critical value associated with a two-tail hypothesis test.

14.6   The test statistic and *p*-value for the slope coefficient

The final steps in testing hypotheses about the slope coefficient depend on the OLS-estimated values for b. These steps have been greatly facilitated by the computer software packages used by economists, so we will only briefly discuss these steps here. As demonstrated in "Using Excel's regression tool in the Data Analysis ToolPak" at the end of this chapter, our software not only generates estimated values for the slope and intercept coefficients, but also computes the standard errors, t-statistics, and p-values required for hypothesis tests of both coefficients. (Readers are advised to pay particular attention to using the p-values generated by Excel.) The software-generated results also include several other statistics that economists use to evaluate the empirical regression results.

Let us now apply what we have learned about hypotheses testing to evaluate the statistical significance of the slope coefficient from the infant mortality model estimated in Chapter 13.

14.7   An example of a hypothesis about the slope coefficient: infant mortality rates

In Section 13.6, we introduced a simple linear regression model in which we presented infant mortality rates (IMR) as a function of the percentage of births attended by skilled medical personnel (ATTEND). We further postulated a negative relationship between IMR and ATTEND based on the economic rationale that, the higher the percentage of births attended by skilled medical personnel in a country, the better those present would be able to address complications with birth, making the birth process safer for both mother and child, thereby reducing infant mortality rates, *ceteris paribus*.

We then estimated the following regression equation:

$$\widehat{IMR} = 120.65 - (1.0519 \times ATTEND)$$

The negative sign on the slope coefficient appears to support our postulated relationship, but we do not yet know if this negative relationship is statistically significant, or if it is negative simply as a matter of chance. To test this, we turn to the hypothesis testing procedures just outlined, by first establishing the null and alternative hypotheses:

$$H_0 : b_{ATTEND} \geq 0 \quad \text{versus} \quad H_a : b_{ATTEND} < 0$$

We again see that the alternative hypothesis is established to reflect the negative relationship postulated by economic theory and that the null hypothesis states the opposite possible relationship between IMR and ATTEND.

Let us now set the level of significance for the hypothesis test at 1 percent ($\alpha = 0.01$), which leads to the following decision rule:

$$\text{reject } H_0 \text{ if } t < t_{152-2}^{0.01} = -2.3515$$

We next use Excel's regression tool to generate the results in Table 14.1. Excel produces what are essentially three groups of statistics for a regression. The first set, called *Regression*

Table 14.1 Infant mortality rate regression results from Excel

Regression Statistics

Multiple R	0.7707
R Square	0.5940
Adjusted R Square	0.5913
Standard error	21.8855
Observations	152

ANOVA

	df	SS	MS	F	Significance F
Regression	1	105,096.26	105,096.26	219.42	3.69E-31
Residual	150	71,846.27	478.98		
Total	151	176,942.53			

	Coefficients	Standard error	t Stat	P-value	P-value/2
Intercept	120.6498	5.8370	20.6697	9.61E-46	
ATTEND	−1.0519	0.0710	−14.8128	3.69E-31	1.85E-31

Statistics, includes the R^2 statistic and number of observations (n). The second set is the analysis of variance (ANOVA) table, which we will discuss further in Part V. The third set of statistics are those pertaining to the regression model's estimated coefficients and thus the focus of the current discussion.

Based on the results in Table 14.1, we see that the test statistic for the IMR–ATTEND model's slope coefficient is −14.8128. Since $t = -14.8128$ is *less* than $t_{125}^{0.01} = -2.3515$, we can reject the null hypothesis with at least 99 percent confidence.

The relevant p-value equals 1.85E-31 (we must divide Excel's two-tail p-value by 2 because we are conducting a one-tail test). It is clearly much smaller than the 1 percent level of significance, which means that we are very confident in rejecting H_0. Thus the empirical evidence strongly suggests that, the greater the percentage of births attended by medical professionals, the lower the infant mortality rate for the 152 countries in the sample.

At this point in the analysis, we might be satisfied with our regression results. First, we estimated that, for a one percentage point increase in the number of births attended by skilled medical personnel, infant mortality rates fell by 1.0519 infant deaths per 1,000 live births. Not only did we obtain the hypothesized sign for b, we also determined that we could be very confident in rejecting the null hypothesis, providing statistical support that the relationship between IMR and ATTEND is negative.

We also observe that the variation in the independent variable, ATTEND, accounted for 59.40 percent of the variation in infant mortality deaths, IMR. While this may be a "good" result for a cross-section model, we can easily think of other variables that might reduce infant mortality rates, such as adequate water and food, sanitation facilities, and prenatal care, all of which would improve the health of the mother before and during pregnancy and the child after birth. Because it is very likely that additional socio-economic variables contribute to the variation in infant mortality rates among the 152 countries surveyed, our model is almost assuredly incomplete. While we will introduce a technique in Part V that allows us to consider these additional factors, let us conclude Chapter 14 by making one additional, and very important, observation regarding hypothesis testing.

14.8   What hypothesis testing does and does not prove

Generating regression results that measure a relationship between economic variables is exciting, and finding that those results are statistically significant is even more satisfying. We do, however, want to remind ourselves what conclusions we can reach from such results, as well as what conclusions would be erroneous.

Hypothesis testing allows economists to assess if the observed relationship between a dependent and an independent variable is statistically reliable or merely one of chance. By establishing null and alternative hypotheses with the goal of rejecting the null, then actually rejecting the null hypothesis means we can conclude it is very unlikely that our sample results would have been observed if the relationship postulated in the null hypothesis was actually true. In other words, by rejecting the null hypothesis, we are confident, for a specified level of significance, that the theoretical relationship stated in the alternative hypothesis is not simply a statistical anomaly. In addition, we can use the *p*-value to state precisely how confident we are in rejecting the null hypothesis.

Hypothesis testing does *not*, however, indicate if the independent variable *causes* the dependent variable to change; it simply indicates that the extent to which the variation in the dependent variable corresponds to the variation in the independent variable. Thus regression analysis does not prove causation, it only assesses the correlation between two variables.

A corollary is that regression analysis *cannot* be used to confirm if the underlying economic theory is correct. That is, economic theory postulates a *causal* relationship between the dependent and independent variables, which regression analysis cannot prove. Thus we must be careful not to attribute causality or proof of theoretical truth to the results obtained from regression analysis.

Summary

Simple linear regression analysis is a powerful tool used by economists to assess the strength and direction of the relationship between two economic variables. Here we have demonstrated how economists evaluate such relationships through the process of hypothesis testing. Appropriate application of this procedure provides us with important information about economic relationships. However, we must remember that hypothesis tests of the simple linear regression model's slope coefficient is a statistical inference, which leaves open the possibility that the statistical evidence of the relationship between the two variables may be totally unrelated in the population from which the sample was drawn.

Concepts introduced

A priori
Best estimator
Consistent estimator
Estimator
One-tail hypothesis test of the slope coefficient
Standard error of the regression (SER)
Standard error of the predicted *Y*-value for each *X* in the regression
Test statistic for the slope coefficient
Two-tail hypothesis test of the slope coefficient
Unbiased estimator

Box 14.1 Using Excel's regression tool in the Data Analysis ToolPak

Once we want to test hypotheses about the simple linear regression model's coefficients, we can no longer estimate the model by fitting a linear trendline to an *XY* (scatter) plot of the dependent and independent variables, because this does not provide us with the standard error of each coefficient. Fortunately, Excel includes a regression tool as part of the Data Analysis ToolPak, which will provide us with the statistics needed to test hypotheses about the coefficients.

To use the regression tool, click on the **Data** tab and then **Data Analysis**. (If **Data Analysis** does not appear on the **Data** ribbon, go to the Office Button and select **Excel Options** at the bottom of the drop-down menu. Next go to **Add-Ins**, and activate the **Analysis ToolPak** and **Analysis ToolPak – VBA** (the latter allows you to run macros in Excel). Click **OK**.)

Under the **Data Analysis** options, click on **Regression** and then **OK**. You will obtain the following **Regression** dialog box:

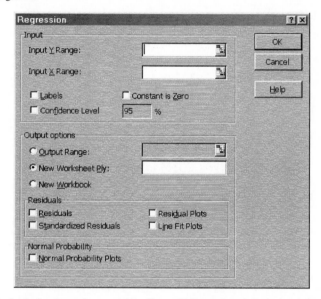

Enter the data for both the dependent variable (**Input Y Range**) and the independent variable (**Input X Range**). Always include a one-cell data label at the top of the column for all variables, and check the **Labels** box.

Under **Output options**, select the one you want. As you begin using the regression tool, it is recommended that you direct Excel to put the output on the same worksheet as your data. To do so, check **Output Range** and then click on the cell where you want the output to appear. That cell will be the top left cell of your output. Click **OK**.

Let us now present the regression output generated by Excel for the GDP–education model:

SUMMARY OUTPUT

Regression Statistics

Multiple *R*	0.5409
R Square	0.2926
Adjusted *R* Square	0.2874
Standard Error	11,135.83
Observations	140

ANOVA

	df	SS	MS	F	Significance F
Regression	1	7,077,101,315.65	7,077,101,315.65	57.0703	5.24038E-12
Residual	138	17,112,924,175.30	124,006,696.92		
Total	139	24,190,025,490.95			

	Coefficients	Standard Error	t Stat	P-value	Lower 95%	Upper 95%	Lower 95%	Upper 95%
Intercept	−7,783.67	2,736.16	−2.8447	0.0051	−13,193.90	−2,373.45	−13,193.90	−2373.45
Average years of total schooling, 15+, 2000	2,665.99	352.90	7.5545	5.24E-12	1,968.19	3,363.78	1,968.19	3363.78

Formatting notes

- Because of the statistical superiority of hypothesis testing over confidence intervals, we will not use the values in the "Lower 95%" or "Upper 95%" columns. You may delete them from the Excel results.
- We will use four decimal places when reporting the following statistics: R Square, Adjusted R Square (in Part V), t Stat, and P-value. Reformat those values if Excel's results do not have four decimal places.

Let us now identify the various results generated by Excel:

SUMMARY OUTPUT

Regression Statistics

Multiple R	0.5409	Correlation coefficient (equation (12.1))
R Square	0.2926	Regression coefficient of determination (equation (13.6))
Adjusted R Square	0.2874	Deferred until Part V
Standard Error	11,135.83	Standard error of the regression (equation (14.7))
Observations	140	Number of observations (n)

ANOVA

	df	SS (Sum of Squares)	Text equations	MS	F	Significance F
Regression	1	7,077,101,315.65	(13.14)	7,077,101,315.65	57.0703	5.24038E-12
Residual	138	17,112,924,175.30	(13.15)	124,006,696.92		
Total	139	24,190,025,490.95	(13.13)			

	Coefficients	Standard Error	t Stat	P-value
Intercept	−7,783.67	2,736.16	−2.8447	0.0051
Average years of total schooling, 15+, 2,000	2,665.99	352.90	7.5545	5.24E-12
Text equations for slope coefficient statistics	(13.5)	(14.6)	(14.1)	(14.2)

Regression Statistics

Multiple *R* is the *correlation coefficient* between GDP and EDUC. Here the value of 0.5409 indicates a strong, positive correlation between GDP and EDUC. It also equals the value obtained for the correlation coefficient (r_{XY}) in equation (12.1).

R Square is the *regression coefficient of determination* for the GDP–EDUC model. Its value of 0.2926 indicates that 29.26 percent of the variation in GDP per capita is accounted for by the variation in adult educational attainment. This statistic is calculated using equation (13.6) (or equation (13.7)).

Standard Error is the *standard error of the regression* defined by equation (14.6). Its primary use is generating the standard errors of the two regression coefficients.

Observations is the number of paired values for *Y* and *X* in the sample.

ANOVA

While we will defer discussion of these statistics until Part V, the SS column contains the values for the three sums of squares described in Section 13.4.

Coefficient results

Coefficients are the estimated values of the regression parameters, or the *regression coefficients*. "Intercept" refers to the intercept term of the model, $\hat{a}$, while "Average years of total schooling, 15+, 2000" is the slope coefficient, $\hat{b}$, for the GDP–EDUC model. Excel uses these values to calculate the numerators of the test statistics for the intercept and slope coefficients.

Standard Error are the estimated standard errors, $se_{\hat{a}}$ and $se_{\hat{b}}$, associated with the model's parameters. Excel uses these values to calculate the denominators of the test statistics for the intercept and slope coefficients.

t Stat is the test statistic for each parameter. In calculating these values, Excel assumes a hypothesized value of zero for the parameter. If we want to test a different hypothesis about the population parameter, we must manually construct the appropriate test statistic.

P-value is the smallest level of significance at which the null hypothesis can be rejected. The *p*-value is also calculated for a hypothesized value of zero for the population parameter. The *p*-value generated by Excel's regression tool is a *two-tail p-value*. To calculate the *p*-value for a *one-tail hypothesis test*, divide Excel's *p*-value by 2. For the GDP–education model, the appropriate *p*-value for the one-tail test is 2.62E-12.

Exercises

1. Return to the three models from Exercise 1 and the marginal productivity model from Exercise 3, both in Chapter 13. Establish the pair of hypotheses for each model based on the underlying economic theoretical relationship along with the decision rule (use $\alpha = 0.05$) for each model.
2. Use Excel's regression tool to estimate the simple linear regression results for each of the four models, and interpret the results.
3. Complete the hypothesis tests from Exercise 1 above. Be sure to state your conclusions using non-technical language.

15 Simple regression analysis

Variable scales and functional forms

In our discussion of the simple regression model thus far, we have assumed a linear relationship between the dependent and independent variables. Inherent in that assumption is that the slope of the regression line (b) is constant across the paired values of Y and X.

But a constant slope may not best describe the relationship between the two variables. For example, suppose economic theory suggests that a constant elasticity between the dependent and independent variables best describes their relationship. Economists often postulate such a relationship for many pairs of economic variables as in the cases of product demand and supply, the Cobb–Douglas production function, and the Solow growth model, to name just a few. If a constant elasticity is theorized, does this mean we can no longer use the ordinary least-squares method for our analysis?

Fortunately, we are able to account for many nonlinear relationships between the dependent and independent variables by using other functional forms. As long as the regression equation remains linear with respect to the model's parameters (a and b), we can still apply the OLS estimation method to our model. But before examining some of these alternative functional forms, let us first examine how rescaling a variable's units of measurement affects the regression model's statistics.

15.1 Rescaling variables and interpreting the regression coefficients

Throughout the text, we have examined economic variables that vary widely in terms of their units of measurement. Some variables, such as those describing a nation's economy, are generally measured in units of billions or trillions of dollars. Other data, especially at the microeconomic level, have been in units as small as individual people, or dollars and cents. Such differences lead us to ask how changing the **scale**, or units of measurement, of a variable in a regression model affects the estimated model's results.

Suppose, for example, that we want to measure the effect of mortgage interest rates on the number of new homes sold between April 1971 and March 2009. We begin by establishing the following theoretical simple linear regression model of the demand for new housing:

NEW HOMES SOLD $= a + (b \times \text{RATES})$

We next collect monthly data to estimate the model. As shown in Table 15.1, the published data for the dependent variable, the number of new homes sold, is measured in thousands of seasonally adjusted units, while the independent variable, 30-year fixed-rate conventional mortgage rates, is measured in percent.

Table 15.1 Excerpt of data on home sales and mortgage rates

Year or month	Published values		Rescaled values	
	HSN1F [monthly data at seasonally adjusted annual rates] (thousands)	*MORTG [average contract rate on commitments for mortgages] (percent per annum)*	*Houses sold [monthly data at seasonally adjusted annual rates]*	*Mortgage rate [average contract rate on fixed-rate first mortgages] (per annum)*
2006-10-01	941	6.36	941,000	0.0636
2006-11-01	1,003	6.24	1,003,000	0.0624
2006-12-01	998	6.14	998,000	0.0614
2007-01-01	872	6.22	872,000	0.0622
2007-02-01	820	6.29	820,000	0.0629
2007-03-01	823	6.16	823,000	0.0616

Data source: FRED, Federal Reserve Economic Data, Federal Reserve Bank of St. Louis (2010b, c), New One-Family Houses Sold (HSN1F) and 30-Year Conventional Mortgage Rate (MORTG).

Using the variables scaled in their original published units, we obtain the following estimated simple linear regression model:

$$\widehat{\text{HOMES}} = 1,129.443 - 44,013 \times \text{RATES}$$

which we will refer to as the original model. Here the slope coefficient indicates that a one percentage point increase in 30-year fixed conventional mortgage rate corresponds to an average 44.013 thousand units decline in the number of new houses sold. While the one percentage point change in the independent variable is readily understood, the 44.013 thousand units change in the dependent variable is less so.

Let us rescale the dependent variable so it represents the actual number of new houses sold by multiplying each month's value by 1,000. (These rescaled values for new houses sold are shown in column 4 of Table 15.1.) Re-estimating the regression model using the rescaled values for the dependent variable yields the following results:

$$\widehat{\text{HOMES}} = 1,129.443 - 44,013 \times \text{RATES}$$

Here the estimated slope coefficient indicates that a one percentage point increase in mortgage rate corresponds to an average decline of 44,013 new houses sold per month.

We observe that multiplying the dependent variable by 1,000 results in a slope coefficient that is 1,000 times the value of the original model's slope coefficient. We can now generalize these results as follows. If we multiply the dependent variable by a constant, c_y, we obtain a slope coefficient that is c_y times the value of the original model's slope coefficient.

What happens to the slope coefficient if, instead, we rescale the independent variable, 30-year conventional mortgage rates? Here we will convert the percentage units to a decimal, or proportion, format by multiplying each rate by 0.01. Now a one-unit change in the independent variable corresponds to a 0.01 mortgage point change.

Re-estimating the regression model estimated using the rescaled mortgage values but published values (those measured in thousands of units) for new homes, we obtain:

$$\widehat{\text{HOMES}} = 1,129.443 - 4,401.3 \times \text{RATES}$$

According to these results, a 0.01 point change in mortgage rate corresponds to an average monthly decline of 4,401 thousand new homes sold. Here, not only is the change in the dependent variable difficult to understand, so is the change in the independent variable, illustrating that rescaling a variable may not always improve our understanding of the underlying economic relationship.

The effect of rescaling only the independent variable (MORTG) in the housing model yields a slope coefficient that is equal to the original model's estimated coefficient times 100, which is the reciprocal of the rescaling factor, $0.01 (= 1/100)$. We can now generalize the result. If we multiply the independent variable by a constant, c_x, we obtain a slope coefficient equal to the original model's slope coefficient divided by that constant, c_x.

We have one final model to consider, one in which the published values of both variables are rescaled. Here we will use the actual number of new homes sold as the dependent variable ($Y \times 1,000$) and the proportion form of the mortgage rate ($X \times (1/100)$) as the independent variable. The estimated model is now:

$$\widehat{HOMES} = 1,129,443 - 4,401,313 \times RATES$$

The slope coefficient indicates that a 0.01 point increase in conventional mortgage rate corresponds to an average monthly decline of 4,401,313 new houses sold.

We observe that the estimated slope coefficient is 100,000 times the value of the coefficient from the original model. As we deduce, this new coefficient value is obtained by multiplying the published data for the dependent variable by 1,000 (c_y) and dividing the published data for the independent variable by the reciprocal of 0.01 ($1/c_x$). More generally, the slope coefficient of the model in which both variables are rescaled equals the original model's slope coefficient multiplied by a factor of c_y/c_x (here $1,000/0.01 = 100,000$).

The effects on the statistics associated with a simple linear regression model for all four possible combinations of rescaled variables are presented in Table 15.2. As we observe, rescaling one, or both, of the variables in the simple linear regression model does change the estimated slope coefficient in the model. Rescaling also changes the value of the slope coefficient's standard error. However, when we solve for the corresponding t-value for each slope coefficient, we obtain the same value for all four models ($t_{\hat{b}} = -15.75$). Thus rescaling one, or both, variables in the simple regression model does not affect the outcome of hypothesis tests of the slope coefficient (assuming that b is hypothesized to be zero).

Table 15.2 Home sales and mortgage rates regression results*

	New houses sold (thousands)	New houses sold
30-year conventional mortgage (percent per annum)	HOMES = $1,129.443 - 44.013 \times$ RATES (26.56) (2.795) $R^2 = 0.3532$	HOMES = $1,129,433 - 44,013 \times$ RATES (26,561) (2,795) $R^2 = 0.3532$
30-year conventional mortgage (per annum)	HOMES = $1,129.443 - 4401.3 \times$ RATES (26.56) (279.5) $R^2 = 0.3532$	HOMES = $1,129,443 - 4,401,313 \times$ RATES (26,561) (2,795) $R^2 = 0.3532$

*Standard errors for estimated coefficients in parentheses.

Data source: FRED, Federal Reserve Economic Data, Federal Reserve Bank of St. Louis (2010b, c), New One-Family Houses Sold (HSN1F) and 30-Year Conventional Mortgage Rate (MORTG).

	No scaling	y scaled by c_y ($y \times c_y$)	x scaled by c_x ($x \times c_x$)	x scaled by c_x and y scaled by c_y ($x \times c_x$) and ($y \times c_y$)
Intercept coefficient, a	a	$a \times c_y$	a	$a \times c_y$
Intercept standard error, se(a)	se(a)	se(a) $\times c_y$	se(a)	se(a) $\times c_y$
Intercept t-statistic, t_a	t_a	None	None	None
Slope coefficient, b	b	$b \times c_y$	b/c_x	$b \times (c_y/c_x)$
Slope standard error, se(b)	se(b)	se(b) $\times c_y$	se(b)$/c_x$	se(b) $\times (c_y/c_x)$
Slope t-statistic, t_b	t_b	None	None	None
R^2	R^2	None	None	None

Figure 15.1 Effects of rescaling the dependent and independent variables by a constant (c).

It can also be shown that rescaling does not affect the R^2 statistic. (We leave the proof to more advanced texts.) This means that the R^2 statistic, as well as the t-statistics, are invariant to changes in the units of the dependent and independent variables, as noted in Figure 15.1.

Having established that rescaling one or both variables does not affect the simple regression model's statistics we use to evaluate hypotheses about the model's coefficients or its overall goodness of fit, we might wonder why economists take the time to rescale economic variables. The answer takes us back to how we can most effectively communicate our statistical results to our audience.

Let us demonstrate using the demand for new houses model, beginning with the independent variable, 30-year fixed-rate mortgages. Most of us are probably more familiar with such rates reported as a percent (as in the published data), rather than as a proportion. This suggests that we use the published data for the independent variable, and report the slope coefficient results as the change in Y for a one percentage point change in mortgage rate. On the other hand, the actual number of new homes sold, rather than that number in thousands of units, is easier for most of us to grasp. This, then, recommends that we use the rescaled values for the dependent variable. Thus the second model (the upper right cell of Table 15.2) produces the most easily understood estimate of the slope coefficient, thereby making it the most effective model for reporting the relationship between home sales and mortgage rates.

15.2 Specifying the regression equation: functional forms

Let us now turn to the regression model itself to examine how it can best describe the relationship between the dependent and independent variables. Mathematically, this means we want to determine what **functional form** is most appropriate for describing that relationship.

The regression models presented in Part IV thus far have had a **linear functional form**. What this means is that the dependent variable, Y, measured in economic units, is a function of the independent variable, X, also measured in economic units:

$$Y = a + bX \tag{15.1}$$

In the linear model, the coefficient, b, measures the slope of the OLS-estimated line. As we have learned, this coefficient indicates by how many units the dependent variable Y changes for a *one unit change in* X:

$$b = \frac{\Delta Y}{\Delta X} \tag{15.2}$$

The slope coefficient is also constant across all pairs of X and Y. But does this most accurately describe the relationship between the dependent and independent variables? In other words, is it reasonable, given the economic context of the model, for the relationship between X and Y to be described by constant unit changes?

To explore this question, let us return to the linear GDP per capita–education model from Chapters 13 and 14. For reasons that will shortly become apparent, we will now refer to that model as a **lin–lin (simple linear) regression model**. As previously shown, the estimated regression model is

$$\widehat{GDP} = -7{,}783.67 + 2{,}665.99 \times EDUC$$

Here the slope coefficient indicates that GDP per capita was, on average, \$2,665.99 (constant 2005 international dollars) higher for each additional year of education completed by the adult population. That is, a one unit increase in education (measured average years of schooling completed by adults) corresponds to a b unit increase in GDP per capita (constant 2005 international dollars).

Now suppose we believe that an additional year of education corresponds not to a dollar change in GDP, but rather to a percentage change in GDP. How do we then estimate this new model? Fortunately, we can still use OLS to estimate the relationship by transforming the dependent variable using natural logarithms. Now the slope coefficient will measure the percentage change in GDP per capita that is associated with an additional year of adult education.

We can transform one (or both) variables in the simple regression model using logarithms without having to abandon OLS as long as the regression model is linear with respect to its coefficients.[1] Mathematically, this means that if the theoretical model is of the general form

$$f(Y) = a + bf(X)$$

then we can use the ordinary least-squares procedure to estimate a and b.

15.3   Semi-log functional forms: the log–lin model

Let us now demonstrate the effects of using a logged dependent variable in the GDP–education model. We first specify the theoretical model for estimating the percentage change in GDP per capita associated with an additional year of adult educational attainment:

$$\ln(GDP) = a + (b \times EDUC)$$

Because the dependent variable, $\ln(GDP)$, is still linear with respect to the intercept and slope coefficients, we can now apply OLS to generate the following estimated model:

$$\ln(\widehat{GDP}) = 5.99 + (0.3635 \times EDUC)$$

(For directions on generating these results in Excel, see " Using Excel's Regression tool with logged variables" at the end of the chapter.) According to the estimated model, GDP per capita is, on average, 36.35 percentage points higher for each additional year of adult educational attainment in the 140 countries. While this is, indeed, an impressive (and statistically significant, with a *p*-value of 1.27E-23) result, we suspect that these results are incomplete because factors other than educational attainment are almost certainly correlated with percentage changes in GDP.

The above equations, generalized below in equation (15.3), utilize what is called a **semi-log functional form**. This name indicates that one of the two variables in the simple regression model has been rescaled using natural logarithms. To further distinguish between the two types of semi-log functional forms, those models in which the dependent variable is measured using a logarithmic scale will be referred to as a **log–lin (simple linear) regression model**

$$\ln Y = a + bX \tag{15.3}$$

The log–lin model differs from the lin–lin regression model in that *b* no longer measures a *constant unit* change in the dependent variable for a unit change in X. Instead, the slope coefficient now measures the *constant percentage change* in Y for a unit change in X since the change in the natural logarithm of a variable approximates the percentage change in that variable:

$$b = \frac{\Delta(\ln(Y))}{\Delta X} \approx \frac{\Delta Y / Y}{\Delta X} = \frac{\% \Delta Y}{\Delta X} \tag{15.4}$$

Because the slope coefficient in a log–lin model measures the *percentage change* in the dependent variable for a unit change in the independent variable, we must multiply $\hat{b}$ by 100 to convert the coefficient to a percentage before we interpret its value.

The log–lin model is often utilized in economics. One popular application of the log–lin form is the human capital earnings model, in which wages are specified as a function of the amount of education, or human capital, an individual worker has attained:

$$\text{WAGES} = a + (b \times \text{EDUC})$$

If we view human capital acquisition as an investment, it can be shown that the slope coefficient in the above model measures the rate of return associated with an additional year invested in schooling (see, for example, Kaufman and Hotchkiss 2006, pp. 382–386). By specifying the dependent variable in logarithmic form (ln(WAGES)) while retaining the original units of measurement (years of education) for the independent variable, *b* will estimate the percentage change in wages associated with an additional year of education.

Let us now estimate and interpret the above earnings model for wives and husbands using the Mroz (1976) dataset. As we observe in Table 15.3, the rate of return to wives for an additional year of education was 6.36 percent, while the return to husbands was slightly higher at 8.74 percent.[2] These results are both easily understood and consistent with the underlying economic theory.

It is often the case that economists transform the dependent (or independent) variable using a logarithmic scale when the variable is measured in monetary units. In the wage model, for example, we would not expect an additional year of education to increase the

Table 15.3 Log–lin wage equations for husbands and wives

Wives' log–lin earns					Husbands' log–lin earns				
Regression Statistics					*Regression Statistics*				
Multiple *R*	0.1201				Multiple *R*	0.4558			
R Square	0.0144				*R* Square	0.2078			
Standard Error	1.2026				Standard Error	0.5160			
Observations	428				Observations	753			
	Coefficients	*Standard Error*	*t Stat*	*One-tail p-value*		*Coefficients*	*Standard Error*	*t Stat*	*One-tail p-value*
Intercept	7.2519	0.3276	22.1399	7.67E-73	Intercept	8.4579	0.0801	105.6503	0.0000
WEDUC	0.0636	0.0255	2.4979	0.064	HEDUC	0.0874	0.0062	14.0339	3.33E-40

Data source: Author's calculations using data from Mroz (1976).

earnings of an individual with only eight years of education by the same dollar amount as an additional year would increase the earnings for a college graduate. Thus the log–lin model's slope coefficient, which measures a constant percentage change in earnings, will more accurately capture the effects of a one-unit increase in education than would the lin–lin model where the slope coefficient measures a constant dollar change in earnings.

Wage models are just one example where the log–lin functional form will more realistically capture the relationship between the dependent and independent variables as compared to the lin–lin model. Another popular application is one we examined in Chapter 8, where we estimated long-run growth rate models. There we saw that a log–lin model generated a slope coefficient that measured the constant percentage change in an economic variable for a given unit change in time.

15.4 Semi-log functional forms: the lin–log model

A different type of semi-log functional form can be used when the underlying economic theory suggests that a unit change in the dependent variable corresponds to a percentage change in the independent variable. This model type, known as a **lin–log (simple linear) regression model**, transforms the independent variable using logarithms but retains the economic units for the dependent variable:

$$Y = a + b \ln X \tag{15.5}$$

where *b* is given by

$$b = \frac{\Delta Y}{\Delta \ln X} \approx \frac{\Delta Y}{\Delta X / X} = \frac{\Delta Y}{\% \Delta X} \tag{15.6}$$

The slope coefficient of the lin–log model measures the constant unit change in the dependent variable for a one percentage point change in the independent variable. Specifically, Y changes by $b/100$ units for a 1 percent change in X, so we must divide $\hat{b}$ by 100 before interpreting its value.

Let us again return to the GDP–education example to examine how a lin–log model specification modifies both the theoretical and estimated regression models:

$$\text{GDP} = a + b\ln(\text{EDUC})$$

and

$$\widehat{\text{GDP}} = -14{,}607.30 + 13{,}849.15\ln(\text{EDUC})$$

Here the estimated model indicates that GDP per capita is, on average, \$138.49 (constant 2005 international dollars) higher for a 1 percent increase in the average years of schooling completed by adults.

While economic theory suggests that additional education corresponds to higher per-capita GDP, the literal interpretation of b may be more difficult to grasp. That is, what is a 1 percent change in educational attainment? Because most of us think about changes in education in discrete terms such as one-year increments, taking the logarithm of the education variable will add little, if anything, to our understanding of education's effects on GDP. Thus, while the results for the lin–log model are statistically correct, the difficulty associated with interpreting those results may recommend that we not use this particular functional form to describe the GDP–education relationship.

In economics, the lin–log model is the least used functional form of those described in this chapter. However, if we hypothesize an economic relationship in which the dependent variable is expected to increase at a decreasing rate in response to changes in the independent variable, as we observe for Engel curves, this function form would be appropriate. One classic application is estimating food expenditure–income Engel curves, where it is assumed that expenditures on food necessities will increase with income, but once income reaches a level where basic needs are met, food expenditures will increase at a decreasing rate.

As this example demonstrates, economists may select a particular functional form for their regression model based on the relationship between the dependent and independent variables suggested by economic theory. And while economists have developed many new regression methods, the practice of "theory guides specification" remains at the heart of linear regression analysis.

15.5   Double-log functional form

We now turn to a fourth, and perhaps the most popular, functional form used in economics, a model in which both the dependent and independent variables are transformed by natural logarithms. Known as the **log–log (simple linear) regression model** or **double-log functional form**, the slope coefficient in such models measures the popular economic concept of elasticity.

Let us first note that the log–log model,

$$\ln Y = a + b\ln X \tag{15.7}$$

is linear with respect to its parameters, a and b, allowing us to estimate those parameters using OLS. In addition, we note that the slope coefficient is constant, as in the lin–lin model, but now b equals the natural log of the dependent variable divided by the natural log of the independent variable. Recalling that the natural log of a variable approximates the percentage change in that variable, b can be written as

$$b = \frac{\ln Y}{\ln X} \approx \frac{\% \Delta Y}{\% \Delta X} \qquad (15.8)$$

Economists are quite familiar with the last term in equation (15.8), since it is the form taken by an **elasticity**. Thus we can interpret the slope coefficient in the double-log linear regression model as the average percentage change in Y for a 1 percent change in X. In addition, because b is already in percentage terms, we do not need to multiply, or divide, b by 100 when interpreting the coefficient.

Let us now run a log–log model using the GDP–education data. After transforming both GDP per capita and adult educational attainment into natural logs, we estimate the following equation:

$$\ln(\text{GDP per capita}) = 4.78 + 2.02 \ln(\text{EDUC})$$

According to these results, a 1 percent increase in average number of years of schooling completed by adults corresponds to a 2.02 percent increase in real GDP per capita. An alternative interpretation is that the elasticity of national output with respect to education is 2.02 percent for the 140 nations in our sample.

Constant elasticity models are frequently found in economics. We know, for example, that constant own-price, cross-price, and income elasticities are key elements in consumer demand models, while wage elasticities can influence labor market outcomes on both the supply and demand side. The Cobb–Douglas production function and the Solow growth model with their assumption of constant returns to scale are two additional popular economic models that assume constant elasticities. Thus any time the underlying economic theory suggests a constant *elasticity* for the dependent variable with respect to the independent variable, the double-log functional form is the appropriate model to use.

15.6 Other functional forms

Figure 15.2 presents the interpretations of the simple linear regression model's slope coefficient for the four functional forms discussed thus far. But these are only some of the functional forms that economists use to estimate relationships using the OLS procedure.

Another popular category is polynomial functional forms, such as the **quadratic functional form**, in which the square of the independent variable is included to capture any increasing or diminishing marginal effects that help to explain the variation in the dependent variable. For example, empirical evidence suggests that earnings in the United States start to decline for workers over the age of 55. To capture this effect in an earnings function, we could include the square of age, in addition to an age variable, in our model ($\text{EARN} = f(\text{AGE}, \text{AGE}^2)$). We might use a different polynomial function, a **cubic functional form**, to estimate the theoretical shape of a firm's marginal cost, where cost is a function of output, output squared, and output cubed ($\text{COST} = f(Q, Q^2, Q^3)$).

Reciprocal relationships are also popular in economic theory, such as that depicted by the Phillips curve, in which the percentage change in wages or prices is an inverse function of the unemployment rate ($\% \Delta \text{WAGE} = f(1/\text{URATE})$). This model, too, can be estimated using OLS. While further discussion of these, and other, forms is left to econometric texts, we do

	Interpretation of b	Coefficient value used
Linear regression model $(Y = a + bX)$	Unit change in Y given a one-unit change in X	b
Lin–log regression model $(Y = a + b\ln X)$	Unit change in Y given a 1 percent change in X	$b/100$
Log–lin regression model $(\ln Y = a + bX)$	Percentage change in Y given a one-unit change in X	$b \times 100$
Double-log regression model $(\ln Y = a + b\ln X)$	Percentage change in Y given a 1 percent change in X	b

Figure 15.2 Interpreting the slope coefficient for four functional forms of a simple regression model.

recognize the wide application of ordinary least squares to empirically estimate economic relationships.

15.7   Selecting the appropriate functional form

How, then, do we select the particular functional form that best describes the relationship between the dependent variable and the independent variable? As previously suggested, economists begin by considering the underlying theoretical relationship between the dependent variable and the independent variable.

If the underlying economic theory does not provide clear direction, we might also consider the empirical relationship between the two variables. As we saw with the earnings–education model, it is probably not reasonable to expect that an additional year of education will have the same impact on the wages for someone who has completed college compared to the earnings for someone who has not graduated from high school. Thus we decided that the log–lin functional form would more accurately reflect the impact of education on earnings.

We also want our estimated results to be easily understood by our audience. For the GDP–education model, we indicated that most people understand what a one-year change in educational attainment means, but would have a difficult time translating a 1 percent change in educational attainment into a meaningful result. Since we always want to communicate all statistical results effectively, this consideration cannot be underestimated.

If the underlying economic theory, the variables' empirical properties, and effective communication considerations do not provide a definitive answer as to the most appropriate functional form, the statistical results for the regression coefficients may provide guidance. However, we cannot use the various model's R^2 statistics to select a "best" model. The reason is that the R^2 statistics are *not* comparable across the four models. To understand why, let us return to the four GDP–education models estimated in this chapter, and interpret each model's R^2 values (Table 15.4).

We see from the interpretations that the sums of squares that determine each R^2 statistic differ across the four models. We observe, for example, that the R^2 statistic for the log–lin model measures the total sum of squares (TSS) using a different dependent variable (ln(GDP)) than the R^2 statistic for the lin–lin model. That model's regression sum of squares (RSS) also differs from the RSS associated with the lin–log and log–log models due to the different expressions on the right-hand side of the regression equations. Thus the log–lin

Table 15.4 Interpretation of the R^2 statistic for the GDP–education simple linear regression models.

Model	R^2 value	Interpretation
Lin–lin	0.2926	29.26 percent of the variation in the dependent variable, GDP per capita, is explained by the variation in the model: $a + (b \times \text{EDUC})$
Log–lin	0.5180	51.80 percent of the variation in the dependent variable, ln(GDP per capita), is explained by the variation in the model: $a + (b \times \text{EDUC})$
Lin–log	0.2466	29.26 percent of the variation in the dependent variable, GDP per capita, is explained by the variation in the model: $a + b$ ln(EDUC)
Log–log	0.5013	50.13 percent of the variation in the dependent variable, ln(GDP per capita), is explained by the variation in the model: $a + b$ ln(EDUC)

	ESS
Linear (lin–lin) regression model	$\sum (Y - \hat{a} - \hat{b}X)^2$
Semi-log (log–lin) regression model	$\sum (\ln Y - \hat{a} - \hat{b}X)^2$
Semi-log (lin–log) regression model	$\sum (Y - \hat{a} - \hat{b}\ln X)^2$
Double-log (log–log) regression model	$\sum (\ln Y - \ln \hat{a} - \hat{b}\ln X)^2$

Figure 15.3 ESS statistics for the linear, semi-log, and double-log functional forms for a simple regression model.

simple regression model's R^2 statistic is not comparable to any of those associated with the other three model specifications.

Such differences are further apparent when we examine each model's error sum of squares (ESS). As demonstrated in Figure 15.3, each ESS is constructed using a different formula. This is why we cannot base our selection of the "best" functional form on the R^2 statistic. We must instead consider the relationships suggested by economic theory, the empirical properties of the variables, the context of our analysis, and the statistical results for the model's coefficients. Relying on the R^2 statistic is thus both incorrect and misleading.

Summary

Chapter 15 has introduced two extensions of the simple linear regression model. The first examined how rescaling one or both of the model's variables affects our regression results, and we noted when such rescaling might be appropriate. We also considered four functional forms for describing the relationship between the dependent and independent variables. These various forms provide us with greater flexibility in accurately modeling the theoretical relationship postulated by economists. We also considered how empirical realities and concerns about effective communication influence our choice of model.

Despite these additions, the simple linear regression model is seriously limited because economists rarely, if ever, believe that the variation in an economic variable is due solely to the variation of only one independent variable. Thus, we now turn to a regression framework in which we include more than one independent variable. These multiple linear regression models are the topic of Part V.

Concepts introduced

Cubic functional form
Double-log functional form
Elasticity
Functional form
Lin–lin (simple linear regression) model
Lin–log (simple linear regression) model
Linear functional form
Log–lin (simple linear regression) model
Log–log (simple linear regression) model
Quadratic functional form
Scale
Semi-log functional form

Box 15.1 Using Excel's Regression Tool with logged variables

We can use Excel's Regression Tool to estimate all of the regression models described in Chapter 15. For those models in which logged variables are present, we must *manually* transform each variable into its natural log:

- Create an obvious name for the variable you wish to transform. Simply appending LN at the beginning of the original variable's label is fine.
- Enter this name at the top of a new column in the same worksheet as the other data needed for the regression.
- Immediately below the new variable's label, enter "=**LN**(*original value for first observation*)." Click **Enter**.
- Copy the formula to calculate the natural log values for all observations of the variable.
- Use these values for the variable in the Regression Tool.

Things to remember

- You cannot take the natural logarithm of either zero or a negative value. If your data contains such values, the LN function will return a #NUM! error, and the Regression Tool will generate an error "Input range contains non-numeric data."
- For data series containing zeros or negative values, consider transforming the variable into an index number. This transformation will reflect the variable's variation, but it will not contain values less than or equal to zero.

Exercises

1. Return to the demand for coffee model from Exercise 1 in Chapter 13, and re-estimate the relationship between the dependent and independent variables using both the semi-log and the double-log functional forms.

(a) Interpret and compare the slope coefficients for all four models. (You may find it useful to consider a ten-point change in the CPI for coffee, rather than the typical one unit change.)
(b) Evaluate for which of the models the slope coefficient is statistically significant in the context of the product demand theory.
(c) Interpret the R^2 statistic for each model.
(d) Based on the results, explain which simple regression model provides the most compelling explanation of the relationship between coffee consumption and the retail price of coffee.

2. Return to the earnings model for registered nurses in 2004 from Exercise 1 in Chapter 13, and re-estimate the relationship between the dependent and independent variables using both the semi-log and the double-log functional forms.

(a) Interpret and compare the slope coefficients for all four models.
(b) Evaluate for which of the models the slope coefficient is statistically significant in the context of the underlying economic theory.
(c) Interpret the R^2 statistic for each model.
(d) Based on the results, explain which simple regression model provides the most compelling explanation of the relationship between nurses' earnings and years of work experience as a nurse.

3. Return to the marginal productivity model described in Exercise 3 from Chapter 13, and explain which of the four functional forms discussed in Chapter 15 would be most appropriate to use given the underlying theory and properties of the variables.

Part V

Relationships between multiple variables

Description and statistical inferences

Part V introduces the multiple linear regression model for estimating the relationships between a dependent variable and more than one independent variable. Such methods allow us to account for the multiple factors that economists theorize may explain variation in the dependent variable, such as the amount of coffee consumed. Not only will that amount depend on the price of coffee, as exemplified in a simple regression model of the demand schedule, but it will also likely be related to disposable income, the price of substitutes (such as tea), the price of complements (such as sugar), and consumer tastes and preferences. Thus, with a multiple regression model, we can examine not only the relationship depicted by a demand schedule, but also those factors which may cause the demand schedule to shift.

Fortunately, the statistical methods developed in Part IV can be extended to the multiple linear regression model. In Chapter 16, we introduce the concept of partial regression coefficients, which we will then estimate and interpret. In Chapter 17, we test hypotheses about these partial regression coefficients, considering how to use that information to refine our regression model, and we discuss new statistics for assessing the multiple regression model's overall goodness of fit. We conclude with Chapter 18, where we introduce dummy variables as a method for including qualitative independent variables in a multiple regression model. This is followed by a brief discussion of some problems that may be associated with a multiple regression model's statistical results.

16 Multiple regression analysis
Estimation and interpretation

16.1 Simple to multiple regression analysis: an introduction

Economists utilize regression analysis to describe the relationship between economic variables, making this statistical method very popular in economic research. Because most economic relationships are rarely confined to only two variables, we now extend the regression tool to the relationship between a dependent and multiple independent variables.

As with simple regression analysis, we use economic theory to guide our choice of the appropriate independent variables to include in the multiple regression model. We must further consider how to represent the mathematical relationship between the dependent variable and each independent variable most accurately, which may also be informed by economic theory. Once we have utilized economic theory to fully specify our multiple regression model, we can then estimate the model and interpret its results.

16.2 The multiple-variable linear regression model

The first step in multiple linear regression analysis is identifying the model to be estimated. Once we have selected the dependent variable (Y), we turn to economic theory to determine what (multiple) factors may be expected to explain the variation in Y. That theory may also inform each factor's expected functional relationship with Y as well:

dependent variable $(Y) = f$ (all theoretically relevant independent variables : $X_1, X_2, \ldots, X_k$)

Once we have identified the k relevant and measurable independent variables, we can write the equation for the **theoretical multiple linear regression model**:

$$Y = b_0 + b_1 X_1 + b_2 X_2 + \cdots + b_k X_k \tag{16.1}$$

Note that we have made a change in notation from the simple regression model. Following econometric conventions, we now designate the intercept term as b_0, and we distinguish the coefficients on each independent variable using the subscripts 1 through k.

This step moves us from the realm of mathematical statistics into that of **econometrics**, which is the application of statistical regression procedures to models that embody economic theory. Thus not all regression analysis is econometric in nature. In terms of equation (16.1), this means that our choice of independent variables is guided by germane economic theory. In fact, we could say that statistical estimation of a regression model not based in economic theory is essentially worthless to economists.

After the crucial step of identifying all theoretically relevant independent variables, we next specify the statistical model we wish to estimate. Since we are virtually certain we cannot generate an estimated "line" that perfectly fits the data,[1] the statistical model must include an error term, e:

$$Y = b_0 + b_1 X_1 + b_2 X_2 + \cdots + b_k X_k + e \tag{16.2}$$

Equation (16.2) is the **statistical multiple linear regression model**, or the model we will estimate empirically using the ordinary least-squares (OLS) method.

After collecting the necessary data and applying OLS, we obtain the **estimated multiple linear regression model** shown in equation (16.3):

$$\widehat{Y} = \hat{b}_0 + \hat{b}_1 X_1 + \hat{b}_2 X_2 + \cdots + \hat{b}_k X_k \tag{16.3}$$

Here $\hat{Y}$ and $\hat{b}_i$ are the values of the dependent variable and the *partial* regression coefficients, respectively, estimated by the OLS technique.

The estimated coefficients in equation (16.3) are called **partial regression coefficients** because each coefficient measures how the dependent variable will vary for a given change in the associated independent variable *while holding constant the influences of the remaining independent variables on Y*. For example, $\hat{b}_2$ measures the isolated effect of X_2 on the variation in Y while holding constant the effects of X_1 and X_3 through X_k on Y.

Technically, the underlying equation for estimating each $\hat{b}$ measures the expected change in the mean value of Y for a unit change in the associated X_i while all other factors that might affect the dependent variable remain the same.[2] That is, when the OLS method estimates each partial regression coefficient, it isolates the effects of that specific independent variable on the dependent variable. This result from the OLS procedure corresponds to the economic condition of *ceteris paribus*, which literally means "all other things being equal." Economists typically invoke the *ceteris paribus* condition in their theoretical analyses of economic phenomena, and the fact that the OLS method isolates the effects of an individual independent variable on the variation in the dependent variable explains why OLS is so popular in econometric analysis.

16.3   Specifying the independent variables and functional form for the multiple regression model

Choosing the relevant independent variables for any multiple regression model is governed first and foremost by economic theory. Here we may rely on one specific economic theory, such as neoclassical consumer demand theory or the Keynesian consumption function, when choosing the model's independent variables. In the former case, for example, demand theory indicates that the price of a good will dictate how many units of the good are demanded (the so-called "law" of demand), so we know to include the good's price as an independent variable in the model. Neoclassical theory also predicts that other factors – such as the price of goods related in consumption as substitutes and/or complements, consumer income, and the number of consumers (the so-called "shift" factors) – can also change the quantity demanded, so we include those factors as well in an econometric model of consumer demand.

It may be the case that multiple economic theories inform our choice of independent variables. To explain the number of hours that an individual will work, economists might rely solely on the labor–leisure model from neoclassical theory to specify the individual labor supply function. Here the worker's wage rate and access to non-labor income (such as a spouse's earnings) are the sole independent variables included. (This theory also includes the difficult-to-quantify tastes and preferences of the worker, so that factor is thus generally omitted from econometric models of labor supply.) Other economists reject the labor–leisure model as incomplete, arguing that, in addition to wages and income, socio-economic factors such as the presence of children, labor force experience, the local unemployment rate, and changes in what constitutes certain standards of living (for example, the goods and services that define a middle-class lifestyle) can influence the decision to participate in the work-force. As a result, this group of economists would include those factors as well in their econometric models.

We must also decide what functional form will best capture the relationship expected between the dependent variable and *each* independent variable. In some models, economic theory may guide our choice. In the case of the consumer demand function, for example, the assumption of constant elasticities between the quantity demanded and such factors as prices and income suggests that we employ the double-log (log–log) functional form. For other models, the functional form may reflect our understanding of how variables actually change in the economy. In the earnings–education model described in Chapter 15, for example, we reasoned that an additional year of education completed would not be likely to change earnings by a constant dollar amount, since different levels of education generate different returns. This then led us to use a log–lin functional form to capture the expected relationship between earnings and education.

Not all econometric models, however, have functional forms that are clearly indicated by economic theory or empirical observation. In such cases, economists often run their regressions using several different functional forms. It is common practice, for example, to estimate both a linear (lin–lin) model and a double-log (log–log) econometric model when interpretations of the partial regression coefficients for each model make economic sense. As we might expect, we would report our results from all econometric models, carefully interpreting the partial regression coefficients obtained for each and discussing the relative merits of each model's results.

16.4 Specifying a multiple regression model for infant mortality rates

Let us now return to the infant mortality rate (IMR) model estimated in Chapter 13, where we theorized that the variation in infant mortality rates across countries could be explained by the variation in births attended by skilled medical personnel (ATTEND). There, using data from 2008, we obtained a statistically significant slope coefficient for the model (1.05), but we also saw that the model accounted for only 59.40 percent of the variation in infant mortality rates. This suggests that we may have omitted other independent variables that might help to explain the differences in IMRs across countries. Let us now consider what additional independent variables might be appropriate to include in an econometric model of infant mortality rates.

How a dependent variable is defined can inform our selection of relevant independent variables for the model. For the current model, infant mortality rates (IMR) represent "the probability of a child born in a specific year/period dying before reaching the age of one" and is measured by the number of infant deaths per 1,000 live births. According to the World Health Organization (WHO), infant mortality rates not only "measure child survival [but] they also reflect the social, economic and environmental conditions in which children (and others in society) live" (World Health Organization Statistical Information System 2010, p. 112). This suggests that a more complete model of IMR might include factors related to the mother's overall health, the quality of the physical environment, and access to resources that can improve the probability of infant survival.

In the simple regression model, we estimated the correlation between IMR and births attended by skilled medical personnel as a percentage of total live births (ATTEND). As defined by the WHO, skilled medical personnel include "doctors, nurses [and] midwives trained in providing life saving obstetric care, including giving the necessary supervision, care and advice to women during pregnancy, childbirth and the post-partum period; to conduct deliveries on their own; and to care for newborns" (World Health Organization Statistical Information System 2010, p. 30). The presence of such personnel will likely improve the survival chances of newborns, leading us to hypothesize a negative relationship between IMR and ATTEND.

What additional factors might contribute to the variation in infant mortality rates across countries? While skilled medical personnel can attend to some problems associated with pregnancy and birth, they cannot necessarily overcome the general poor health of women of childbearing age. As a result, we will include another independent variable that measures differences in women's health across countries, the prevalence of anemia among pregnant women (ANEMIA). Measured by the World Bank as the percentage of pregnant women whose hemoglobin level is less than 110 grams per liter at sea level, anemia during pregnancy increases the risk of pregnancy complications, including an increased risk of preterm delivery and low birthweight, as well as a higher risk of stillbirth or newborn death. (See, for example, Allen (2000) for further discussion of the relationship.) Consequently, we would expect to observe higher infant mortality rates in countries with higher percentages of anemic pregnant women. Mathematically, this means we expect a positive correlation between IMR and ANEMIA.

Another set of factors that might explain the variation in infant mortality rates relate to the general health of a country's population as manifested in the physical environment in which people live. Access to safe water, sanitation facilities, and adequate food, for example, all improve general health in a country, and all have been shown to be positively correlated with lower infant mortality rates. (For further discussion, see, for example, "Goal 4: Reducing child mortality" in the Millennium Development Goals (United Nations Children's Fund 2010), and the WHO's "Newborns: reducing mortality" factsheet (World Health Organization 2010a).) For our model, we will select one of these factors, access to safe water (H_2O), which is measured as the percentage of a country's population with access to improved water sources, as a third independent variable in our econometric model. We note that this choice was governed by data availability because we have more observations for access to safe water for the countries in our sample than observations available for access to sanitation facilities and the number of average calories consumed per person. This demonstrates how considerations other than economic theory can influence specification of the econometric model.

Variations in infant mortality rates may also be the result of considerable differences in access to economic resources between countries. This differential could be measured using a nation's poverty rate or Gini coefficient, but data availability from the WHO's Global Health Observatory (World Health Organization 2010b) leads us to measure this factor using gross national income (GNI). Specifically, we will use GNI per capita to eliminate any effects that the size of a country's population might have on infant mortality rates, and we will measure the variable in current U.S. dollars using the Atlas method[3] so the currency units are comparable across countries.

We now ask if it is reasonable to expect a constant unit change in infant mortality rates across all countries for a given dollar change. That is, would a $100 increase in GNI per capita in a country such as Guinea-Bissau, with annual per-capita income of $250, change infant deaths by the same number as that same dollar amount would affect infant deaths in a country such as Libya, whose annual GNI per capita equals \$12,380? While that is not likely to be the case, it may be reasonable that the same percentage increase in GNI per capita will change infant deaths by a constant number across countries. Thus we will use a lin–log functional form on GNI per capita, which means that the coefficient on the ln(GNI) variable will be interpreted as the change in the number of infant deaths per 1,000 live births for a 1 percent change in GNI per capita.

Let us now assume, for discussion purposes, that these four independent variables will capture the various socio-economic factors that contribute to the variation in infant mortality rates across the countries in our sample. This means that we are ready to specify the theoretical and statistical econometric models of the IMR function, respectively:

$$IMR = b_b + b_{ATTEND} \times ATTEND + b_{ANEMIA} \times ANEMIA + b_{H_2O} \times H_2O + b_{GNI} \times \ln(GNI) \tag{16.4}$$

$$IMR = b_b + b_{ATTEND} \times ATTEND + b_{ANEMIA} \times ANEMIA + b_{H_2O} \times H_2O + b_{GNI} \times \ln(GNI) + e \tag{16.5}$$

Procedurally, the next step is to hypothesize the signs expected on each partial regression coefficient and state the appropriate decision rule, but we will defer doing so until Chapter 17. For now, let us assume we have performed these steps, which makes our next task the collection of appropriate data for the five variables in the model. While this step may, in the abstract, seem straightforward, the reality is that identifying and then locating appropriate data is an important and time-consuming aspect of conducting econometric analyses.

Let us now revisit some of the considerations associated with specifying this particular multiple linear regression model. While economic theory was the primary guide for our variable and functional form choices, our ability to locate complete datasets for certain variables may also determine which variables we can, and which we cannot, include in the econometric model. As noted for the IMR model, the lack of data on access to sanitation facilities and the number of average calories consumed per person, two variables we expect to contribute to a population's general health, led us to omit those variables from the econometric model.

This demonstrates a common tradeoff encountered in econometric analysis. If we were to include those two variables in the model, we would reduce the number of observations available for the statistical estimation of the model's parameters, which reduces the robustness of our statistical results. If we do not include the two variables (assuming, in essence, that the access to water (H_2O) variable will sufficiently capture the effects of improved general health on IMR), we run the risk of introducing what is called omitted-variable bias into the model. While neither solution is ideal, we are rarely able to estimate an econometric model that contains all the variables recommended by economic theory. Thus here, as is often the case, we will proceed under the assumption that the IMR model defined by equation (16.4) is complete, leading us to estimate the model in equation (16.5) accordingly.

16.5 Estimating a multiple regression model for infant mortality rates

Applying the OLS method to estimate equation (16.5) yields the following model:

$$\widehat{IMR} = 148.35 - 0.2482 \times ATTEND + 0.3813 \times ANEMIA$$
$$- 0.8529 \times H_2O - 4.2894 \times \ln(GNI)$$

What do these results tell us about infant mortality rates in 2008? To interpret the results, we consider each estimated regression coefficient's effect on IMR in isolation from the other factors in the model. This means we interpret the OLS-estimated coefficients[4] as follows.

- b_{ATTEND}: a one percentage point increase in the number of births attended by skilled medical personnel corresponds to an average decrease in infant mortality of 0.25 infant deaths per 1,000 live births, *ceteris paribus*.
- b_{ANEMIA}: a one point increase in the percentage of pregnant women with anemia corresponds to an average increase in infant mortality of 0.38 deaths per 1,000 live births, *ceteris paribus*.
- b_{H_2O}: a one point increase in the percentage of the population with access to improved water sources (H_2O) corresponds to an average decrease in IMR of 0.85 deaths per 1,000 live births, *ceteris paribus*.
- b_{GNI}: a one percent increase in GNI per capita corresponds to an average reduction of 0.0429 infant deaths per 1,000 live births, *ceteris paribus* (recall that we must divide the estimated coefficient by 100 when using a lin–log functional form).

Including the phrase *ceteris paribus* as part of our interpretation of each partial regression coefficient reminds us that the OLS procedure isolates the effect of that independent variable on the dependent variable. Thus, the coefficient for ATTEND measures the change in infant deaths per 1,000 live births for a unit change in ATTEND (the percentage of total births attended by skilled medical personnel) while holding constant any effects that the prevalence of anemia among pregnant women, access to improved water sources, and GNI per capita might have on the variation in infant mortality rates.

We further observe that the sign on each partial regression coefficient is consistent with the underlying economic reasoning. While we will formally test their statistical significance in Chapter 17, we can now discuss the **economic significance** of each partial regression coefficient. That is, we can examine the relative size of each independent variable's estimated effect on infant mortality rates.

We observe that the largest change in infant mortality rates is associated with improved access to safe water sources. According to the model, if the percentage of a country's population with such access increases by one point, the corresponding drop in infant mortality rates is almost one death (0.85, to be precise) per 1,000 live births.

The second largest effect is associated with the ANEMIA variable, where a one point decrease in the percentage of women with anemia corresponds to a 0.38 drop in infant deaths. We can use this result to determine by what percentage in the prevalence of pregnant women with anemia would be required to reduce infant deaths by one child per 1,000 live births. Taking the reciprocal of the coefficient, we find that the percentage of pregnant women with anemia would need to fall by 2.6 percent to reduce the number of infant deaths by one for every 1,000 live births.

An increase in the presence of skilled medical professionals corresponds to an even smaller decline in infant mortality. This results translates to needing a 4 percent increase in the number of births attended by skilled medical personnel to reduce IMR by one death.

Given these three environmental factors, the model's results suggest that improving access to safe water sources would have the greatest effect on reducing infant mortality. However, as with simple linear regression analysis, we cannot use these results (even if they are statistically significant) to prove that any of these factors actually *cause* infant mortality rates to change. Thus, when presenting our results, we want to caution our audience about what they do, and do not, tell us about infant mortality.

Let us turn to the final result, the coefficient on the GNI per capita variable. The 0.043 value indicates that a 1 percent increase in GNI corresponds to a very small change in infant mortality rates. In fact, according to the estimated coefficient, GNI per capita would need to increase by over 23 percentage points before infant mortality would decline by even one death per 1,000 live births. Such an increase is virtually economically impossible.

How do we respond to the size of the GNI coefficient? Assuming that the small value is not the result of statistical problems, we might be tempted to conclude that differences in GNI are simply not correlated with differences in infant mortality rates for this particular sample of countries in 2008. But such a conclusion is not consistent with the statistical evidence, where the pairwise correlation between the two variables is -0.76 percent. Thus a more likely explanation is that GNI per capita may not be the most appropriate variable for measuring overall resource availability across countries.[5]

While the language in our discussion of the partial regression coefficients from the IMR model might suggest a causal relationship between the dependent variable and each independent variable, let us reiterate: causality is *not* something we can prove from our econometric results (or those associated any other statistical inference method). Indeed, regression analysis is really a method for assessing the correlation between pairs of variables and cannot be used to determine causation. In addition, we cannot use our statistical results (even statistically significant ones) to conclude that the *theoretical* relationships described by the multiple regression model are, in fact, true. Thus we must never use the results from an econometric analysis to claim that the underlying economic theory is correct.

Summary

Multiple regression analysis takes us into the more realistic world of complex economic relationships. However, the statistical procedures for generating multiple regression results critically depend on what economic theory and empirical observation tell us about those relationships. This means we must completely specify the theoretical multiple regression model to include all economically relevant independent variables as well as the functional forms that best describe the relationships between variables if we want to obtain meaningful regression results. Not doing so ensures that the model's results, even if statistically significant, contribute little to our understanding of economic phenomena, which is of course our motivation for engaging in such analysis.

Concepts introduced

Econometrics
Economic significance
Estimated multiple linear regression model
Partial regression coefficient
Statistical multiple linear regression model
Theoretical multiple linear regression model

Box 16.1 Where can I find…? Integrated Public Use Microdata Series (IPUMS)

Collecting data for multiple regression analyses can lead us to a wide range of locations, raising the possibility of data incompatibility between the various sites. The IPUMS project eliminates those potential problems by harmonizing, or identically coding, the variables in each data series across time, thereby eliminating a very time-consuming step of empirical analysis.

IPUMS is the acronym for the **Integrated Public Use Microdata Series**. The seven series available as of December 2010 are microdata, which IPUMS defines as follows:

> IPUMS is not a collection of compiled statistics; it is composed of **microdata**. Each record is a person, with all characteristics numerically coded. In most samples persons are organized into households, making it possible to study the characteristics of people in the context of their families or other co-residents. Because the data are individuals and not tables, researchers must use a statistical package to analyze the millions of records in the database. A data extraction system enables users to select only the samples and variables they require. (Available online from <http://cps.ipums.org/cps-action/faq>.)

The IPUMS project represents a series of data collection, harmonization, and dissemination projects housed at the Minnesota Population Center (MPC). Accessible free of charge online, IPUMS provides a valuable resource for over 50,000 researchers, policy-makers, and students worldwide.

Data available through IPUMS

Data is available online through IPUMS at <http://www.ipums.org>. The following descriptions were obtained from the links to each IPUMS data series as listed.

IPUMS-International <https://international.ipums.org/international/>

IPUMS-International is an integrated series of census microdata samples from 1960 to the present. As of December 2010, IPUMS-I describes approximately 326 million persons recorded in 159 censuses taken from 55 countries: Argentina, Armenia, Austria, Belarus, Bolivia, Brazil, Cambodia, Canada, Chile, China, Colombia, Costa Rica, Cuba, Ecuador, Egypt, France, Ghana, Greece, Guinea, Hungary, India, Iraq, Israel, Italy, Jordan, Kenya, Kyrgyz Republic, Malaysia, Mali, Mexico, Mongolia, Nepal, Netherlands, Pakistan, Palestine, Panama, Peru, Philippines, Portugal, Puerto Rico, Romania, Rwanda, Saint Lucia, Senegal, Slovenia, South Africa, Spain, Switzerland, Tanzania, Thailand, Uganda, United Kingdom, United States, Venezuela, and Vietnam.

The data series in IPUMS-I includes information on a broad range of population characteristics, including fertility, nuptiality, life-course transitions, migration, labor-force participation, occupational structure, education, ethnicity, and household composition. The information available in each sample varies according to the questions asked in that year and by differences in post-enumeration processing.

IPUMS-I 2010 [Online] Available at
<https://international.ipums.org/international/overview.shtml>

IPUMS-USA <http://usa.ipums.org/usa/>

IPUMS-USA consists of more than 50 high-precision samples of the American population drawn from 15 federal censuses and from the American Community Surveys (ACS) of 2000–2009. Some of these samples have existed for years, and others were created specifically for this database. These samples, which draw on every surviving census from 1850 to 2000, and the 2000–2009 ACS samples, collectively constitute our richest source of quantitative information on long-term changes in the American population. However, because different investigators created these samples at different times, they employed a wide variety of record layouts, coding schemes, and documentation. This has complicated efforts to use them to study change over time. The IPUMS assigns uniform codes across all the samples and brings relevant documentation into a coherent form to facilitate analysis of social and economic change.

IPUMS-USA 2010 [Online] Available at <http://cps.ipums.org/cps-action/faq>

IPUMS-CPS <http://cps.ipums.org/cps/>

IPUMS-CPS is an integrated set of data from 49 years (1962 to 2010) of the March Current Population Survey (CPS). The CPS is a monthly U.S. household survey conducted jointly by the U.S. Census Bureau and the U.S. Bureau of Labor Statistics. Initiated in the 1940s in the wake of the Great Depression, the survey was designed to measure unemployment. A battery of labor force and demographic questions, known as the "basic monthly survey," is asked every month. Over time, supplemental inquiries on special topics have been added for particular months. Among these supplemental surveys, the March Annual Demographic File and Income Supplement (hereafter referred to as the March CPS) is the most widely used by social scientists and policy-makers, and it provides the data for IPUMS-CPS.

IPUMS-CPS 2010 [Online] Available at <http://cps.ipums.org/cps-action/faq#ques0>

Other data available from the Minnesota Population Center

North Atlantic Population Project (NAPP) <http://www.nappdata.org/napp/>

The North Atlantic Population Project (NAPP) is a machine-readable database of the complete censuses of Canada (1881), Great Britain (1881), Norway (1865, 1900), Sweden (1900), the United States (1880) and Iceland (1870, 1880, and 1901 – forthcoming). These nine censuses comprise our richest source of information on the population of the North Atlantic world in the late nineteenth century, and they have only recently become available for social science research. Samples of census data are also available for Canada (1871, 1901), Great Britain (1851), the German state of Mecklenburg-Schwerin (1819), Norway (1875), and the United States (1850, 1860, 1870, 1900, 1910), which support cross-temporal analyses.

The NAPP database has enabled linking of individuals between census years for longitudinal analysis. NAPP includes samples of the United States that link 1880 to seven other census years, and six linked samples for Norway that link males and couples between the 1865, 1875, and 1900 censuses.

NAPP 2010 [Online] Available at <http://www.nappdata.org/napp/intro.shtml>

National Historical Geographic Information System (NHGIS) <http://www.nhgis.org/>

NHGIS provides U.S. aggregate census data and GIS-compatible boundary files for tracts and counties between 1790 and 2000.

Integrated Health Interview Series (IHIS) <http://www.ihis.us/ihis/>

The Integrated Health Interview Series (IHIS) provides integrated data and documentation from the U.S. National Health Interview Surveys (NHIS) from 1969 to the present.
Over 1,000 integrated variables are currently available, and linking keys enable users to include other, not yet integrated, variables from the NHIS files.

American Time Use Survey (ATUS) Extract Builder <http://www.atusdata.org/index.shtml>

ATUS-X is a project dedicated to making it easy for researchers to use data from the American Time Use Survey (ATUS). The ATUS is an ongoing time diary study, started in 2003, that is funded by the U.S. Bureau of Labor Statistics (BLS) and fielded by the U.S. Census Bureau. It provides detailed information about the activities in which respondents engage, together with extensive information about the characteristics of those respondents and other members of their households.

ATUS-X 2010 [Online] Available at <http://www.atusdata.org/aboutbuilder.shtml>

Box 16.2 Using Excel's Regression Tool in multiple regression analysis

We can use Excel to estimate multiple regression models using the **Regression Data Analysis** tool. Although similar to performing simple regression analysis with Excel, we note some important differences when conducting multiple regression analyses.

After entering the label and data for the model's dependent variable in the **Input Y Range** (and checking the **Labels** box), enter the labels and the range of data for the multiple independent variables in the **Input X Range**.

- The data for the entire range of independent variables *must* be in *contiguous columns* (columns adjacent to one another). If not, you will receive the following error message: "Input range must be a contiguous reference".
- If you do not include descriptive labels in the first cell of the columns containing the independent variable data, your regression results will look like the following:

	Coefficients
Intercept	148.3480
X Variable 1	−0.2482
X Variable 2	0.3813
X Variable 3	−0.8529
X Variable 4	−4.2894

Such generic results will complicate your interpretation of the regression's estimates, especially when you conduct multiple runs of the model.
- You can incorporate *only 16 independent variables* in Excel's regression tool. For econometric models that include more than 16 independent variables, use software dedicated to statistical analysis. Three software packages frequently used by economists are SAS, SPSS, and STATA. Check the internet for more information on each.

Exercises

1. Return to Exercise 1 in Chapter 13, and identify at least three additional independent variables that economic theory would suggest for each of the four models. Be sure to explain the rationale behind each independent variable chosen.
2. The following example estimates the demand for bus travel across 40 cities in the United States in 1988. The first variable, BUSTRAVL, is the dependent variable, and the remaining six are the independent variables. Units of measurement are in parentheses:

BUSTRAVL: Demand for urban transportation by bus
(thousands of passenger hours)
FARE: Bus fare (dollars)
GASPRICE: Price of a gallon of gasoline (dollars)
INCOME: Average income per capita (dollars)
POP: Population of city (thousands)
DENSITY: Density of population (persons per square mile)
LANDAREA: Land area of the city (square miles)

After running the regression, the following estimated model was obtained:

$$\widehat{BUSTRAVL} = 2{,}744.68 - 238.65 \times FARE + 522.11 \times GASPRICE$$
$$- 0.19 \times INCOME + 1.71 \times POP + 0.12$$
$$\times DENSITY - 1.16 \times LANDAREA$$

Interpret each partial regression coefficient in the estimated model.

3. A new model for bus travel in 1988 is estimated using the double-log functional form. Interpret each partial regression coefficient in this estimated model:

$$\ln(\widehat{BUSTRAVL}) = 44.72 + 0.48 \times \ln(FARE) - 1.73 \times \ln(GASPRICE)$$
$$- 4.85 \times \ln(INCOME) + 1.69 \times \ln(POP) + 0.28$$
$$\times \ln(DENSITY) - 0.82 \times \ln(LANDAREA)$$

4. The Keynesian consumption function is a popular macroeconomic model to estimate econometrically. Go to a source for national data in a specific country, and collect appropriate data for the following variables: personal consumption expenditures (dependent variable), personal disposable income, a low-risk interest rate (in the U.S., the rate on three-month Treasury Bills is often used), and a measure of consumer confidence (in the U.S., the Michigan Index of Consumer Sentiment is available from FRED®).

(a) First estimate the original relationship identified by Keynes between consumer spending and personal income using the simple regression model. Interpret the coefficients for the average propensity to consumer (a) and the marginal propensity to consume (b).

(b) Add the two other independent variables, and re-estimate the model. Interpret the coefficients.

(c) Compare the marginal propensities to consume for the two models. Discuss if it appears that the addition of explanatory variables to the second model alters the marginal propensity to consumer.

5. The following labor supply model for married women with children was estimated using the data from Mroz (1976):

$$\widehat{WHRS} = -2{,}764.73 - 47.64 \times WW + 466.48 \times \ln(FAMINC) - 261.28 \times KL6$$
$$+ 37.40 \times AX - 20.06 \times WA$$

where WHRS is the wife's hours of work in 1975, WW is the wife's average hourly earnings (in 1975 dollars), FAMINC is the family income (in 1975 dollars), KL6 is the number of children less than six years old in the household, AX is the actual years of wife's previous labor market experience, and WA is the wife's age.

Interpret the partial regression coefficients estimated for the model, and explain what the overall results indicate about the factors that influenced these women's labor supply decision.

17 Multiple regression analysis

Hypothesis tests for partial regression coefficients and overall goodness of fit

Let us now evaluate the statistical significance of the partial OLS-estimated regression coefficients. As with the simple regression framework, we will test the model's coefficients by evaluating hypotheses that reflect the expected economic relationship between the dependent variable and each independent variable. We also introduce two new statistics to evaluate the overall goodness of fit, or statistical significance, of the full multiple regression model.

17.1 General procedures for testing the significance of the partial regression coefficients (b_i)

Testing hypotheses of the partial regression coefficients in the multiple regression model is analogous to testing hypotheses about a simple regression model's slope coefficient. Here we will describe a similar multi-step procedure, noting along the way the general differences associated with multiple regression analysis. This discussion will be followed by an analysis of the econometric results for the infant mortality rate model estimated in Chapter 16.

The first step is establishing the hypotheses to test, but now we will test a pair of hypotheses associated with each partial regression coefficient. Following economic convention, we again begin by specifying the alternative hypothesis H_a for each partial regression coefficient to reflect the theoretically expected relationship between the dependent variable and each independent variable, writing the null hypothesis as "not H_a." Rejecting the null hypothesis for a specific coefficient is thus akin to obtaining statistical support for the relationship posited by economic theory.

We next establish a level of significance, α, to use for each partial regression coefficient's hypothesis test. In multiple regression analysis, it is common practice to select the same level of significance, typically 1 percent or 5 percent, for the hypothesis tests of all partial regression coefficients. Again, the low levels of significance are chosen to minimize the probability of committing a Type I error, which now means that we would incorrectly reject a true null hypothesis about each partial regression coefficient.

Establishing the decision rule for each partial regression coefficient introduces an important modification to the simple linear regression procedure, a change in the degrees of freedom. In a multiple regression model with k independent variables, the degrees of freedom equal $n - k - 1$. Similar to the reasoning used with the simple regression model, we must now calculate the means of Y and all k independent variables ($X_1, X_2, \ldots, X_k$) to estimate the partial regression coefficients, thereby losing the variability associated with $(k-1)$ observations. (We can also use this formula to generate the $n - 2$ degrees of freedom associated with the simple regression model where $k = 1$.)

	Null hypothesis	Alternative hypothesis	Decision rule
One-tail tests			
Statistically significant *positive* relationship between Y and X_i	$H_0: b_i \leq 0$	$H_a: b_i > 0$	Reject H_0 if $t_{b_i} > t_{n-k-1}^{\alpha}$
Statistically significant *negative* relationship between Y and X_i	$H_0: b_i \geq 0$	$H_a: b_i < 0$	Reject H_0 if $t_{b_i} < -t_{n-k-1}^{\alpha}$
Two-tail tests			
Statistically significant but theoretically ambiguous or unknown relationship between Y and X_i	$H_0: b_i = 0$	$H_a: b_i \neq 0$	Reject H_0 if $|t_{b_i}| > t_{n-k-1}^{\alpha}$

Figure 17.1 Possible hypothesis tests and corresponding decision rules for partial regression coefficients.

Figure 17.1 presents the three pairs of hypotheses that we might test for each partial regression coefficient (b_i). As with the slope coefficient in the simple regression model, we select the test based on the relationship between Y and X_i that is predicted by economic theory. If we have postulated a directional relationship between the two variables, we conduct a one-tail test on the partial regression coefficient. If, however, the relationship is theoretically ambiguous or one to be empirically determined, we conduct a two-tail test.

The three steps just articulated can, and should, be completed prior to estimating the actual regression model. Doing so will insure that economic theory, not the statistical results, guide the hypotheses specified for each partial regression coefficient.

The next step is estimating the regression model, followed by calculating the test statistic associated with each partial regression coefficient. As with the simple regression model, the test statistic is a ratio that contains the estimated regression coefficient (b_i) and its estimated standard error (se_{b_i}) along with the hypothesized value for the population parameter (β_i). Because our regression coefficients are now partial, we modify the formula for the test statistic by adding a subscript to denote which independent variable is our focus:

$$t_{b_i} = \frac{b_i - \beta_i}{se_{b_i}} \tag{17.1}$$

The formulas for calculating the partial regression coefficient and its standard error are considerably more complex than those for the simple regression model because they now must account for the additional independent variables in the model. Fortunately, we need not generate these results by hand; the statistical software used to generate regression results will calculate not only the estimated value for b_i, but also its standard deviation as well as the value for the test statistic when β_i is hypothesized to have a value of zero.

Our statistical software will also generate the *p*-value associated with each partial regression coefficient. We can again compare the estimated *p*-value with the level of significance previously established for making a decision about the null hypothesis. As shown in Figure 17.2, if the *p*-value is less than 0.05 or 0.01 (depending on the level of significance), we conclude that we can reject the null hypothesis.

The *p*-value also allows us to determine precisely how confident we are in our decision to reject the null hypothesis for each partial regression coefficient. Subtracting the *p*-value from a value of one, we can state the degree of confidence we have in rejecting H_0.

Table 17.1 Hypothesis tests for partial regression coefficients in the infant mortality rate model

Theorized relationship	Economic rationale	Null hypothesis	Alternative hypothesis
Negative relationship between IMR and ATTEND	The more skilled medical personnel available per 1000 live births, the more likely complications can be addressed, thereby reducing infant mortality rates	$H_0: \beta_{\text{ATTEND}} \geq 0$	$H_a: \beta_{\text{ATTEND}} < 0$
Positive relationship between IMR and ANEMIA	The greater the prevalence of anemia among pregnant women, the greater the likelihood of pregnancy complications, stillbirths, and newborn birth complications, thereby increasing infant mortality rates	$H_0: \beta_{\text{ANEMIA}} \leq 0$	$H_a: \beta_{\text{ANEMIA}} > 0$
Negative relationship between IMR and H_2O	The greater the percentage of a country's population with access to improved water sources, the healthier the overall environment in which births occur, thus reducing the probability of infant mortality	$H_0: \beta_{H_2O} \geq 0$	$H_a: \beta_{H_2O} < 0$
Negative relationship between IMR and GNI	The higher GNI per capita, the greater access each person has, on average, to economic resources that can be used to reduce infant deaths, thus lowering the risk of infant mortality	$H_0: \beta_{\text{GNI}} \geq 0$	$H_a: \beta_{\text{GNI}} < 0$

17.2   Testing the significance of partial regression coefficients (b_i): IMR model

We now return to the infant mortality rate model to test the null hypotheses for each partial regression coefficient. While we articulated the expected theoretical relationships between IMR and each independent variable in Chapter 16, let us now formalize those relationships by stating the hypotheses we will test in Table 17.1. As we observe, we will conduct one-tail tests for all four partial regression coefficients.

We next establish a 1 percent level of significance for our hypothesis tests to minimize the likelihood of incorrectly rejecting a true null hypothesis. We also determine that we have

Table 17.2 Decision rules for partial regression coefficients in the infant mortality rate model

Reject H_0 if $t_{\text{ATTEND}} < -t_{147}^{0.01} = -2.3520$
Reject H_0 if $t_{\text{ANEMIA}} > t_{147}^{0.01} = 2.3520$
Reject H_0 if $t_{H_2O} < -t_{147}^{0.01} = -2.3520$
Reject H_0 if $t_{\text{GNI}} < -t_{147}^{0.01} = -2.3520$

147 degrees of freedom for this model, obtained from the 152 observations and four independent variables ($k = 4$). Thus the $t_{147}^{0.01}$ critical value equals ± 2.3520. Table 17.2 presents the resulting decision rules for the four partial regression coefficients.

We now use the OLS method to estimate the IMR econometric model, obtaining the statistical results for the partial regression coefficients shown in Table 17.3. We observe that all partial regression coefficients have the hypothesized sign. If such results had not obtained, we immediately know that we will *not* be able to reject the null hypothesis for the offending partial regression coefficient(s). Thus we will place the corresponding independent variable(s) on a list for possible removal from the model after we have finished our statistical testing.

Because all four estimated coefficients did have the hypothesized sign, we now evaluate the decision rules established in Table 17.2. Specifically, we will determine if the statistical evidence permits us to reject the null hypothesis regarding the relationship between infant mortality rates and each independent variable.

Let us begin with the statistical results for b_{ATTEND}, the sole independent variable considered in the simple regression model. We first observe that the estimated partial regression coefficient on ATTEND (-0.25) is noticeably smaller in value compared to the simple regression slope estimate, which equaled -1.05. This change suggests that the independent

Table 17.3 Estimates of partial regression coefficients and related statistics from the infant mortality rate model

	Coefficients	Standard error	t-statistic	p-value	p-value/2	$[1 - (p/2)] \times 100$
Intercept	148.3480	17.5729	8.4419	0.0000	0.0000	n/a
Number of skilled medical personnel per 1000 live births (ATTEND)	−0.2482	0.0922	−2.6926	0.0079	0.0040	99.60
Prevalence of anemia among pregnant women in percent (ANEMIA)	0.3813	0.1412	2.7001	0.0077	0.0039	99.61
Percentage of population with access to improved water source (H₂O)	−0.8529	0.1284	−6.6422	0.0000	0.0000	100.00
ln(GNI per capita) in current U.S. dollars (GNI)	−4.2894	1.6443	−2.6087	0.0100	0.0050	99.50

variables added to the multiple regression model account for some of the effects on infant mortality that were statistically attributed to ATTEND in the simple regression model.

The b_{ATTEND} slope coefficient was also a statistically significant result at the 1 percent level in the simple regression model, but does that hold in the multiple regression model? We observe that the standard error for b_{ATTEND} equals 0.0922, yielding a test statistic for b_{ATTEND} of -2.6929. Because -2.6929 is less than the critical t-value of -2.3520, we can reject the null hypothesis that b_{ATTEND} is either greater than or equal to zero with at least 99 percent confidence. Alternatively, we can say that we are at least 99 percent confident that infant mortality rates and the number of skilled medical personnel are negatively related in our sample.

The associated p-value of 0.0040 confirms this conclusion, since it is smaller than the 0.01 level of significance. We can also use this value to state precisely how confident we are in our decision to reject the null hypothesis. Using the calculation in the last column of Table 17.3, we determine that we are 99.60 percent confident that the negative relationship between IMR and ATTEND is statistically significant and not merely one of chance.

We next consider the statistical results for b_{ANEMIA}. Using the estimated value for b_{ANEMIA} (0.38) and the standard error of 0.1412, we obtain a test statistic for b_{ANEMIA} of 2.7001, which is greater than the critical t-value, 2.3520. Thus we can, with at least 99 percent confidence, reject the null hypothesis that b_{ANEMIA} is either less than or equal to zero, concluding that we are at least 99 percent confident that infant mortality rates and the percentage of pregnant women with anemia are positively related across the countries in our sample. The associated p-value of 0.0039 is also less than the 0.01 level of significance, confirming our conclusion about the null hypothesis. We can also say that we are 99.61 percent confident there is a statistically significant positive relationship between IMR and ANEMIA in this sample.

The test statistic for the access to safe water (H_2O) coefficient of -6.6422 was obtained by dividing the coefficient estimate (-0.85) by its standard error (0.1284), and it is noticeably smaller than the critical value of -2.3520. Thus we reject the null hypothesis that the H_2O coefficient is statistically greater than or equal to zero with at least 99 percent confidence. The p-value of 0.000 (technically, 2.80E-10) indicates that we are very confident in our decision to reject the null hypothesis, leading us to conclude that we are almost 100 percent confident that the negative relationship between IMR and H_2O is a statistically significant one.

We now turn to the statistical results for the coefficient for the GNI per capita variable. Here the test statistic of -2.6087 (obtained by dividing the estimated coefficient of -4.2894 by the standard error of 1.6443) is less than the critical value of -2.3520, which means that we can, with at least 99 percent confidence, reject the null hypothesis that the natural log of GNI per capita is either greater than or equal to zero. The p-value of 0.005 is also less than α (0.01) and indicates that we are 99.50 percent confident of a statistically significant negative relationship between IMR and GNI per capita.

The hypothesis tests on all four partial regression coefficients suggest that each of the four independent variables are statistically related to infant mortality rates in ways consistent with the underlying economic rationale. While this might suggest that we have specified a complete econometric model for explaining the variation in infant mortality rates, we still do not know how much of the variation in infant mortality rates is explained by this particular multiple regression model, nor do we know if the overall model is statistically significant. We now turn to statistics that we will permit us to assess both questions.

17.3 Evaluating the overall goodness of fit

In the simple regression model, our statistical results included the coefficient of determination, R^2. We used this statistic to assess how much of the variation in the dependent variable could be "explained" by the variation in the single independent variable. In other words, we used the coefficient of determination to assess the goodness of fit for the model.

When we conduct multiple regression analysis, we would also like to evaluate how well the model "fits" the data. But now that we have more than one independent variable in our regression equation, it is no longer appropriate for us to use the R^2 statistic. Instead, to assess how much of the variation in the dependent variable is "explained" by the regression model, we use a new statistic, the adjusted R^2.

Perhaps the best way to understand why this new statistic is necessary is to re-examine the formula for the unadjusted R^2 statistic discussed in Chapter 13. There we saw that the unadjusted R^2 can be calculated by subtracting the ratio of the regression model's error sum of squares (ESS) divided by the total sum of squares (TSS) from a value of one:

$$\text{unadjusted } R^2 = 1 - \frac{\text{ESS}}{\text{TSS}} \tag{17.2}$$

Let us now introduce the formula for the **adjusted R^2**:

$$\text{adjusted } R^2 = 1 - \left[\frac{\text{ESS}/(n-k-1)}{\text{TSS}/(n-1)} \right] \tag{17.3}$$

We observe that the adjusted R^2 statistic includes not only the error and total sums of squares found in the unadjusted R^2, but also the degrees of freedom associated with each deviation. Specifically, the numerator in the bracketed term is the ESS divided by $(n-k-1)$, the number of degrees of freedom associated with the regression model's error sum of squares, while the denominator divides TSS by $(n-1)$, the number of degrees of freedom associated with the total sum of squares.

As with the unadjusted R^2 statistic, the adjusted R^2 statistic, or $\bar{R}^2$ (R-bar squared), ranges in value from (approximately) 0 to 1. It can be shown that if the multiple regression model does not explain any of the variation in the dependent variable, $\bar{R}^2$ approaches a value of zero,[1] while a multiple regression model that explains the entire variation in the dependent variable (meaning ESS = 0) will have an $\bar{R}^2$ equal to one. This tells us that the closer $\bar{R}^2$ is to one in value, the better the overall model fits the data.

The numerator and denominator in the bracketed term of equation (17.3) are referred to as the **mean square errors** (MSE) associated with the error sum of squares and total sum of squares, respectively, and are obtained by dividing each sum of squares by its respective degrees of freedom. We can thus rewrite equation (17.3) as

$$\text{adjusted } R^2 = 1 - \left[\frac{\text{error MSE}}{\text{total MSE}} \right] \tag{17.4}$$

While we will not explore the technical details, the error and total mean square errors used in the adjusted R^2 statistic represent unbiased estimators of the error and total variances, respectively.[2] This is one reason why the $\bar{R}^2$ statistic, rather than R^2, is used to assess the goodness of fit for multiple regression models.

Let us now consider what happens to the two sums of squares, ESS and TSS, when we move from a simple regression model to a model with two independent variables. We first recall that the total sum of squares (TSS) is calculated by the formula, $\sum (Y_i - \overline{Y})^2$. Since neither Y_i nor $\overline{Y}$ change when we move from a simple regression model to a multiple regression model, the value of TSS will not change if we modify the right-hand side of the regression model.

That is not the case with the error sum of squares, which is defined by the following equation:

$$\sum (Y_i - \hat{Y}_i)^2 = \sum (Y_i - \hat{b}_0 - \hat{b}_1 X_1 - \hat{b}_2 X_2)^2 = \sum e_i^2$$

Assuming that the independent variable we add is relevant for understanding the variation in Y, the model's residuals, and thus ESS, will almost certainly decline.[3] If we now refer back to equation (17.2), we see that adding another relevant independent variable will then increase the model's unadjusted R^2 statistic.

While this appears to be a desirable result, we have not yet factored in the cost associated with adding another independent variable to the model. Because we now must estimate the sample mean for the new independent variable (here, X_2) to obtain the two partial regression coefficient estimates for the new model, this additional computation costs us an additional degree of freedom, such that we now have only $n - 3$ degrees of freedom instead of the $n - 2$ associated with the simple regression model. Consequently, every time we increase the number of independent variables in the regression model, the degrees of freedom associated with the error sum of squares necessarily declines. This means that we cannot predict, *a priori*, if the adjusted R^2 statistic will decrease or increase when we add another relevant independent variable to the regression model.

Because we want to account for both the benefits (a lower ESS) and costs (a lost degree of freedom) associated with adding an independent variable to the model, economists use the adjusted R^2 statistic, rather than the unadjusted R^2, to evaluate the overall fit of a multiple regression model. Let us now return to the infant mortality model to see if adding three more independent variables to the simple regression model increases the model's adjusted R^2 statistic.

17.4   Interpreting the infant mortality model's overall goodness of fit

Table 17.4 presents the unadjusted and adjusted R^2 statistics for both the simple and multiple regression models, along with each model's sums of squares and mean square errors. As we observe, the unadjusted R^2 for the multiple regression model is higher than the unadjusted R^2 for the simple regression model. We are not surprised by this result, since the residual sum of squares (ESS) from the multiple regression model is almost half the size as that for the simple regression model. We further note, as mentioned above, that the TSS is the same value for both models.

Now let us compare the two goodness-of-fit statistics associated with the multiple regression model. If we do not account for the three fewer degrees of freedom associated with the new independent variables, our model explains 77.77 percent of the variation in infant mortality rates. Once we do account for the lost variability, the adjusted R^2 statistic indicates that the model accounts for a slightly lower percentage, 77.16 percent, of the variation in infant mortality rates. Thus the multiple regression model does explain more of the variability in infant mortality rates than did the simple regression model.

Table 17.4 Coefficients of determination for the simple and multiple regression IMR models

	IMR $= a + b \times$ ATTEND	IMR $= b_0 + b_{\text{ATTEND}} \times$ ATTEND $+ b_{\text{ANEMIA}} \times$ ANEMIA $+ b_{\text{H}_2\text{O}} \times \text{H}_2\text{O} + b_{\text{GNI}} \times$ GNI
Unadjusted R^2	0.5940	0.7777
Adjusted R^2	n/a	0.7716
Sum of squares		
Residual	71,846.27	39,342.27
Total	176,942.53	176,942.53
Degrees of freedom		
Residual	150	147
Total	151	151
Mean square error		
Residual	478.98	267.63
Total	1,171.80	1,171.80

However, the IMR multiple regression model accounts for only about three-quarters of the variation in infant mortality rates. As a result, we might be tempted to add more independent variables to the model in an effort to reduce the adjusted R^2 statistic even more, but doing so is inconsistent with econometric analysis. Remember that we fully specify the econometric model *before* we estimate it. Thus if we let the statistical results govern our choice of independent variables, we engage in inappropriate econometric practices, leading some to sneer that "economists can prove anything with statistics" and others to dismiss the results as measurement without theory.

While the multiple regression results appear to be statistically superior to those from the simple regression model, we have not yet assessed the overall significance of the regression model. We now turn to a statistical test for addressing this concern, the F-test of overall significance. Here we will evaluate the null hypothesis that the group of partial regression coefficients estimated by the model jointly equal zero.

17.5 Testing joint hypotheses about the regression model's overall significance: basic concepts and procedures

Evaluating the statistical significance of the overall multiple regression model means we want to determine if the adjusted R^2 value is statistically different from zero. If $\bar{R}^2$ is not statistically different from zero, then we must conclude that the overall model does not explain any of the variation in the dependent variable. In more technical terms, assessing a regression model's overall significance evaluates if the group of independent variables in the model *jointly* (that is, taken together) account for any of the variation in the dependent variable.

Let us consider the case in which the overall regression model explains none of the variation in the dependent variation. In that case, the entire variation in the dependent variable is explained by the model's residuals, and none of the variation in Y is explained by the *group* of independent variables in the model. Economists often interpret this result as indicating that none of the independent variables add explanatory power to the multiple regression model.

To operationalize the test of the multiple regression model's overall statistical significance, we will evaluate a pair of hypotheses about the partial regression coefficients ($b_1, b_2, \ldots, b_k$) taken together. (Note that the test only includes the partial regression coefficients; we do *not* include the intercept estimate, b_0.) This means that we will test the following pair of hypotheses:

$$H_0 : b_1 = b_2 = \cdots = b_k \quad \text{versus} \quad H_a : \text{at least one } b_i \neq 0$$

If we cannot reject this null hypothesis, it means we must conclude that all partial regression coefficients equal zero. That is, it indicates that the group of independent variables in the model do not help explain the variation in the model's dependent variable.

Rejecting the null hypothesis does not, however, mean that *all* independent variables contribute to the model's explanatory power. As we see in the alternative hypothesis, rejecting the null hypothesis means only that *at least one* partial regression coefficient is not equal to zero. To determine which ones are not equal to zero, we must perform another series of hypothesis tests to evaluate which partial regression coefficients are statistically different from zero. (We will discuss this step further in Section 17.7.)

Let us now complete the steps for testing the multiple regression model's overall significance. After selecting a level of significance for the test, again generally 1 percent or 5 percent, we next establish a decision rule for evaluating the statistical significance of the adjusted R^2 statistic. Before doing so, we must introduce a new statistic that reflects the distributions of the variances of the sums of squares associated with the OLS method. It can be shown that these variances are normally distributed, and, as you may recall from your statistics course, a ratio of normally distributed variances will follow an F-distribution. This means that we will use an **F-statistic** to test the regression model's overall significance:

$$F\text{-stat} = \frac{\text{RSS}/k}{\text{ESS}/(n-k-1)} = \frac{[\sum(\hat{Y} - \bar{Y})^2]/k}{(\sum e^2)/(n-k-1)} = \frac{\text{regression MSE}}{\text{error MSE}} \qquad (17.5)$$

Let us now note several features of the F-statistic.

- The F-statistic is the ratio of two variances, or mean square errors:
  - the regression deviation (sum of squares) (RSS) divided by the number of independent variables (k), the degrees of freedom associated with the regression deviation. This statistic is also known as the **regression mean square error**;
  - the error deviation (sum of squares) (ESS) divided by ($n - k - 1$), the degrees of freedom for the model's residuals. This statistic is also known as the **error mean square error**.

- The F-statistic has two sets of degrees of freedom associated with it. The k degrees of freedom in the statistic's numerator correspond to the number of independent variables in the model, while the ($n - k - 1$) degrees of freedom in the statistic's denominator correspond to the number of "free" observations available after estimating the sample means for the dependent and all independent variables. We use both degrees of freedom to obtain both the critical F-value and the test statistic for F.
- If the regression coefficients *jointly* equal zero, the regression mean square error (regression MSE) will equal zero. This then yields an F-test statistic of zero, which indicates that the regression model does not explain any of the variation in the dependent variable.
- The F-distribution has no upper limit since the underlying probability distribution is asymptotic. This means that we cannot conclude that one regression model is "more significant" than another because it has a higher F-statistic.

- We can construct an F-statistic for the simple regression model, but generally do not. When we only have one independent variable in the regression model, we can use the two-tail statistical test on the model's slope coefficient to evaluate the model's overall significance. If we cannot reject the null hypothesis that the slope coefficient equals zero, we have obtained the same result as we would from the F-statistic.

Let us now formally establish the decision rule for the null hypothesis regarding the multiple regression model's overall significance:

$$\text{reject } H_0 \text{ if } F > F^{\alpha}_{k,n-k-1}$$

where F is the test statistic we will obtain from the estimated regression model, and $F^{\alpha}_{k,n-k-1}$ is the critical value. (We can obtain the critical value using Excel's FINV function. See "Using Excel to determine the critical value associated with the F-distribution" at the end of the chapter.)

We next construct the F-test statistic, the necessary components for which are found in the **analysis of variance (ANOVA)** results. Generated by statistical software packages as part of the regression results, the ANOVA table typically includes the following statistics:

- the three sums of squares (SS) associated with the regression model, i.e. RSS, ESS, and TSS;
- the degrees of freedom associated with the three sums of squares, i.e. k degrees of freedom associated with the regression sum of squares, $(n - k - 1)$ degrees of freedom associated with the error sum of squares, and $(n - 1)$ degrees of freedom associated with the total sum of squares;
- the mean square errors (MSE) for the regression and error sums of squares, which, as noted, are calculated by taking the corresponding sums of squares values and dividing each by the appropriate degrees of freedom;
- the F-test statistic generated by equation (17.5), i.e.

$$F = \frac{\text{regression MSE}}{\text{error MSE}}$$

- the p-value associated with the F-test statistic.

Once the F-test statistic is calculated, we can evaluate the decision rule about the multiple regression model's overall significance. Figure 17.3 presents the two possible outcomes of the F-test, along with the general interpretation of each set of statistics.

If we can reject the null hypothesis regarding the model's overall significance, we conclude that at least one of the independent variables adds explanatory power to the model. If we cannot reject the null hypothesis regarding the model's overall significance, we conclude that none of the independent variables help explain the variation in the dependent variable. Thus we must return to the drawing board, and reconsider what factors might affect the variation in the dependent variable.

	Result	Decision	Interpretation of decision	p-value test*	Interpretation of p-value
One-tail test for *positive* relationship					
Statistically significant *positive* relationship between Y and X_i	$t_{b_i} > t^\alpha_{n-k-1}$	Reject H_0	We are at least $[(1-\alpha) \times 100]$ percent confident that we can reject the null hypothesis that b_i is less than or equal to 0	$\alpha > p/2$	We are $[(p/2) \times 100]$ percent confident that we can reject H_0
Statistically *insignificant positive* relationship between Y and X_i	$t_{b_i} \leq t^\alpha_{n-k-1}$	Cannot reject H_0	We cannot, with $[(1-\alpha) \times 100]$ percent confidence, reject H_0	$\alpha \leq p/2$	
One-tail test for *negative* relationship					
Statistically significant *negative* relationship between Y and X_i	$t_{b_i} < t^\alpha_{n-k-1}$	Reject H_0	We are at least $[(1-\alpha) \times 100]$ percent confident that we can reject the null hypothesis that b_i is greater than or equal to 0	$\alpha > p/2$	We are $[(p/2) \times 100]$ percent confident that we can reject H_0
Statistically *insignificant negative* relationship between Y and X_i	$t_{b_i} \geq t^\alpha_{n-k-1}$	Cannot reject H_0	We cannot, with $[(1-\alpha) \times 100]$ percent confidence, reject H_0	$\alpha \leq p/2$	
Two-tail test for relationship					
Statistically significant relationship between Y and X_i	$|t_{b_i}| > t^\alpha_{n-k-1}$	Reject H_0	We are at least $[(1-\alpha) \times 100]$ percent confident that we can reject the null hypothesis that b_i is equal to 0	$\alpha > p$	We are $[p \times 100]$ percent confident that we can reject H_0
Statistically *insignificant* relationship between Y and X_i	$|t_{b_i}| \leq t^\alpha_{n-k-1}$	Cannot reject H_0	We cannot, with $[(1-\alpha) \times 100]$ percent confidence, reject H_0	$\alpha \leq p$	

Figure 17.2 Possible outcomes of hypothesis tests for partial regression coefficients.

*Most software packages, including Excel, generate a two-tail p-value. When conducting a one-tail test, we must divide the p-value by 2 before comparing it with α.

Table 17.5 ANOVA results for the IMR multiple regression model

ANOVA	df	SS	MS	F	Significance F
Regression	4	137,600.26	34,400.06	128.53	5.90E-47
Residual	147	39,342.27	267.63		
Total	151	176,942.53			

17.6 Testing joint hypothesis about the IMR multiple regression model's overall significance

We now return to the IMR multiple regression model to evaluate its overall significance. For this model, the null hypothesis to be tested is

$$H_0 : b_{\text{ATTEND}} = b_{\text{ANEMIA}} = b_{\text{H}_2\text{O}} = b_{\text{GNI}} = 0$$

while the alternative hypothesis is

$$H_a : \text{at least one } b_i \neq 0$$

The decision rule for the test at the 1 percent level of significance is

$$\text{reject } H_0 \text{ if } F > F_{4,147}^{0.01} = 3.4494$$

As Excel's ANOVA results presented in Table 17.5 show, the test statistic for F equals 128.53, a value clearly greater than the critical F-value of 3.4494. This means we can, with at least 99 percent confidence, reject the null hypothesis that all four of the partial regression coefficients equal zero. The corresponding p-value of 5.90E-47 confirms this decision, indicating that we are virtually 100 percent confident that the overall IMR multiple regression model is statistically significant.

Let us now consider what it means to have rejected the null hypothesis that $b_{\text{ATTEND}} = b_{\text{ANEMIA}} = b_{\text{H}_2\text{O}} = b_{\text{GNI}} = 0$. In statistical terms, rejecting the null hypothesis means that *at least one* partial regression coefficient (b_i) is *not* equal to zero. In economic terms, it means that *at least one* independent variable adds explanatory power to the model. However, we have not yet determined which of the four independent variables contribute to the explanatory power of the regression model; it could be just one, two, three, or all four variables. This leads us to one more round of hypothesis tests to identify which independent variables are statistically related to the dependent variable.

17.7 Is each independent variable statistically related to the dependent variable?

To determine which independent variables have a statistical relationship with the dependent variable, we utilize a two-tail hypothesis test to evaluate if the partial regression coefficients are statistically different from zero. Procedurally, we first establish the following pair of hypotheses for each coefficient:

$$H_0 : b_i = 0 \quad \text{versus} \quad H_a : b_i \neq 0$$

Table 17.6 Results for evaluating the statistical relationship between infant mortality rates and each independent variable

	$\hat{b}_i$	t-statistic	Decision rule	Can we reject H_0: $b_i = 0$ at the 1 percent level?	p-value	How confident in rejecting H_0? $(1 - p\text{-value}) \times 100\ (\%)$
b_{ATTEND}	−0.2482	−2.6926	Reject H_0 if $|t| > 2.6097^*$	Yes	0.0079	99.2087
b_{ANEMIA}	−0.8529	−6.6422	Reject H_0 if $|t| > 2.6097$	Yes	0.0000	100.0000
$b_{\text{H}_2\text{O}}$	0.3813	2.7001	Reject H_0 if $|t| > 2.6097$	Yes	0.0077	99.2255
b_{GNI}	−4.2894	−2.6087	Reject H_0 if $|t| > 2.6097$	No	0.0100	98.9973

Data source: *The critical value used in each decision rule corresponds to $t_{147}^{0.01}$.

If we can reject H_0, we conclude that there is a statistically significant relationship between Y and X_i, which means the independent variable does contribute to explaining the variation in the dependent variable. In terms of the model's specification, this evidence directs us to retain that independent variable in the model.

Table 17.6 presents the necessary statistics for evaluating the two-tail tests for each partial regression coefficient in the IMR model. According to our results, we can reject, at the 1 percent level, the null hypotheses associated with three of the four independent variables in the IMR model – the number of skilled medical personnel (ATTEND), the percentage of pregnant women with anemia (ANEMIA), and the percentage of the population with improved water access (H₂O). However, we are only 98.9973 percent confident that the natural log of GNI per capita helps to explain the variation in IMR.

You might wonder why we do not first conduct two-tail tests on the partial regression coefficients. The reason? One-tail tests reflect the economic theory underlying the econometric model, which is always our primary consideration. By first evaluating if the theoretical relationships are statistically significant, we emphasize the relationships between the dependent variable and each independent variable over how well the model fits the data.

What is our next step? The statistical results suggest that we remove the statistically insignificant independent variable, ln(GNI), since it does not, at the 1 percent level, contribute to the model's explanatory power. We then rerun the regression model without the offending variable, and compare those results with the original model to determine if one model is "better" both statistically and economically than the other. But before we do so for the infant mortality example, let us first discuss the general procedures associated with determining a "best" multiple regression model.

17.8   The "best" statistical multiple regression model

Choosing the multiple regression model that provides the most compelling statistical evidence for explaining the variation in the dependent variable requires us to complete several steps in a particular order. We always begin by examining the sign obtained on each partial regression coefficient to see if it reflects the underlying economic theory. We next evaluate each coefficient's t-statistic and p-value to determine its statistical significance in the context of that theory. We then evaluate the overall significance of each model using the F-statistic

	Result	Decision	Interpretation of decision	p-value test*	Interpretation of p-value
Statistically significant relationship between Y and the joint set of independent variables	$F > F^{\alpha}_{k,n-k-1}$	Reject H_0	We are at least α percent confident that we can reject the null hypothesis, which indicates that the adjusted R^2 is statistically different from zero	$p > \alpha$	We are ($p \times 100$) percent confident that we can reject H_0
Statistically *insignificant* relationship between Y and the joint set of independent variables	$F \leq F^{\alpha}_{k,n-k-1}$	Cannot reject H_0	We cannot, with α percent confidence, reject H_0	$p \leq \alpha$	We are *only* ($p \times 100$) percent confident that we *can* reject H_0

Figure 17.3 Possible outcomes of hypothesis tests for the overall significance of the multiple regression model.

and its *p*-value. The last round of tests examines if each partial regression coefficient adds explanatory power to the regression model.

As indicated in Figure 17.4, the most important consideration when evaluating any econometric model is whether the partial regression coefficients exhibit the signs posited by economic theory. Thus, for each model, we first compare the sign obtained on each partial regression coefficient with the sign hypothesized by the economic theory justifying that independent variable's inclusion in the model. Not obtaining a hypothesized sign raises an immediate red flag that could ultimately lead us to remove that variable from the model.

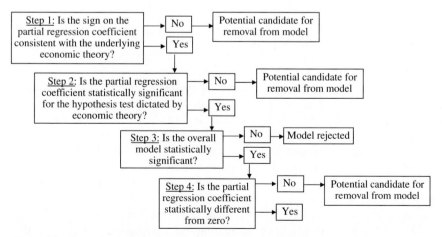

Figure 17.4 Statistical tests and possible outcomes for the multiple linear regression model.

For those partial regression coefficients that do exhibit the hypothesized signs, we evaluate their statistical significance with respect to the underlying economic theory. That is, we conduct the one-tail hypothesis tests suggested by the theoretical relationship between the dependent and independent variables. If we can reject the null hypothesis for a partial regression coefficient, we have obtained statistical evidence that the relationship between the two variables is not merely one of chance. If we cannot reject the null hypothesis, we flag the variable as an additional candidate for possible removal from the model.

We next test each model's overall significance by conducting an F-test. If we cannot reject the null hypothesis, then our econometric model is, in essence, of no value, and we must restart our statistical analysis of the dependent variable. If, however, we can reject the null hypothesis regarding the joint significance of the independent variables, we know that at least one of the independent variables helps explain the variation in the dependent variable. We next proceed to a two-tail test on each partial regression coefficient to determine if it is statistically different from zero. If we cannot reject the null hypothesis associated with a two-tail test, we have identified an additional independent variable that we might remove from the model.

If we reach this point and the various models deliver comparable statistical results, *and* assuming the dependent variables in the various models have the same functional form (because neither R^2 nor adjusted R^2 are comparable across different functional forms), we can then compare each model's overall goodness-of-fit statistic, the adjusted R^2. That is, if, and only if, the models under consideration have demonstrated similarly "good" statistical results, can we use the values of $\bar{R}^2$ to select the "best" multiple regression model. As our discussion suggests, given the various hypothesis tests that come before this step, this statistic is perhaps the least valuable for assessing our regression results. Consequently, we would be wise to avoid the tendency to get too excited (or to be too disappointed) by the value of the adjusted $\bar{R}^2$ statistic that we obtain; nor do we want to rely too heavily on its usefulness when assessing our econometric results.

The reality of determining the "best" econometric model is rarely as straightforward as this section implies, typically involving numerous regression runs. As we shall now see for the infant mortality models, it is also common that each of the regression models we compare has both "good" and "less than good" results associated with it. Thus determining the "best" model is truly more of an art than a science.

17.9   The "better" statistical infant mortality model?

We pick up the infant mortality rate model where we left off in Section 17.6. There, we saw that the GNI variable was not, at the 1 percent level, statistically different from zero. This result leads us to omit GNI as an independent variable and rerun the regression model. To distinguish between the two models, we will refer to the multiple regression model with the four independent variables (ATTEND, ANEMIA, H_2O, and GNI) as the *original model*, and the model without the GNI variable as the *GNI-less model*.

Table 17.7 presents the results for the GNI-less model, followed by the results for the original model. We first note that the partial regression coefficients (and standard errors) for ATTEND, ANEMIA, and H_2O in the GNI-less model do not have the same values as in the original model. This demonstrates an important point, which we must remember as we evaluate the statistical results for different regression models: Any time we modify a regression equation by removing or adding another independent variable, all the estimated regression results are likely to change. Consequently, we must always re-evaluate *all* results, beginning with the coefficient signs through the *p*-values for the two-tail tests on the regression coefficients, when we modify an econometric model.

Table 17.7 also presents the hypothesis-related results from the two models. Following the steps outlined above, we first consider the signs on the partial regression coefficients to determine if they are consistent with the relationship hypothesized by economic theory. Here we find all partial regression coefficients have the theoretically expected signs. Based solely on this evidence, all independent variables will be retained in each model.

We next determine if each partial regression coefficient is statistically significant with respect to the hypothesized theoretical relationship. According to the decision rule for the *t*-statistic, we find that all coefficients in both models are statistically significant at the 1 percent level, which is confirmed by the *p*-test.

Let us now evaluate the overall significance of each model. As we see, the *F*-test statistics for both models (162.73 for the GNI-less model and 128.53 for the original model) are considerably larger than the corresponding critical *F*-values ($\alpha = 0.01$) of 3.9167 and 3.4494, respectively. As the associated *p*-values confirm, we are very confident that at least one independent variable adds explanatory power to each model. Thus we can conclude at this point that both models provide equally "good" statistical results.

We pause here to note the following. When deciding which regression model is "best," we cannot compare the relative size of the *t*-statistics or *p*-values for the same partial regression coefficient or the *F*-statistic across models to make that determination. For example, here we cannot conclude that the GNI-less IMR model is better than the original model because its *F*-statistic is larger, since the statistics underlying the *F*-statistic's construction are not comparable. Thus the only regression results we can compare across models are the estimated coefficient values and, assuming each model's dependent variable has the same functional form, the adjusted R^2 statistic.

The final set of hypothesis tests evaluate which partial regression coefficients are statistically different from zero. We have already determined that, in the original model, three of the four coefficients from the original model were statistically different from zero at the 1 percent level, but the fourth, GNI, was not. In the GNI-less model, we determine that all three partial regression coefficients are statistically different from zero at the 1 percent level. Based solely on the two-tail hypothesis tests of the partial regression coefficients, these results suggest that the GNI-less model is superior to the original model.

The last step is to compare the adjusted R^2 statistics for the two IMR models, since the statistical results obtained thus far are comparable. We also note that this comparison is feasible because the dependent variable, IMR, has the same functional form in the two models. Turning to the adjusted R^2 statistics, we observe that $\bar{R}^2$ for the GNI-less model indicates that the model explains 76.26 percent of the variation in infant mortality rates, while the $\bar{R}^2$ indicates that the original model accounts for 77.16 percent of the variation in infant mortality rates. Based solely on the $\bar{R}^2$ statistics, the original model explains more of the variation in infant mortality rates than the GNI-less model, suggesting that the original model is statistically "better" than the GNI-less model.

The various statistical results obtained for the two IMR models do not give us a clear winner. According to the partial regression coefficient results, the GNI-less model is "better" than the original model, but the adjusted R^2 statistic is higher for the original model. Given that the GNI coefficient in the original model *is* statistically different from zero at just over the 1 percent level (1.003 percent, to be precise), most economists would conclude that the original model is the better of the two, particularly since all four partial regression coefficients are statistically different from zero at a 5 percent level of significance. While

Table 17.7 Evaluating the regression results for two IMR multiple regression models

GNI-less model: $\text{IMR} = b_0 + b_{\text{ATTEND}} \times \text{ATTEND} + b_{\text{ANEMIA}} \times \text{ANEMIA} + b_{\text{H}_2\text{O}} \times \text{H}_2\text{O}$

Regression statistics

Multiple R	0.8760	Critical t-value for one-tail test of coefficients $= \pm 2.3518$
R Square	0.7674	Critical t-value for two-tail test of coefficients $= 2.6095$
Adjusted R Square	0.7626	
Standard error	16.6773	
Observations	152	

ANOVA

	df	SS	MS	F	Significance F
Regression	3	135,778.91	45,259.64	162.738	1.1670E-46
Residual	148	41,163.62	278.13		
Total	151	176,942.53			

	Coefficients	Standard error	t-statistic	p-value	$[1 - (p/2)] \times 100$	$(1 - p) \times 100$
Intercept	116.2264	12.7812	9.0936	0.0000		
ATTEND	−0.3384	0.0871	−3.8848	0.0002	99.9923	99.9846
ANEMIA	0.5749	0.1225	4.6931	0.0000	99.9997	99.9994
H_2O	−0.8825	0.1304	−6.7689	0.0000	100.0000	100.0000

Original model: $\text{IMR} = b_0 + b_{\text{ATTEND}} \times \text{ATTEND} + b_{\text{ANEMIA}} \times \text{ANEMIA} + b_{\text{H}_2\text{O}} \times \text{H}_2\text{O} + b_{\text{GNI}} \times \text{GNI}$

Regression statistics

Multiple R	0.8818	Critical t-value for one-tail test of coefficients $= \pm 2.3520$
R Square	0.7777	Critical t-value for two-tail test of coefficients $= 2.6097$
Adjusted R Square	0.7716	
Standard error	16.3595	
Observations	152	

ANOVA

	df	SS	MS	F	Significance F
Regression	4	137,600.26	34,400.06	128.53	5.9034E-47
Residual	147	39,342.27	267.63		
Total	151	176,942.53			

	Coefficients	Standard error	t-statistic	p-value	$[1 - (p/2)] \times 100$	$(1 - p) \times 100$
Intercept	148.3480	17.5729	8.4419	0.0000		
ATTEND	−0.2482	0.0922	−2.6926	0.0079	99.6044	99.2087
ANEMIA	0.3813	0.1412	2.7001	0.0077	99.6127	99.2255
H_2O	−0.8529	0.1284	−6.6422	0.0000	100.0000	100.0000
GNI	−4.2894	1.6443	−2.6087	0.0100	99.4986	98.9973

Table 17.7 Continued

	Original model	GNI-less model
1. Do the partial regression coefficients exhibit the signs hypothesized by the underlying economic theory?		
ATTEND	Yes	Yes
ANEMIA	Yes	Yes
H_2O	Yes	Yes
ln(GNI)	Yes	n/a

	Original model	GNI-less model
2. Are the partial regression coefficients statistically significant at the 1 percent level with respect to the underlying economic theory?		
ATTEND	Yes	Yes
ANEMIA	Yes	Yes
H_2O	Yes	Yes
lnGNI	Yes	n/a
3. Is the overall model statistically significant?		
	Yes	Yes
4. Are the partial regression coefficients statistically different from zero at the 1 percent level?		
ATTEND	Yes	Yes
ANEMIA	Yes	Yes
H_2O	Yes	Yes
ln(GNI)	No	n/a
5. What percentage of the variation in the dependent variable is explained by the regression model?		
	77.16	76.26

we must not return and change our choice of α, economists finesse results such as ours by reporting that the GNI coefficient is statistically significant at the 5 percent level but not at the 1 percent level.

The original model is also more consistent with the factors that economic theory suggests will explain the variation in infant mortality rates, providing further support for the original model as the "better" econometric model. However, neither model has been evaluated for the numerous statistical problems that can arise in multiple regression analysis, so we shall reserve final judgment until Chapter 18, where we examine some of those possibilities.

In practice, economists typically report all plausible results from their various regression runs, explaining, as we did above, how the statistical results compare. These runs may include alternative functional forms or different independent variables considered in an effort to obtain the model that most fully explains the variation in the dependent variable. Such practice is not inconsistent with the multiple regression steps outlined here. For example, there may be multiple independent variables that capture a particular relationship with the dependent variable, so running regressions with those alternative independent variables may be useful for understanding why the dependent variable varies across elements in the sample. A similar argument could be made for alternative functional forms in those cases where economic theory might suggest it. In the end, however, we must not let the statistical regression results *alone* govern our analysis of economic phenomena.

17.10   Closing observations about the IMR multiple regression results

Several final comments are in order here. When conducting an econometric analysis, we must never forget to provide an economic interpretation of the statistical findings. While it is easy to get caught up in the statistical considerations, we must not lose sight of why we are estimating the multiple regression model in the first place. That is, we must interpret the results in the context of the *economic* analysis at hand and comment on their economic significance as well as statistical significance.

We also saw that the adjusted R^2 statistics for the two IMR models indicate that neither model fully accounts for the variation in infant mortality rates across the 152 countries in the sample. One reason is related to data limitations. Thus it is possible (actually, it is likely) that we have omitted relevant variables from the model, even though we strive to fully specify the model prior to running the regression model. As we shall discuss briefly in Chapter 18, omitting relevant independent variables from the regression model will bias the partial regression coefficients and may lead us to erroneous conclusions about the statistical relationships between the dependent and independent variables.

Another possible explanation for the "low" adjusted R^2 statistic in the IMR models may be the result that they are cross-section models. The nature of cross-section data is such that unobservable and immeasurable differences obtain across elements in the sample. For example, we cannot reasonably expect to capture all factors that affect the variation in infant mortality across countries as economically, politically, and culturally diverse as Argentina, Botswana, and Thailand. Thus we generally expect the $\bar{R}^2$ statistics for cross-section models to be lower in value than those for time-series models.

We have also not yet evaluated our regression results for the myriad statistical problems that may occur in linear multiple regression analysis. Any of these problems could both alter our conclusions and make our statistical results less reliable. We will introduce some of these problems in Chapter 18, leaving further considerations to more advanced texts. As this very brief discussion suggests, many factors may contribute to the statistical results obtained through multiple regression analysis. Thus effective econometric analysis, like all empirical economic analysis, requires us to exercise informed judgment in all steps of the multiple regression process and to clearly and completely communicate that process, and its results, to our audience. Even when the results are not statistically significant or compatible with economic theory, we can learn something about the empirical manifestation of economic phenomena.

Summary

Multiple regression is a powerful statistical method. Widely employed by economists, it is an exciting tool to use because it allows us to make connections between economic theory and empirical observation. However, we have merely scratched the surface of how to apply this technique. As you can imagine, there are many statistical issues that can reduce the validity of our results. There are also many nonlinear models available for measured variables, and there are models for assessing relationships between a qualitative dependent variable and various independent variables. As this suggests, we can make numerous modifications to the multiple regression model so that it describes more completely the underlying economic relationships.

Time is such that we can only discuss a few of these modifications and concerns. In Chapter 18, we examine how economists use dummy variables to account for independent attribute variables that may help explain the dependent variable's variation. We will also briefly discuss the cause and effects of several major statistical problems that can arise in multiple regression analysis. We first examine multicollinearity, a problem associated with high correlation between independent variables, and then discuss omitted variable bias. We also examine patterns in the model's residuals to see if either serial correlation or heteroskedasticity are present, since both reduce the reliability of our hypothesis tests of the partial regression coefficients. While necessarily not complete, Chapter 18 will provide a glimpse of both the power and limitations associated with the very popular statistical method of multiple regression analysis.

Concepts introduced

Adjusted R^2
Analysis of variance (ANOVA)
Error mean square error
F-statistic
Mean square error
Regression mean square error

Box 17.1 Using Excel to determine the critical value associated with the F-distribution

Excel has a built-in function that determines the critical F-value used to evaluate the multiple regression model's overall significance.

In Excel 2007, that function is the **FINV** function and has the following syntax:

FINV(*probability*, $k, n - k - 1$)

where *probability* equals the selected level of significance between 0.00 and 1.00; k is the degrees of freedom associated with the regression deviation; and $n - k - 1$ is the degrees of freedom associated with the residuals.

In Excel 2010, the appropriate function is **F.INV.RT**, and its syntax is identical:

F.INV.RT(*probability*, $k, n - k - 1$).

We will, as with the other critical values used in hypothesis testing, report the critical value for the F-statistic to four decimal places.

Exercises

1. Return to the bus travel demand model from Exercise 2 in Chapter 16, and evaluate the statistical significance of these results for the linear model.

Regression statistics	
Multiple R	0.9597
R Square	0.9210
Adjusted R Square	0.9067
Standard error	742.9113
Observations	40

ANOVA

	df	SS	MS	F	Significance F
Regression	6	212,411,004	35,401,834	64.143	0.0000
Residual	33	18,213,267	551,917		
Total	39	230,624,271			

	Coefficients	Standard error	t Stat	p-value
Intercept	2744.68	2641.67	1.0390	0.3064
FARE	−238.65	451.73	−0.5283	0.6008
GASPRICE	522.11	2658.23	0.1964	0.8455
INCOME	−0.19	0.06	−3.0013	0.0051
POP	1.71	0.23	7.3972	0.0000
DENSITY	0.12	0.06	1.9543	0.0592
LANDAREA	−1.16	1.80	−0.6409	0.5260

(a) Establish the hypotheses for each partial regression coefficient. Five of the six independent variables have obvious directional relationships with the dependent variable; the sixth, INCOME, is ambiguous. Explain why we may not, *a priori*, know the expected sign on INCOME and how that ambiguity is addressed in the pair of hypotheses to be tested.

(b) Establish the decision rules for each partial regression coefficient and then test the hypotheses. Once you have reached each conclusion, describe in detail (that is, go beyond stating that you either reject or cannot reject the null hypothesis) what that decision tells us about the relationship between bus travel and the independent variable.

(c) Interpret the appropriate goodness-of-fit statistic for the model.

(d) Evaluate the model's overall significance.

(e) Given all your analysis, explain what your next step would be.

2. The double-log model for bus travel in Exercise 3 in Chapter 16 generated the following regression results.

SUMMARY OUTPUT

Regression statistics

Multiple R	0.8105
R Square	0.6570
Adjusted R Square	0.5946
Standard error	0.7370
Observations	40

ANOVA

	df	SS	MS	F	Significance F
Regression	6	34.3312	5.7219	10.5339	0.0000
Residual	33	17.9252	0.5432		
Total	39	52.2564			

	Coefficients	Standard error	t Stat	p-value
Intercept	44.72	20.750	2.1550	0.039
lnFARE	0.48	0.425	1.1200	0.271
lnGASPRICE	−1.73	2.495	−0.6946	0.492
lnINCOME	−4.85	1.047	−4.6332	0.000
lnPOP	1.69	2.696	0.6257	0.536
lnDENSITY	0.28	2.663	0.1035	0.918
lnLANDAREA	−0.82	2.713	−0.3010	0.765

(a) Discuss if, and then how, the hypotheses for each partial regression coefficient change from those established for the linear bus travel demand model.
(b) Establish the decision rules for each partial regression coefficient and then test the hypotheses. Once you have reached each conclusion, describe in detail (that is, go beyond stating that you either reject or cannot reject the null hypothesis) what that decision tells us about the relationship between bus travel and the independent variable.
(c) Interpret the appropriate goodness-of-fit statistic for the model.
(d) Evaluate the model's overall significance.
(e) Given all your analysis, explain what your next step would be.

3. Given your findings for Exercises 1 and 2 above, explain which results you can compare between the two models and which results cannot be compared.
4. Evaluate the statistical significance of the regression results for the two Keynesian consumption functions estimated in Exercise 4 in Chapter 16. Which of the two models is "better" from a statistical perspective? Which is "better" from a theoretical perspective?

18 Multiple regression analysis

Dummy variables and statistical problems

Multiple regression analysis is a powerful statistical tool that can be applied to a wide variety of economic issues. In this final chapter, we introduce a method for including independent qualitative, or attribute, variables in the multiple regression model. Such variables, known as dummy variables, allow us to account for such factors as race-ethnicity or geographical location that may help explain the variation in a dependent variable.

We conclude our discussion of multiple regression analysis by introducing several statistical problems that can arise in multiple regression models. We provide an intuitive explanation of the sources of such problems, and we examine basic detection methods for identifying their presence in the model.

18.1 Independent dummy variables: basic concepts and considerations

In the regression models discussed thus far, we have included only quantitative, or measured, independent variables. But some of the variation in a dependent variable may be related to differences in attributes across elements in the sample. For example, labor economists typically include an attribute variable for sex when estimating earnings functions to account for the empirical fact that men's earnings are, on average, higher than women's even when all quantifiable productivity-related factors between men and women are the same.

We may also find that a dependent variable varies in response to a particular identifiable event or policy change. For example, when estimating the long-run growth pattern of federal government spending on defense in Chapter 8, the level of defense spending appeared to be higher during times of war compared to periods of peace. To determine if the average amount of government expenditures during wartime was *statistically* greater than during peacetime, we could include an attribute variable for war in the model to assess this.

A popular method for incorporating both types of effects in a regression model is to use what are called **dummy variables**. Dummy variables can account for differences in the dependent variable that are separate from the effects captured by the measured independent variables. In terms of the regression "line," those differences may be manifested in the line's intercept term, the line's slope coefficient, or both components.

Let us now turn to a multiple regression model for estimating the earnings of men and women. Here we will introduce a dummy variable that distinguishes between the two groups based on the single attribute of gender. We will further assume this difference affects only the earnings function's intercept term.

To explain the variation in earnings between husbands and wives in the Mroz (1976) sample, we consider the productivity-enhancing factors of the amount of education a worker has attained (EDUC) and age (AGE) as a proxy for years of work experience, along

EARNS

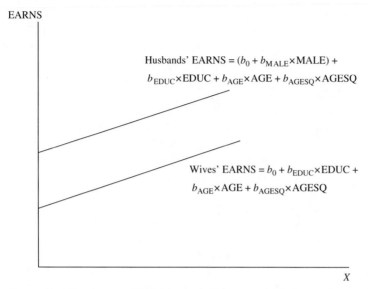

Husbands' EARNS $= (b_0 + b_{MALE} \times MALE) +$
$b_{EDUC} \times EDUC + b_{AGE} \times AGE + b_{AGESQ} \times AGESQ$

Wives' EARNS $= b_0 + b_{EDUC} \times EDUC +$
$b_{AGE} \times AGE + b_{AGESQ} \times AGESQ$

X

Figure 18.1 Earnings model with intercept dummy variable for gender.

with the square of age (AGESQ) to account for the observed decline in earnings after age 55. However, even after accounting for such differences in productivity-related factors, empirical studies have shown that men's and women's earnings are, on average, still unequal. To capture this effect, we introduce a dummy variable, MALE, to distinguish between the men and women in the sample. In terms of the regression "line," this means that we expect the earnings function for husbands to lie above that for wives, as depicted in Figure 18.1.

But we need not run separate regressions for husbands and wives. That is, we can use the MALE dummy variable to capture the different intercepts between husbands and wives by specifying the following statistical model:

$$\ln(EARNS) = b_0 + b_{MALE} \times MALE + b_{EDUC} \times EDUC + (b_{AGE} \times AGE)$$
$$+ b_{AGESQ} \times AGESQ + e$$

where MALE $= 1$ for husbands in the sample and MALE $= 0$ for wives.

Using Mroz's data, we obtain the following OLS-estimated earnings function:

$$\ln(\widehat{EARNS}) = 4.20 + 0.0840 \times EDUC + 0.1250 \times AGE - (0.0013 \times AGESQ)$$
$$+ 1.4893 \times MALE$$

Each regression coefficient has the sign predicted by economic theory, all coefficients and the F-statistic are statistically significant at the 1 percent level, and the model has an adjusted R^2 of 0.4562.[1] We also observe that an additional year of education corresponds to a 8.4 percent increase in annual earnings, while the marginal effect of age *at the sample median age of 44 years* corresponds to a 0.67 percent increase in annual earnings.[2]

What does the coefficient on the MALE dummy variable indicate? Here we must modify how we interpret the partial regression coefficient, since it now measures the *percentage difference*[3] in men's and women's earnings, *ceteris paribus*. Thus $\hat{b}_{\text{MALE}} = 1.4893$ tells us that men's earnings were, on average, 148.93 percent higher than those earned by women after accounting for the other three factors, EDUC, AGE, and AGESQ.

Let us now consider the characteristics of the MALE dummy variable. We first observe that we have two sexes in the model, men and women. Thus to define the MALE variable, we distinguish between the two groups using what is technically called a binary dummy variable. A **binary dummy variable** has a value of either one or zero, reflecting the two possible outcomes that might obtain. It will be assigned a value of one if the observation exhibits the specified characteristic, and we refer to that group of elements as the **reference group** or **benchmark group**. If an observation does not exhibit the characteristic, the dummy variable will be assigned a value of zero, with that group being labeled the **comparison group**. For the earnings function, we (arbitrarily) made men the reference group, so that MALE is assigned a value of one for all men in the sample. This then means that the MALE variable for all women in the sample is assigned a value of zero.

The second characteristic relates to which component of the regression "line," intercept or slope, we expect to differ across the two groups. Here we use an **intercept dummy variable** because we expect the entire regression line for men's earnings to lie above the regression line for women's earnings. In general, if we expect the regression model's intercept to differ between the reference group and the comparison group, we would utilize an intercept dummy variable.

In the earnings model, we also implicitly assumed that a person's gender does not interact with either the education or the age of individuals to affect earnings. If we *did* believe that the interaction between a person's gender and years of education attained influenced earnings (suppose, for example, we had evidence that husbands in the sample had chosen more lucrative college majors than their wives), we could incorporate what is known as a **slope dummy variable** to account for this differential. Constructed by multiplying the quantitative variable for education by the attribute variable for sex, the coefficient on such a variable would estimate the difference in earnings due to the interaction between education and gender.

After estimating a regression model with a binary intercept dummy variable, we can perform a statistical significance test on the dummy variable's coefficient. If we expect the intercept for the reference group to vary in a particular direction compared to that for the comparison group, we hypothesize the sign on the coefficient. If we have no expectations of direction, believing only that there may be a difference between the two groups, then we perform a two-tail hypothesis test. In the case of the earnings function, we do expect that men's earnings are higher than women's, so we test the following pair of hypotheses:

$$H_0 : b_{\text{MALE}} \leq 0 \quad \text{versus} \quad H_a : b_{\text{MALE}} > 0$$

Given the $\hat{b}_{\text{MALE}}$ coefficient's *p*-value of 1.95E-141, we are very confident that b_{MALE} is statistically greater than zero. This demonstrates that, for this sample, men's earnings were, on average, higher than women's even after accounting for differences in education and age/experience.

	Binary intercept dummy variable (D_i)	Quantitative variable (X_i)
Variable's inclusion in the model is based on …	… a previously observed empirical economic fact or known event	… economic theory
Data for the independent variable is …	… constructed by the researcher	… collected from a reliable statistical source or through reliable survey results.
Values for the independent variable are …	… either 0 or 1	… defined by the variable's range and may be either discrete or continuous.
The estimated partial regression coefficient, when the dependent variable is in its original units, measures …	… by how many units Y *differs* between the reference group and the comparison group	… by how many units Y changes for a unit change in X_i
When a directional relationship is hypothesized, the alternative hypothesis on the partial regression coefficient is …	… $b_{\text{DUMMY}} > 0$ or $b_{\text{DUMMY}} < 0$	… $b_i > 0$ or $b_i < 0$
The alternative hypothesis for a two-tail test on the partial regression coefficient is …	… $b_{\text{DUMMY}} \neq 0$, which tests if there is a statistically significant difference in the dummy coefficient for the reference and comparison groups	… $b_i \neq 0$, which tests if there is a statistically significant relationship between Y and X_i

Figure 18.2 Independent binary intercept dummy variables and independent quantitative variables.

Figure 18.2 summarizes key differences between an independent binary intercept dummy variable and a quantitative independent variable. Let us mention three additional points about binary intercept dummy variables:

- Dummy variables, unlike measured quantitative variables, are constructed by the researcher. We first determine the reference and comparison groups, and then we manually construct the dummy variable.
- When we include a binary dummy variable in the regression model, we include a dummy variable *only* for the reference group. In the above example, this meant that we included only one variable, MALE, to distinguish between the men and women in the sample. If we had included a second dummy variable for women, FEMALE, the two dummy variables, MALE and FEMALE, would be perfect linear combinations of each other. Such perfect collinearity violates one of the key assumptions associated with the OLS method, and the resulting problem, known as the **dummy variable trap**, will prevent us from estimating the regression model.
- We can use more than one dummy variable in a multiple regression model if we want to account for multiple attributes, such as sex and race. When using multiple binary dummy

variables, we must make sure that we do not inadvertently fall into the dummy variable trap. (See Exercise 1 for an example.)

Let us now return to the infant mortality rate model, where we add a binary intercept dummy variable and determine if its inclusion enhances the explanatory power of the multiple linear regression model.

18.2   Independent dummy variables: infant mortality rates and sub-Saharan Africa

It is well documented that infant mortality rates are higher in certain regions of the world due to extreme long-term poverty, the prevalence of communicable diseases such as malaria and HIV, political strife, and inadequate physical infrastructures and health-care services. Countries in sub-Saharan Africa,[4] in particular, while having seen noticeable improvements in infant mortality rates since 1970, still lag behind countries not only in northern Africa but also in the rest of the world. In 2008, the average infant mortality rate for the sub-Saharan African countries in our sample was 71.51 per 1,000 live births compared to an average IMR average of 15.39 for the remaining 108 countries in the sample.

This reality suggests that we might want to include an independent variable in the IMR model that accounts for this difference between the two groups of countries. Because we cannot include some variables, such as those for communicable diseases, due to data limitations, and others, such as the difficult-to-measure political strife, let us instead include a binary intercept dummy variable in our model for the countries in sub-Saharan Africa.

Technically, we can use this binary intercept dummy variable to determine if the average infant mortality rate in sub-Saharan Africa is statistically different from that in the remaining countries even after controlling for the four factors we identified in Chapter 16. If it is, the dummy variable's value will measure the average difference in the number of infant deaths per 1,000 live births between the two regions. Econometrically, this dummy variable will help account for those unavailable and immeasurable independent variables.

Here we will establish the countries in sub-Saharan Africa as our reference group. This means the constructed dummy variable for all sub-Saharan African countries (SSAFR) will be assigned a value of one, while the value of the SSAFR dummy variable for countries in the comparison group will be set to zero.

We next establish the pair of hypotheses for testing the partial regression coefficient on the sub-Saharan Africa dummy variable. Because we expect the infant mortality rate for countries in sub-Saharan Africa to be, on average, higher than the IMR rates for the other countries in the sample, we will conduct a one-tail hypothesis test of a positive difference between the two groups' infant mortality rates:

$$H_0 : b_{\text{SSAFR}} \leq 0 \quad \text{versus} \quad H_0 : b_{\text{SSAFR}} > 0$$

The corresponding decision rule is:

$$\text{reject } H_0 \text{ if } t_{\text{SSAFR}} > t_{n-k-1=146}^{0.01} = 2.3522$$

Here we will again use the same 1 percent level of significance as we did with the other partial regression coefficients in Chapter 17. However, the critical t-value used for all coefficients does change to reflect the additional independent variable in the model, such that we now have 146 (152 − 5 − 1) degrees of freedom.

Table 18.1 Multiple regression results for IMR model with sub-saharan Africa dummy variable

Regression statistics	
Multiple R	0.9089
R Square	0.8262
Adjusted R Square	0.8202
Standard error	14.5147
Observations	152

ANOVA

	df	SS	MS	F	Significance F
Regression	5	146,183.61	29,236.72	138.7748	1.2302E-53
Residual	146	30,758.92	210.68		
Total	151	176,942.53			

	Coefficients	Standard error	t Stat	p-value	$[1-(p/2)] \times 100$	$(1-p) \times 100$
Intercept	121.7133	16.1400	7.5411	4.56E-12		
ATTEND	−0.2266	0.0818	−2.7690	0.0064	99.68	99.36
ANEMIA	0.2635	0.1267	2.0801	0.0393	98.04	96.07
H_2O	−0.6947	0.1166	−5.9587	1.82E-08	100.00	100.00
ln(GNI)	−3.0336	1.4721	−2.0608	0.0411	97.95	95.89
SSAFR	21.6452	3.3911	6.3829	2.17E-09	100.00	100.00

We are now ready to estimate the following statistical model:

$$\text{IMR} = b_b + b_{\text{ATTEND}} \times \text{ATTEND} + b_{\text{ANEMIA}} \times \text{ANEMIA} + b_{H_2O} \times H_2O$$
$$+ b_{\text{GNI}} \times \ln(\text{GNI}) + b_{\text{SSAFR}} \times \text{SSAFR} + e \tag{18.1}$$

Table 18.1 presents the regression results obtained from Excel. What does the coefficient on the sub-Saharan Africa dummy variable tell us? The partial regression coefficient's value of 21.65 means that, on average, there were 21.65 more infant deaths per 1,000 live births in sub-Saharan Africa as compared to the remaining countries in the sample, *ceteris paribus*. That is, even after accounting for differences in the number of births attended by skilled medical personnel, the prevalence of anemia among pregnant women, access to improved water sources, and GNI per capita, countries in sub-Saharan Africa still averaged almost 22 more infant deaths per 1,000 births than the countries in the comparison group. We also see that the dummy coefficient is statistically greater than zero given its *p*-value of 1.09E-09.

The sub-Saharan Africa dummy variable's presence in the model has also changed the coefficient values for the other four independent variables. This suggests that the new independent variable is accounting for some of the variation in infant mortality that was previously (statistically) attributed to the other four independent variables. Such results are typical. As discussed earlier, any modification to an econometric model is likely to change the model's estimated parameters. In this case, none of the signs on the coefficients change, but the two coefficients on the ANEMIA and GNI independent variables are no longer significant at the 1 percent level.

Evaluating the model's other results, we see that the overall regression model is statistically significant ($F = 138.7748 > F_{k=5;146}^{0.01} = 3.1451$ and a *p*-value of 1.23E-53) at the

1 percent level. In addition, while only three of the independent variables, ATTEND, H_2O, and SSAFR, are now statistically different from zero at the 1 percent level, all five are statistically different from zero at the 5 percent level ($t_{146}^{0.05} = \pm 1.9763$). Lastly, the regression model with the sub-Saharan Africa dummy variable accounts for 82.02 percent of the variation in infant mortality rates, which is higher than the $\bar{R}^2 = 77.16$ percent from Chapter 17's original model.

Have we improved the IMR econometric model by including an independent variable that distinguishes between the two regions? Taken as a whole, the evidence supports the model containing the sub-Saharan Africa dummy variable. The statistical significance of the SSAFR coefficient, and that model's higher $\bar{R}^2$, both recommend the new model since that model explains more of the variation in infant mortality rates. Thus most economists would agree that the loss of statistical significance for two previously significant variables at the 1 percent level (but not at the 5 percent level) is outweighed by the gains from adding the sub-Saharan Africa dummy variable to the model.

The above reasoning again illustrates that econometrics is more art than science. Because our statistical evidence does not give us a clear-cut "winner" between the two multiple regression models (an outcome quite common in econometric analysis), we must weigh the benefits and costs associated with each model, and then use our informed judgment to identify the "best" model.

But before making such a determination, we must evaluate the estimated models for possible statistical problems, and correct those that we can. In other words, economists typically resolve any statistical problems before choosing the "best" model(s) and reporting their regression results.

We now turn to a brief discussion of several statistical issues that often arise in linear multiple regression models: multicollinearity, omitted variable bias, heteroskedasticity, and serial correlation. Here we will provide an intuitive explanation of each problem, leaving the technical details to dedicated econometric texts. We will also mention basic detection methods for identifying such problems.

18.3 Multicollinearity

The first statistical issue can arise when two or more independent variables are very highly correlated with each other. We previously alluded to this problem with the earnings function where we did not include two dummy variables for gender. Because FEMALE = $(1 - \text{MALE})$, FEMALE is an exact linear combination of MALE. This means that the two independent variables are perfectly collinear with each other, and including both variables in the model gives rise to the dummy variable trap.

What do we mean by **perfect collinearity**? In technical terms, it means that one independent variable is an exact *linear* function of one or more other independent variables in the model.[5] In terms of the OLS method, perfect collinearity means we cannot estimate the partial regression coefficients for those variables because the OLS coefficient estimator cannot distinguish between the differential effects of the perfectly collinear variables on the dependent variable. In other words, when two (or more) independent variables are perfectly collinear, we cannot hold the effect of one independent variable on the dependent variable constant while estimating the coefficient for the other independent variable(s).

Avoiding perfect collinearity in multiple regression analysis is largely a matter of paying attention to the independent variables included in the model. In addition to avoiding the dummy variable trap, we want to avoid using two or more independent variables that measure the same effect using different units of measurement. For example, we would not want to include real personal disposable income measured in both dollars as well as in billions of dollars in a Keynesian consumption function model. We also must not incorporate independent variables that are defined by an additive identity, such as real and nominal interest rates in a demand-for-money model since real interest rates are defined as nominal interest rates minus inflation.

Perfect multicollinearity is rare in economic analysis because it is easily identified by the researcher. However, imperfect collinearity, known as **multicollinearity**, does often obtain. Multicollinearity occurs when two or more independent variables are more highly collinear with one another than they are with the dependent variable. The presence of multicollinearity can lead to "wrong" signs on the partial regression coefficients, large standard errors (and thus low and possibly statistically insignificant t-statistics) for the partial regression coefficients, and very high $\bar{R}^2$ statistics.

Methods for detecting multicollinearity are largely beyond the scope of this text, but we will briefly examine one basic approach for a certain type of multicollinearity. If we expect multicollinearity to be a problem between any *pair* of independent variables, we can compare the correlation between the two independent variables against the correlations between each independent variable and the dependent variable. If the two independent variables are more closely correlated with each than each one is with the dependent variable, multicollinearity may generate undesirable regression results.

Suppose, for example, we want to account for a country's commitment to the health of its population in the IMR model, since we would expect a lower infant mortality rate for countries that devote more resources to maintaining their populations' physical well-being. If we include a variable for health, we then include the following pair of hypotheses and decision rule in our testing:

$$H_0 : b_{\text{HLTH}} \geq 0 \quad \text{versus} \quad H_a : b_{\text{HLTH}} < 0$$

$$\text{reject } H_0 \text{ if } t_{\text{HLTH}} < t_{152-6-1}^{0.01} = 2.6102$$

Collecting data on per-capita expenditures on health measured at the average exchange rate in U.S. dollars, we specify this new independent variable, HLTH, using natural logs, and estimate the following statistical model:

$$\widehat{\text{IMR}} = b_0 + b_{\text{ATTEND}} \times \text{ATTEND} + b_{\text{ANEMIA}} \times \text{ANEMIA} + b_{\text{H}_2\text{O}} \times \text{H}_2\text{O}$$

$$+ b_{\text{GNI}} \times \ln(\text{GNI}) + b_{\text{SSAFR}} \times \text{SSAFR} + b_{\text{HLTH}} \times \text{HLTH} + e \qquad (18.2)$$

Applying OLS, we obtain the Excel results shown in Table 18.2. We immediately see that we do not obtain the hypothesized sign on HLTH. (According to the regression results, a \$1 increase in per-capita expenditures on health corresponds to an *increase* of almost four infant deaths per 1,000 births.) We also observe that b_{GNI} no longer exhibits the positive sign postulated by economic theory (and that we obtained in previous runs), and that this coefficient is no longer statistically significant at any acceptable level. We further observe that the coefficients on ATTEND and ANEMIA are no longer statistically significant at the 1 percent level (b_{ATTEND} is statistically different from zero only at the 5 percent level, while b_{ANEMIA} is statistically different from zero only at a 10 percent level). We do, however,

Table 18.2 Multiple regression model with HLTH independent variable

Regression statistics

Multiple R	0.9143
R Square	0.8360
Adjusted R Square	0.8292
Standard error	14.1466
Observations	152

ANOVA

	df	SS	MS	F	Significance F
Regression	6	147,924.22	24,654.04	123.19	2.29E-54
Residual	145	29,018.31	200.13		
Total	151	176,942.53			

	Coefficients	Standard error	t Stat	p-value	$(1 - p) \times 100$
Intercept	80.0501	21.1431	3.7861	0.0002	
H_2O	−0.6503	0.1146	−5.6737	0.0000	100.0000
ATTEND	−0.1667	0.0823	−2.0249	0.0447	95.5289
ANEMIA	0.2264	0.1241	1.8249	0.0701	92.9921
lnGNI per capita	**−0.9470**	1.5997	**−0.5920**	**0.5548**	**44.5196**
SSAFR	19.2714	3.4017	5.6652	0.0000	100.0000
lnHLTH	**3.9563**	1.3415	2.9492	0.0037	99.6284

obtain a higher adjusted R^2 (0.8292) for this model versus 0.8202 on the model with the sub-Saharan dummy. This may seem promising until we realize that all of these results exhibit the telltale signs of multicollinearity.

The results on HLTH and GNI in particular raise a red flag about potential multicollinearity in the model. Neither coefficient has the hypothesized sign, and $\hat{b}_{GNI}$ is not statistically significant. Given the definition of multicollinearity, such results are not surprising since the amount of GNI per capita is likely to be linearly related to a country's per-capita expenditure on health.

To formally test if our suspicions are correct, let us now compare the correlation between per-capita GNI and per-capita health expenditures as a percentage of GNI with the correlation between each variable and IMR. As we see in Table 18.3, the correlation coefficient between the two variables, −0.8292,[6] falls in the strong correlation range from Figure 12.1. This correlation is also larger, in absolute value,[7] than either the correlation between IMR and GNI ($r_{IMR, GNI} = -0.7632$) or that between IMR and HLTH ($r_{IMR, HLTH} = 0.8154$). These results suggest that multicollinearity between GNI and HLTH may be an issue in our model.

Given this information, what is our next step, assuming we want to retain both independent variables in the model? The two most commonly recommended responses are to increase the

Table 18.3 Pairwise correlations between IMR, GNI, and HLTH

$r_{GNI, HLTH} = -0.8292$
$r_{IMR, HLTH} = 0.8154$
$r_{IMR, GNI} = -0.7632$

sample size or to do nothing. Because multicollinearity is really a sample problem, the first response is generally considered more desirable but it is also typically unfeasible. Data limitations often constrain the independent variables we can include in our multiple regression models, as was the case for the IMR model. In addition, economists are rarely able to expand the sample size because we typically use data collected by others. Thus the cost of drawing a new, larger sample may be prohibitive.

As a result, the second option, "do nothing," is the one usually chosen, particularly since economists do not agree that concerns about multicollinearity are warranted (see, for example, Wooldridge 2009, pp. 95–99). The reason? Many economic variables are closely correlated with one another, which could cause multicollinearity if such variables are chosen as the model's independent variables. If we have strong theoretical reasons for including a group of highly collinear independent variables in the regression model, we may then choose to accept (and acknowledge) the presence of multicollinearity despite any statistical problems that might occur. Since there is also no standard for judging what constitutes a serious multicollinearity problem, nor is there often much economists can do about its presence, some econometricians view the so-called solutions to multicollinearity as ineffective, thereby further supporting the "do nothing" option.

How, then, do we address the (apparent) multicollinearity in the infant mortality model? Given the results obtained for the various models estimated in Part V, those associated with equation (18.1) are statistically stronger than those for equation (18.2).[8] In addition, GNI is a broader measure than HLTH of the general availability of resources in a country that could be devoted to improving prenatal and natal environments. Given the statistical and theoretical issues considered thus far, we will choose equation (18.1) as the "best" IMR model.

18.4 Model misspecification and omitted variable bias

One key assumption of the ordinary least-squares method is that the regression model be fully and correctly specified. If it is not, the model is said to be misspecified, and OLS will not generate statistically unbiased partial regression coefficients.

In economics, our primary concern is to identify and then include all relevant independent variables as indicated by economic theory.[9] Omitting such variables will cause the regression model to be **underspecified**, with the partial regression coefficients exhibiting **omitted variable bias**. Technically, this means the expected value of those partial regression coefficients that are affected by the omitted variable(s) will not equal the true population parameters. In economic terms, it means our coefficient estimates will not provide an accurate measure of the relationship between the dependent variable and affected independent variables.

To understand the effects of omitted variable bias, let us reconsider two of the IMR models from Chapter 17: the original model containing four independent variables (ATTEND, ANEMIA, H_2O, and GNI), and the GNI-less model, where we omitted the GNI variable. As we showed in Table 17.7, all values on the partial regression coefficients changed when we removed GNI from the model. But did they change in a way consistent with omitted variable bias?

Let us discuss one method, outlined in Table 18.4, for assessing the potential bias associated with theoretically relevant omitted variables. While we will not address measuring the size of the bias, we will examine how we might determine its direction using the three following factors: the strength of the correlation between the omitted variable and the remaining independent variables; the direction of that correlation; and the expected correlation between the omitted variable and the dependent variable. As we begin, let us also note that the

Table 18.4 Assessing omitted variable bias in the IMR multiple regression model

Step 1: Are the remaining independent variables correlated with the omitted variable?	
Correlation coefficient ($r_{b_{\text{OMIT}}b_i}$)	*Statistically different from zero?* ($t_{152-2}^{0.01} = 2.6090$)
$\hat{b}_{\text{ATTEND}}$ $r_{\text{GNI, ATTEND}} = 0.7393$	Yes, since $t_{\text{GNI, ATTEND}} = 13.4462$
$\hat{b}_{\text{H}_2\text{O}}$ $r_{\text{GNI, H}_2\text{O}} = 0.7579$	Yes, since $t_{\text{GNI, H}_2\text{O}} = 14.2282$
$\hat{b}_{\text{ANEMIA}}$ $r_{\text{GNI, ANEMIA}} = -0.7693$	Yes, since $t_{\text{GNI, ANEMIA}} = -14.7462$

Step 2: Do we obtain the expected sign on the correlation coefficient between each remaining independent variable and the omitted variable?			
	Expected correlation with GNI?	*Expected sign on r?*	*Statistically greater/less than zero?* $t_{152-2}^{0.01} = 2.3515$
$\hat{b}_{\text{ATTEND}}$	Positive	Yes	Yes, since $t_{\text{GNI, ATTEND}} = 13.4462$
$\hat{b}_{\text{H}_2\text{O}}$	Positive	Yes	Yes, since $t_{\text{GNI, H}_2\text{O}} = 14.2282$
$\hat{b}_{\text{ANEMIA}}$	Negative	Yes	Yes, since $t_{\text{GNI, ANEMIA}} = -14.7462$

Step 3: What is the expected relationship between the omitted variable and the dependent variable, and is it statistically significant?		
Expected correlation between X_{OMIT} *and Y?*	*Expected sign on r?*	*Statistically greater/less than zero?* $t_{152-2}^{0.01} = 2.3515$
Negative	Yes ($\hat{r}_{\text{GNI, IMR}} = -0.7575$)	Yes $t_{\text{GNI, IMR}} = -14.2098$

Step 4: What is the direction of the expected bias regarding equation (18.3)? The direction of expected bias in $\hat{b}_i = $ sign on $b_{\text{omitted variable}} \times$ sign on $r_{b_i * b_{\text{OMIT}}}$.	
Direction of expected bias in $\hat{b}_i$?	
$\hat{b}_{\text{ATTEND}}$	Negative: $(-) \times (+) = -$
$\hat{b}_{\text{H}_2\text{O}}$	Negative: $(-) \times (+) = -$
$\hat{b}_{\text{ANEMIA}}$	Positive: $(-) \times (-) = +$

Step 5: Do we observe the predicted change in the remaining partial regression coefficients when we omit a relevant independent variable from the model?				
	Original model model	*GNI-less model*	*Difference in value*	*Consistent with predicted bias?*
$\hat{b}_{\text{ATTEND}}$	−0.2482	−0.3384	−0.0902	Yes
$\hat{b}_{\text{H}_2\text{O}}$	−0.8529	−0.8825	−0.0297	Yes
$\hat{b}_{\text{ANEMIA}}$	0.3813	0.5749	0.1936	Yes

following method can only be used if we know which independent variable(s) has been omitted from the model.

We first determine if each of the remaining independent variables are correlated with GNI. (This step is important, since such correlation is required for the bias to affect the model's coefficients.) In the infant mortality rate model, it is likely that the number of births attended by skilled medical personnel, access to safe water sources, and the prevalence of anemia among pregnant women are all related, to varying degrees, to the per-person resources available in a country. This leads us to anticipate that each of the remaining independent variables will be correlated with GNI, which is confirmed by the pairwise correlation coefficients presented in step 1 of Table 18.4. There we also observe that, for each pair of variables, the correlation falls into the strong range ($r > 0.50$) defined in Table 12.1, and all correlation coefficients are statistically different from zero.

In step 2, we consider the direction of the correlation between each remaining independent variable and the omitted GNI variable. Economic reasoning leads us to expect that higher GNI per capita will likely correspond to more health professionals being available and better access to safe water, but a lower prevalence of anemia among pregnant women who would have more potential resources available for prenatal care. As we observe, we do obtain all expected signs on the correlation coefficients, and each is statistically significant at the 1 percent level.

The third consideration is the expected correlation between the omitted variable and the dependent variable. As we recall, economic theory led us to hypothesize a negative relationship between GNI and the dependent variable, IMR, such that we expect $\hat{r}_{GNI, IMR}$ to be negative. As shown in step 3, this correlation coefficient is both negative and statistically significant.

Having established that the omitted variable is correlated with the remaining independent variables, we can use equation (18.3) to determine the direction of the expected bias in a partial regression coefficient for the remaining variable:

$$\text{direction of expected bias in } \hat{b}_i = \text{sign on } b_{\text{omitted variable}} \times \text{sign on } r_{b_i * b_{\text{OMIT}}} \qquad (18.3)$$

As we see in step 4 of Table 18.4, the expected sign of the omitted variable bias is positive for the ATTEND and H_2O variables and negative for the ANEMIA variable. We also observe, in step 5, that all three actual outcomes are consistent with how the coefficients on the remaining variables are predicted to change when we omit GNI from the model. Thus, based on our method, omitting GNI from the IMR regression model biases the remaining partial regression coefficients. As a result, we choose to retain the GNI variable in the IMR model. (We also recall that, when GNI is included, the regression model's adjusted R^2 statistic rises, further supporting GNI's inclusion.)

Given our earlier discussion about the high correlation between many economic variables, we might correctly expect that omitted variable bias will be present in many of our multiple regression models. As a rule, economists accept this bias if it does not change the expected relationship between the dependent variable and each remaining independent variable. And even when the expected bias does alter those relationships, we may not be able to correct for it if the variable was omitted because of data constraints or because the omitted variable is unmeasurable.

Omitting variables from the multiple regression model, as we now know, is sometimes a function of data being unavailable for our analysis. The lack of data for theoretically relevant variables may then lead us to less than desirable choices about those variables we can actually include in the econometric model. In the case of the infant mortality rate model, for example, we noted that data constraints prevented us from including such potentially important independent variables as the percentage of women receiving adequate prenatal care and the prevalence of communicable diseases. While we might consider using equation (18.3) to identify the direction of any potential bias, the lack of data for those variables precludes us from applying the detailed procedure outlined above.

Another reason why we may omit important independent variables from the regression model is that those variables are not measurable. A good example occurs for the earnings function estimated in Section 18.2. While education and experience are predicted to improve an individual worker's productivity, neither variable will necessarily capture a person's capacity to learn or the level of work commitment brought to a job, both of which would increase worker productivity. But it is unlikely we can find a dataset that contains measures of such factors, again leaving us to omit relevant variables from the econometric model.

Given the repercussions of omitting relevant independent variables from our model, which not only biases the remaining coefficients, but also includes the loss of the omitted variable's coefficient, we always strive to specify our multiple regression model as fully as possible when undertaking an econometric analysis. However, because omitted variables are likely to be an unavoidable fact in econometric analysis, economists typically discuss all potentially relevant independent variables when specifying the theoretical model to be estimated, and then follow that discussion with an explanation of which variables were omitted and the reason(s) why.

18.5   "Misbehaved" regression residuals: heteroskedasticity and serial correlation

Another pair of key assumptions underlying the OLS method pertain to the regression model's residuals. Specifically, the OLS technique assumes both that the variance of the model's residuals is equal across all observations and that no correlation exists between a model's residuals. If the first condition does not hold, then the model is said to exhibit heteroskedasticity, while violation of the second condition results in serial correlation. We now provide a brief description of each problem and demonstrate how we might use scatter diagrams to examine if either problem obtains in our models.

Heteroskedasticity is largely a cross-section problem that results from differences in the variation across economic elements of different sizes. Its name is derived from the OLS assumption that the regression model's residuals have a constant, or **homoskedastic**, variance across all observations. If the variances of the residuals are not homoskedastic, they are said to be **heteroskedastic**.

Typically, heteroskedasticity results from a wide dispersion between the minimum and maximum values of the dependent variable, which in turn leads to larger residuals for some observations. For example, suppose we wanted to estimate family expenditures on food. We would expect to see relatively little variation in the dollar amount spent on food in families with low levels of income, since their primary concern will be meeting basic nutritional needs. However, as family income increases, families of gourmands may devote increasing portions of their budgets to food expenditures, while other families, whose tastes run to activities other than eating, may be more frugal. Thus we would observe greater variation in food expenditures for families with high levels of income compared to families with low income levels. If this does, in fact, obtain, then we might expect to have heteroskedastic errors in an OLS-estimated regression model of food expenditures across families.

Heteroskedasticity does not bias the partial regression coefficients, so it will not affect the estimated coefficient values, nor does it affect our interpretation of the adjusted R^2 statistic. However, its presence does bias the variances of the model's partial regression coefficients. While it is not possible to know, *a priori*, if the bias will be positive (leading OLS to overestimate the coefficients' true variances) or negative (such that OLS underestimates the coefficients' true variances), we do know that hypothesis tests based on the t- and F-distributions are no longer statistically reliable. This, in turn, may result in misleading conclusions about the statistical significance of both the model and its coefficients.

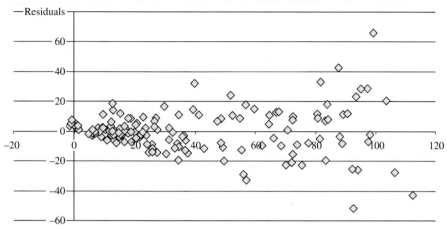

$$\widehat{\text{IMR}} = \hat{b}_0 + \hat{b}_{\text{ATTEND}} * \text{ATTEND} + \hat{b}_{\text{ANEMIA}} * \text{ANEMIA} + \hat{b}_{\text{H}_2\text{O}} * \text{H}_2\text{O} + \hat{b}_{\text{GNI}} \ln \text{GNI} + \hat{b}_{\text{SSAFR}} * \text{SSAFR}$$

Chart 18.1 Scatterplot of residuals (Y-axis) and predicted IMR (X-axis).

We can acquire an approximate idea of heteroskedasticity's presence by plotting the regression model's residuals against the estimated, or predicted, values of the dependent variable ($\widehat{Y}$). If the residuals exhibit a pattern that is either expanding or contracting for larger values of the predicted dependent variable, exhibiting a fan shape, this suggests that the model's residuals are heteroskedastic.

Chart 18.1 (p. 417) plots the residuals for the infant mortality model estimated by equation (18.1) against $\widehat{\text{IMR}}$. It clearly demonstrates that the IMR model's residuals are more widely dispersed at higher predicted values of infant deaths per 1,000 live births. This, then, suggests that heteroskedasticity is present in the infant mortality rate model's residuals, and if confirmed by formal testing, may lead us to different conclusions regarding the statistical significance of the regression coefficients or the model's overall significance.

Software dedicated to statistical analysis can easily correct for heteroskedasticity, but that is not the case with Excel. Consequently, we can only mention the possible presence of heteroskedasticity if it is suggested by the scatter diagram, and indicate that our decisions regarding the null hypotheses may be biased. In the case of the IMR model, for example, we would include Chart 18.1 in our discussion, and indicate that our conclusions based on the OLS-estimated coefficients may be unreliable.

A second problem associated with "misbehaved" residuals occurs when the residuals are serially correlated. Also known as **autocorrelation**, the problem of **serial correlation** arises primarily in time-series models when a particular period's residual is in some way dependent on the residuals from recent previous periods. Because serial correlation is best demonstrated graphically, we now turn to a well-known time-series model in macroeconomics, the Keynesian consumption function, for evidence of autocorrelated residuals.

The Keynesian consumption function theorizes what economic factors might lead to changes in personal consumption expenditures over time. While the theory's originator, John Maynard Keynes, focused primarily on the role of income, subsequent economists have included additional variables in their econometric Keynesian consumption functions. For our purposes, we will estimate a Keynesian consumption function for the United States in which real monthly per-capita personal consumption expenditures (PCE) are a function of real per-capita personal disposable income (DPI), Three-Month Treasury Bill rates (3MTBILL), and real per-capita consumer credit outstanding (CREDIT). Here we expect a positive relationship between PCE and DPI and between PCE and CREDIT, but a negative one between PCE and 3MTBILL, which we use to measure the opportunity cost of spending over saving. Thus the statistical function to be estimated is

$$PCE_t = b_0 + b_{DPI} \times DPI_t + b_{3MTBILL} \times 3MTBILL_t + b_{CREDIT} \times CREDIT_t + e_t$$

Using U.S. data for the period of January 1982 through April 2004, we obtain the following estimated Keynesian consumption function:

$$\widehat{PCE}_t = -3795.44 + 1.00 \times DPI_t - 62.36 \times 3MTBILL_t + 0.46 \times CREDIT_t$$

Table 18.5 presents the complete regression results. Let us interpret the coefficient results. We first observe that a $100 increase in real per-capita personal disposable income corresponds to a $100 increase, on average, in real per-capita personal consumption expenditures, *ceteris paribus*, while a $100 increase in real per-capita consumer credit outstanding corresponds to a $46 increase, on average, in real per-capita personal consumption expenditures, *ceteris paribus*. We also see that, when the opportunity cost of spending, as measured by the lowest-risk savings instrument, Three-Month Treasury Bills, increases by one percentage point, real per-capita personal consumption expenditures decrease, on average, by $62.36, *ceteris paribus*. We further observe that all three coefficients are statistically significant at the 1 percent level, while the model itself is also statistically significant and accounts

Table 18.5 Multiple regression results for Keynesian consumption function

$PCE = b_0 + b_{DPI} \times DPI + b_{3MTBILL} \times 3MTBILL + b_{CREDIT} \times CREDIT$

Regression statistics

Multiple R	0.9970
R Square	0.9941
Adjusted R Square	0.9940
Standard error	299.2873
Observations	268

ANOVA

	df	SS	MS	F	Significance F
Regression	3	3,960,522,479.37	1,320,174,159.79	14,738.55	0.0000
Residual	264	23,647,238.00	89,572.87		
Total	267	3,984,169,717.37			

	Coefficients	Standard error	t Stat	p-value
Intercept	−3795.44	449.7454	−8.4391	0.0000
Real per-capita DPI	1.00	0.0243	41.0954	0.0000
Three-month treasury bill rates	−62.36	11.3796	−5.4796	0.0000
Real per-capita consumer credit outstanding	0.46	0.0995	4.6154	0.0000

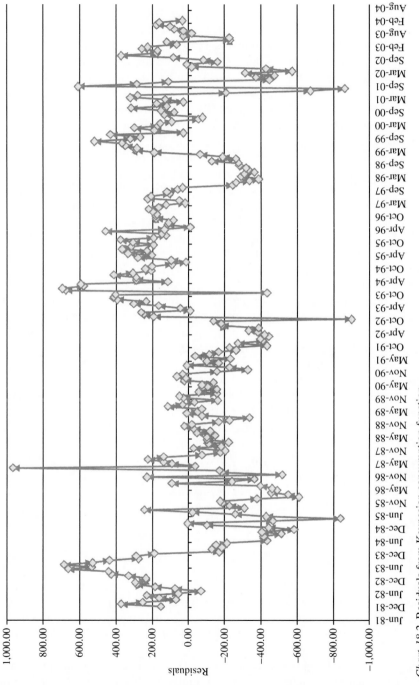

Chart 18.2 Residuals from Keynesian consumption function.

for 99.40 percent of the variation in real per-capita personal consumption expenditures. However, the high value of the adjusted R^2 statistic can be a signal that serial correlation is present in the model.

Let us now examine the Keynesian consumption function's residuals, which we plot against the sample's observations in chronological order. To detect possible serial correlation, we look for strong correlation between successive values of the residuals. As we observe in Chart 18.2 (p. 419), the Keynesian consumption function's residuals clearly exhibit a distinctive pattern over consecutive time periods. We see, for example, that the model's residuals decline in varying degrees each successive month between August 1983 and July 1984. Thus if we knew the residual for, say, December 1983, we could "predict" that the residual for January 1984 will be lower in value. Such a result violates the assumption of a random error term that is crucial to the OLS method, and it has the effect of biasing the OLS formulas for estimating the variances of the regression coefficients.

What are the consequences of serial correlation? Like heteroskedasticity, serial correlation does not bias the partial regression coefficients, so its presence will not affect the estimated coefficient values. Serial correlation does, as noted, bias the coefficient's variances, such that they are no longer efficient (minimum variance) estimates. With serial correlation, this bias is often (but not always) negative, which leads OLS to underestimate the variances, and thus the standard errors, of the regression coefficients. In such cases, the t-statistics generated by the OLS method will be overestimated when serial correlation is present, and that may lead us to commit a Type I error by incorrectly rejecting a true null hypothesis about the partial regression coefficient.

Formal tests for serial correlation are beyond the scope of this text. Thus, in our analyses, we will utilize the scatter diagram of our model's residuals to see if it suggests the presence of serial correlation. If it appears that autocorrelation is present, we report this possibility in our findings as we discuss our (uncorrected) regression results.

Because serial correlation is largely a time-series problem, it can be attributed, in part, to the fact that many measured economic variables exhibit similar patterns of change over time. This tendency among variables not only can lead to serially correlated residuals, but also has the further effect of reducing the model's error sum of squares, which in turn can raise the model's adjusted R^2 statistic. Thus the higher adjusted R^2 statistics that typically accompany time-series regression models may be less reliable measures of a model's overall goodness of fit than we would like. That is, some of the value of $\bar{R}^2$ may be attributable to the general tendency of how economic variables change over time rather than to how well the regression model captures the variation in the dependent variable. This possibility further recommends that we not place too much stock in the $\bar{R}^2$ values associated with multiple regression models.

Summary

The introduction of linear multiple regression analysis in Part V is just that – an introduction to a widely used statistical method in economics. But even though there is a great deal more to be learned about linear multiple regression analysis (and other econometric techniques), Part V does examine the basic features that all users of multiple regression analysis must master.

Chapter 16 introduced the key concepts associated with estimating and interpreting partial regression coefficients in the multiple regression framework. Economists undertake multiple regression analysis for the purpose of learning more about economic relationships, so it is critical that we fully specify our econometric models, which are grounded in economic theory, and then interpret what the estimated model actually tells us about the economy.

Chapter 17 explained how to assess the statistical significance of the partial regression coefficients so we might determine if our empirical evidence is statistically sound. We also introduced a method for evaluating the overall significance of the regression model so we might determine if the variation in the dependent variable is statistically related to the model's independent variables. We also discussed some of the considerations associated with selecting the "best" econometric model.

In Chapter 18, we have examined how attribute independent variables, in a very basic form, can be included in the multiple regression model. As we might imagine, the concept of the so-called dummy variable can be extended to account for more than binary cases as well as for slope differences and interactions between independent variables.

Chapter 18 also introduced some of the problems that can arise in multiple regression analysis as a result of violations of key assumptions associated with the ordinary least-squares estimation method. We leave it to the reader to pursue further work to learn more about these, and other more advanced, topics in econometrics.

Concepts introduced

Autocorrelation
Benchmark group
Binary dummy variable
Comparison group
Dummy variables
Dummy variable trap
Intercept dummy variable
Heteroskedasticity
Homoskedasticity
Multicollinearity
Omitted variable bias
Perfect collinearity
Reference group
Serial correlation
Slope dummy variable
Underspecified (regression model)

Exercises

1. In the late 1990s, Hirsch and Schumacher (1998) conducted a cross-sectional analysis of the effect of unions on earnings among three occupations in the health-care sector, including 38,555 registered nurses. Their model was specified as follows:

$$\ln(W) = b_0 + b_S \times S + b_E \times \text{EXP} + b_{E2} \times (\text{EXP}^2/100) + b_{\text{UNION}}$$
$$\times \text{UNION} + b_{\text{BLACK}} \times \text{BLACK} + b_M \times \text{MALE} + b_{\text{MAR}}$$
$$\times \text{MARRIED} + b_D \times \text{DSW} + b_H \times \text{HOSPITAL} + b_P$$
$$\times \text{PUBLIC} + b_{\text{MET}} \times \text{METRO} + b_{\text{PT}} \times \text{PT} + e$$

where $W \equiv$ nurse's annual earnings; BLACK $= 1$ if nurse was African American, and $= 0$ if nurse was white; DSW $= 1$ if nurse was divorced, separated, or widowed, and $= 0$ if nurse was single; EXP $\equiv$ years of potential experience, and $(EXP^2/100)$ is a term to account for diminishing earnings as one ages; HOSPITAL $= 1$ if nurse worked in a hospital, and $= 0$ if nurse worked elsewhere; MALE $= 1$ if nurse was male, and $= 0$ if nurse was female; MARRIED $= 1$ if nurse was currently married, and $= 0$ if nurse was not currently married; METRO $= 1$ if nurse worked in a metropolitan area, and $= 0$ if nurse worked outside a metropolitan area; PT $= 1$ if nurse worked part-time, and $= 0$ if nurse worked full-time; PUBLIC $= 1$ if nurse worked in the public sector, and $= 0$ if nurse worked in the private sector; $S \equiv$ years of schooling; and UNION $= 1$ if nurse was a member of a union, and $= 0$ if nurse was not a union member.

(a) Identify which of the independent variables are dummy variables and which are quantitative variables.

(b) Hirsch and Schumacher obtained the following estimates for the partial regression coefficients and their standard errors.

Results from Hirsch and Schumacher's earnings model for registered nurses

Independent variable	Coefficient value	Standard error
BLACK	−0.095	0.007
DSW	0.015	0.006
EXP	0.013	0.001
$(EXP^2/100)$	−0.027	0.001
HOSPITAL	0.159	0.004
MALE	0.028	0.008
MARRIED	0.015	0.005
METRO	0.102	0.004
PT	0.007	0.004
PUBLIC	0.019	0.004
S	0.033	0.001
UNION	0.032	0.005
$\bar{R}^2 = 0.187$		
$n = 38{,}555$		

Interpret each of the partial regression coefficients.

(c) Evaluate which of the partial regression coefficients are statistically different from zero.

(d) Union members typically earn what is called a *union wage premium* over earnings of non-union workers. Given these results, discuss if the unionized registered nurses in this sample did, in fact, earn a union wage premium.

2. Chart 8.11 depicted the long-run growth trend in U.S. defense consumption expenditures between 1929 and 1990 and included the years of World War II to demonstrate an episodic variation in the data. Go to the U.S. Bureau of Economic Analysis (at <www.bea.gov>), and retrieve the data from NIPA Table 3.10.5. (Real data is not available for this variable.)

(a) Estimate the long-run growth rate of defense expenditures.

(b) Construct a WAR binary intercept dummy variable for the years 1941–1945, and re-estimate the long-run growth rate.

(c) Compare the results, and discuss which of the two models provides a better explanation of long-run changes in nominal defense expenditures.

3. Assume we can locate a dataset of worker characteristics and earnings that includes an acceptable measure of each worker's intellectual capabilities. Explain how omitting this factor as an independent variable in an earnings function might affect the coefficients on a worker's level of education and a worker's years of experience (age).

4. Evaluate the Keynesian consumption function estimated in Exercise 4 in Chapter 16 for autocorrelation. Explain what the scatter diagram tells you about the reliability of the OLS-generated results.

Notes

1 The role of statistics in economics

1 The National Bureau of Economic Research (NBER) is the official dater of the U.S. business cycle. Not all economists agree with either date, citing the upward trend in the unemployment rate as evidence.

2 The U.S. Bureau of Labor Statistics constructs five alternative measures of labor underutilization that capture various aspects not measured by the official unemployment rate; see its Local Area Unemployment Statistics (U.S. Bureau of Labor Statistics 2010g).

3 Price stability is sufficiently important that the U.S. Congress included it as one of the U.S.'s primary macroeconomic goals, two others of which are full employment and economic growth. See the *Economic Report of the President* (1947) for more information on how the three macroeconomic goals were established in the United States.

4 In the long run, however, the causal "trigger" of fewer workers available may lead to less output and slower long-term economic growth, which suggests a positive relationship between the two variables.

5 Recall that *ceteris paribus* means "all other things held constant." For these results, this means that the estimated slope is calculated based on the assumption that no other factors potentially affecting the relationship between state economic growth and unemployment have changed.

2 Visual presentations of economic data

1 This is an updated version of chart 3 in Fuglie *et al.* (2007).

2 Basic methods for estimating a variable's secular trend are discussed in Chapter 8. Chart 2.3 uses a more advanced method, the Hodrick–Prescott filter, following Fuglie *et al.* (2007).

3 Economists do not agree about which factors influenced these results. See, for example, Bernanke (2010) and Elliott and Baily (2009).

4 Industries are classified using the North American Industrial Classification System (NAICS). According to the NAICS Frequently Asked Questions (U.S. Census Bureau 2010e), "NAICS is a two- through six-digit hierarchical classification system, offering five levels of detail. Each digit in the code is part of a series of progressively narrower categories, and the more digits in the code signify greater classification detail. The first two digits designate the economic sector, the third digit designates the subsector [not used for educational services], the fourth digit designates the industry group, the fifth digit designates the NAICS industry, and the sixth digit designates the national industry."

5 An excellent example of this distortion is found in the work of William Playfair, a Scottish political economist who is generally credited with being the first to use charts to illustrate quantitative information. In "Chart of the National Debt of Britain from the Revolution to the End of the War with America" – see Friendly (2010) for the actual image – Playfair presented the (nominal) dollar value of British debt along a tall vertical axis and time on a shortened horizontal axis, which suggested a huge increase in Britain's national debt. Playfair then argued that Britain's government spending on foreign wars was out of control. To his credit, he corrected this distortion in subsequent work.

3 Observations and frequency distributions

1 As we shall discuss in Part III, we can also use a simple random sample of families.
2 "White," "black," and "Asian" are three race categories historically used by the U.S. Census Bureau. Beginning with the 2000 Decennial Census, the race categories were expanded to account for multiracial individuals.
3 Scholars disagree about what differentiates the interval and ratio scales. See, for example, Balnaves and Caputi (2001) or Cliff (1996) for more on the debate about measurement scales developed by Stanley Smith Stevens (1946). However, scholars do generally agree that the scale distinction for quantitative variables is less an issue in the social sciences than in the physical sciences, so we will not engage further in the debate.
4 Monthly housing costs average the monthly costs of all of the applicable following items: electricity, gas, fuel oil, other fuels (e.g. wood, coal, kerosene, etc.), garbage and trash, water and sewage, real estate taxes, property insurance, condominium fees, home-owner's association fees, mobile home park fees, land or site rent, other required mobile home fees, rent, mortgage payments, home equity loan payments, other charges included in mortgage payments, and routine maintenance.
5 The data was obtained from DataFerrett. See Box 3.1 "Where can I find . . . ? DataFerrett" at the end of the chapter for more details.
6 Beginning with the 2000 Decennial Census, respondents were allowed to select more than one race. The "Race" categories denoting "Only" refer to those respondents who selected only one race.

4 Measures of central tendency

1 Here we will interpret the median as demarcating those values that are less than or equal to the median value and those that are greater than the median value. Others might interpret the median as indicating which values are less than the median and which are greater than or equal to the median. Such semantic differences rarely affect our analysis.
2 The author drew a sample of two-or-more-person families from the March 2008 CPS from Data-Ferrett and found that only 11 families out of 57,269 (or 0.02 percent) consisted of 12 or more members.
3 The procedures for determining the median of a frequency distribution will be discussed in Section 4.6. As for the mode of a frequency distribution, it is sufficient for most economic analyses to simply identify the modal class.
4 The definition of shelter employed is that used by Abby Duly at the Division of Consumer Expenditure (CE) Surveys, Bureau of Labor Statistics. Duly includes expenditures on shelter and utilities, but omits expenditures on "other components used in the CE survey, such as household furnishings and equipment," which she argues "are not, in fact, basic goods" (Duly 2003, p. 5).
5 According to the Consumer Expenditure Survey Glossary (U.S. Bureau of Labor Statistics 2010a), household operations consist of personal services and other household expenses. Personal services "includes baby-sitting; day care, nursery school, and preschool tuition; care of the elderly, invalids and handicapped; adult day care; and domestic and other duties." Other household expenses "includes housekeeping services, gardening and lawn care services, coin-operated laundry and dry-cleaning (non-clothing), termite and pest control products and services, home security systems service fees, moving, storage, and freight expenses, repair of household appliances and other household equipment, repair of computer systems for home use, computer information services, reupholstering and furniture repair, rental and repair of lawn and gardening tools, and rental of other household equipment."
6 The following states are included in the U.S. Census Bureau's delineation of the Midwest region: Illinois, Indiana, Iowa, Kansas, Michigan, Minnesota, Missouri, Nebraska, North Dakota, Ohio, South Dakota, and Wisconsin.
7 To facilitate the analysis, we will assume that prices for the items under household operations are comparable across the region.
8 We are using hypothetical data since the actual terms of an ARM are quite complex. We also do not need to know the loan amount to calculate the average interest rate.
9 The procedure described in Section 4.6 is one of several that we might use. We could, for example, round the percentile rank to the nearest whole unit, which here would be the ninth person. While we might think the interpolation procedure provides more precise estimates, scholars do not agree

that it is necessarily a superior procedure, nor do they agree as to which of the various procedures available is statistically "the best."

10 Excel's PERCENTILE function locates the ith percentile using the following rank equation:

$$R_{P_i} = 1 + \left[\frac{P_i}{100} \times (N-1) \right]$$

5 Measures of dispersion

1 In 2004, the median monthly housing costs for owner-occupied units in Sacramento, CA, was $1,206, and 36.8 percent of these homeowners incurred monthly costs of $1,500 and higher (U.S. Department of Housing and Urban Development and U.S. Census Bureau 2005a, p. 70). In the Seattle–Everett metropolitan area, median costs were $1,240, and 38.0 percent of home-owners had monthly costs equal to $1,500 and higher (U.S. Department of Housing and Urban Development 2005b, p. 70).

2 While not a complete analysis, we can note that the U.S. automobile industry experienced difficult times throughout the 2000s. Early in the decade, domestic manufacturers faced growing labor costs, especially related to health care because of an aging labor force and legacy retirees who were guaranteed coverage, as well as rising gasoline prices and the concurrent drop in demand for sports utility vehicles and light trucks. Their problems were compounded by the recent recession and internal financial problems, although by the end of the decade, Ford Motor Company (alone) saw sales and profits improving.

By comparison, Japanese-owned automobile makers fared better during the 2000s due in part to greater fuel efficiency and lower labor costs, although they too were affected by the recent economic recession (albeit at a later date than domestic producers). In addition, starting in November 2009, Toyota Motor Sales, which saw sales and market share continue to grow through much of the 2000s, was plagued by safety recalls for its vehicles. This suggests that Toyota's common stock prices might exhibit greater volatility when compared with Honda, with both companies' stock prices being less volatile than Ford's.

3 It is not clear in Bernat's article if he multiplied the coefficients of variation by 100 per convention; see chart 1 (Bernat 2001) for his results.

4 See, for example, the World Bank Institute (2005, Chapter 6) and Bellù and Liberati (2006) for details on the Gini coefficient's limitations, along with limitations associated with other measures of inequality used by economists.

6 Measuring changes in price and quantity

1 Economists and statisticians continue to seek the most superlative index. The "most superlative" might sound redundant, but it is a term introduced by Diewert (1976) to indicate a class of indices that could account for shifting consumption patterns in response to price changes. One superlative index, the Fisher ideal index, will be discussed in Section 6.5.

2 Hardware and software developers, among others, initially used two digits to signify years. As the year 2000 approached, programmers had to address how their products would handle the transition to four-digit years. This led many businesses and government agencies to upgrade their information technology to avoid problems.

7 Descriptions of stability

1 Recall the relationship between logarithmic and exponential functions. The exponential function is $y = e^x$. Taking the natural logarithm of y yields x; i.e. $\ln(y) = x$. If we now want to find the value of y, it must be true that $y = e^x = e^{\ln y}$. Therefore, the value of y can be found from $\ln(y)$ simply by finding the exponential of $\ln(y)$.

2 The properties of logarithms means we can also write equation (7.9) as

$$r = \left(\frac{\ln(P_t) - \ln(P_0)}{t} \right) \times 100$$

8 Patterns of long-term change

1 We will not derive the normal equations here. Readers interested in the derivation are directed to introductory econometrics textbooks for the proof.

2 Dated by the National Bureau of Economic Research (NBER), a recession in the United States is measured from the previous peak (the high point of the business cycle) to the trough (the cycle's low point) and occurs when real GDP declines for two consecutive quarters. (As of July 2003, NBER used a more complex, and controversial, yardstick; see <www.nber.org/cycles/recessions.html#faq> for details.) NBER (2010) provides the dates and duration of U.S. business cycles since December 1854, allowing economists to incorporate this information in their analyses of long-run growth for historical time series. See <http://www.nber.org/cycles/main.html> for more information.

9 Basic concepts in statistical inference

1 The mean was obtained from DeNavas-Walt *et al.* (2009, p. 29), while the standard deviation was calculated by the author based on the 2008 statistics.

2 Applying the Empirical Rule to the statistics indicates that some U.S. households in 2008 had incomes less than zero. Because households can report negative incomes in the Current Population Survey, our empirical results are not inconsistent with the underlying data.

10 Statistical estimation

1 We will not address limitations associated with the non-sampling errors, as they are beyond the scope of the text.

2 The slight difference in values here is the result of rounding all values to the nearest dollar.

3 "Student" was the name used by statistician William S. Gosset, who discovered the *t*-distribution while working for Guinness Brewing. He published his findings, "Probable error of a mean" in 1908 anonymously per Guinness' request.

4 This formula assumes that $\overline{X}_1$ and $\overline{X}_2$ are disjoint subsets of the population, which holds in our example because no person can be a member of both age groups. If one group is a subset of the other, the appropriate standard error of the difference between two non-disjoint estimates is

$$\mathrm{SE}(\overline{X}_2 - \overline{X}_1) = \sqrt{\mathrm{Var}(\overline{X}_1) + \mathrm{Var}(\overline{X}_2) - 2r[\mathrm{Var}(\overline{X}_1) + \mathrm{Var}(\overline{X}_2)]}$$

where r is the correlation coefficient between $\overline{X}_1$ and $\overline{X}_2$.

11 Statistical hypothesis testing of a mean

1 Freddie Mac is a government-sponsored business established by the U.S. Congress "to provide liquidity, stability, and affordability to the U.S. housing market." While it does not provide mortgages directly to home-owners, it provides funding to mortgage originators through the secondary loan markets in support of home-ownership and affordable housing (Freddie Mac 2010). For more details on the Home Possible® mortgages program, see <http://www.freddiemac.com/homepossible/>.

2 In practice, it is unlikely that eligibility for affordable mortgages would be determined at the regional level. However, in order to use actual economic data in the example, we will assume that the policy is written as stated.

3 Here we use median family income as the average. This is consistent with the income eligibility statistics used by Freddie Mac.

4 Both statistics were obtained from the 2007 American Community Survey (U.S. Census Bureau 2007d).

12 Correlation analysis

1 According to the World Bank, "GDP per capita based on purchasing power parity (PPP) is gross domestic product converted to international dollars using purchasing power parity rates. An international dollar has the same purchasing power over GDP as the U.S. dollar has in the United States. GDP at purchaser's prices is the sum of gross value added by all resident producers in the

economyplus any product taxes and minus any subsidies not included in the value of the products. It is calculated without making deductions for depreciation of fabricated assets or for depletion and degradation of natural resources. Data are in constant 2005 international dollars." See Notes for the variable in the *World Development Indicators* database.

2 Even if we had ten billion degrees of freedom (i.e., $n - 2 = 10,000,000,000$) and set α equal to 0.99999999999, the critical value of t equals 0.0009768. Thus the critical t-value is unlikely to ever equal zero.

3 Technically we should draw the new sample for the same year. However, the limited availability of multiple samples that measure GDP per capita and educational attainment prevents us from doing so. In addition, while we would prefer to draw a sample closer to the original year of 2000, educational attainment data is only available every five years from the World Bank.

13 Simple linear regression analysis

1 Economists distinguish between *simple regression analysis*, which assesses the relationship between the dependent variable and one independent variable, and *multiple regression analysis*, in which multiple independent variables are used in the regression model. The latter method is the focus of Part V.

2 A fourth relationship, pooled cross-section data, uses the same variables, but not the same elements, over multiple time periods. The econometric methods required for working this type of data, as well as for panel data, are beyond the scope of this text.

3 The procedure involves solving two simultaneous equations, the so-called *normal equations*, for a and b:

$$\sum Y = na + b \sum X \quad \text{and} \quad \sum XY = a \sum X + b \sum X^2$$

The solution of these equations for the unknown values of a and b will thus specify the linear function that minimizes the squared deviations measured on the Y-axis for the particular set of observations. Their derivation can be found in most introductory econometrics texts.

14 Simple regression analysis

1 The mathematical model assumes that, for all values of X:

- the mean value of the population error term, u, is zero;
- the variance of u is constant;
- observations of the error term are uncorrelated with each other;
- the population error term u is normally distributed.

If any of these assumptions is violated, our regression results may be biased and/or no longer best. Such problems will be addressed in Part V.

2 The model used included non-wage income for all individuals. The regression results were generated by the author using 1976 Panel Study of Income Dynamics data collected by Mroz (1976).

3 In the following two equations, the critical t-value at the 1 percent level for both slope coefficients equals 2.5824, and the estimated t-statistics are 12.9044 for the wives' model and -6.6797 for the husbands' model. Thus we are at least 99 percent confident that both coefficients are statistically different from zero.

15 Simple regression analysis

1 "Linear with respect to the *estimated* coefficients" means that we can define the dependent variable, Y, as a function of a and b, but not, for example, as a function of b^2 or $\ln(a)$. This assumption is also key for the ordinary least-squares technique to generate the statistically "best" estimates for a regression model's parameters.

2 We see that both coefficients are statistically greater than zero at a 99 percent level of confidence, but also note the very low R^2 statistics for both models. This suggests that we have omitted other factors, such as work experience and the presence of children, that might help to explain the variation in earnings for members of married couples.

16 Multiple regression analysis

1 Even though we are now estimating the relationship between three or more variables and no longer working in two dimensions, we will, for discussion purposes, continue to refer to the regression "line."
2 We will not present the much more complex formulas for estimating the partial regression coefficients here. For our purposes, what is important is understanding what each coefficient measures and how to interpret it in the context of the estimated model.
3 According to the World Development Indicators of the World Bank (2010b), the Atlas method "applies a conversion factor that averages the exchange rate for a given year and the two preceding years, adjusted for differences in rates of inflation between the country" for the purpose of smoothing fluctuations in prices and exchange rates.
4 Note that we did not interpret the estimated intercept term. To attach economic meaning to that coefficient, we must assume that the means of all four independent variables are zero, which is neither economically realistic nor consistent with the actual data values.
5 The literature on infant mortality suggests that factors such as war and political instability, unequal distribution of wealth, HIV/AIDS, and communicable diseases all contribute to increased deaths among children. Adequate data series for these variables, however, are not available for the number of countries in the sample.

17 Multiple regression analysis

1 The degrees of freedom in the bracketed term of equation (17.2) mean that $\bar{R}^2$ will not equal zero. It may also occur that, in cases where the unadjusted R^2 statistic is small in value, $\bar{R}^2$ will be negative in value.
2 The following table illustrates the relationship between the variations and variances in the context of regression models.

Unbiased estimators of the multiple regression model's variations and variances

	Variation	Variance
Total	$TSS = \sum(Y_i - \bar{Y})^2$	$\sum(Y_i - \bar{Y})^2/(n-1)$
Regression	$RSS = \sum(\hat{Y}_i - \bar{Y})^2$	$\sum(\hat{Y}_i - \bar{Y})^2/k$
Error	$ESS = \sum(Y_i - \hat{Y}_i)^2 = \sum e_i^2$	$\sum(Y_i - \hat{Y}_i)^2/(n-k-1)$

3 It is possible that adding another relevant independent variable will not change the sum of the squared residuals. It will not, however, increase $\sum e_i^2$.

18 Multiple regression analysis

1 The complete regression results from Excel are below:

$$\ln(\text{EARNS}) = b_0 + b_{\text{AGE}} \times \text{AGE} + b_{\text{AGESQ}} \times \text{AGESQ} + b_{\text{EDUC}} \times \text{EDUC} + b_{\text{MALE}} \times \text{MALE}$$

Regression statistics

Multiple R	0.6768
R Square	0.4580
Adjusted R Square	0.4562
Standard error	0.8275
Observations	1181

ANOVA

	df	SS	MS	F	Significance F
Regression	4	680.44	170.11	248.44	1.0454E-154
Residual	1176	805.21	0.68		
Total	1180	1485.65			

	Coefficients	Standard error	t Stat	p-value	1 − p/2
Intercept	4.20	0.7234	5.8001	0.0000	
EDUC	0.0840	0.0088	9.5690	0.0000	100.0000
AGE	0.1250	0.0329	3.7985	0.0002	99.9924
AGESQ	−0.0013	0.0004	−3.6258	0.0003	99.9850
MALE	1.4893	0.0510	29.2038	0.0000	100.0000

2 The marginal effect of experience at 44 years of age is determined by the following equation:

$$\left.\frac{\partial \ln \text{EARNS}}{\partial \text{AGE}}\right|_{\text{MED AGE}=44} = \hat{b}_{\text{AGE}} + [2\hat{b}_{\text{AGESQ}} \times \text{AGE}_{\text{MED}}]$$
$$= 0.1250 + [2 \times (-0.0013) \times 44] = 0.0067$$

We must multiply this result by 100 to convert it into percentages.

3 For the earnings model, the coefficient measures the *percentage* difference because the dependent variable is ln(EARNS).

4 The sub-Saharan countries included in this statistical analysis are Angola, Benin, Botswana, Burkina Faso, Burundi, Cameroon, Cape Verde, Central African Republic, Chad, Comoros, Republic of the Congo, Democratic Republic of the Congo, Cote d'Ivoire, Equatorial Guinea, Eritrea, Ethiopia, Gabon, The Gambia, Ghana, Guinea, Guinea-Bissau, Kenya, Lesotho, Liberia, Madagascar, Malawi, Mali, Mauritius, Mozambique, Namibia, Niger, Nigeria, Rwanda, Sao Tome and Principe, Senegal, Sierra Leone, Swaziland, Tanzania, Togo, Uganda, Zambia, and Zimbabwe. Following the categories in UNICEF's *The State of Africa's Children 2008*, the statistical analysis also included Djibouti and Sudan in the group. The Seychelles, Somali, and South Africa were excluded from the analysis due to incomplete data.

5 We note that an independent variable *can* be a nonlinear function of other independent variables in the multiple regression model.

6 The negative relationship between the two independent variables may be unexpected, but it might be explained as follows. Countries with higher standards of living, as measured by GNI per capita, may be able to devote a smaller percentage of GNI to per-capita health expenditures as compared to countries with low levels of GNI per capita.

7 In detecting multicollinearity, it is the strength of the correlation, rather than its direction, that is crucial. Consequently, we use the absolute values of the correlation coefficients in our determinations.

8 Another model in which the GNI variable was removed and HLTH was included produced the following results.

Regression statistics	
Multiple R	0.9141
R Square	0.8356
Adjusted R Square	0.8300
Standard error	14.1151
Observations	152

ANOVA

	df	SS	MS	F	Significance F
Regression	5	147,854.09	29,570.82	148.42	2.12E-55
Residual	146	29,088.44	199.24		
Total	151	176,942.53			

	Coefficients	Standard error	t Stat	p-value	$[1-(p/2)] \times 100$
Intercept	70.4606	13.5564	5.1976	0.0000	
lnHLTH	4.3075	1.2005	3.5882	0.0005	99.9773
ANEMIA	0.2556	0.1136	2.2501	0.0259	98.7032
ATTEND	−0.1769	0.0803	−2.2018	0.0292	98.5377
H_2O	−0.6498	0.1144	−5.6823	0.0000	100.0000
SSAFR	19.2952	3.3939	5.6853	0.0000	100.0000

Here again we do not obtain the expected sign on HLTH, which suggests we remove HLTH from the model. Doing so yields the following results.

Regression statistics

Multiple R	0.9061
R Square	0.8211
Adjusted R Square	0.8162
Standard error	14.6742
Observations	152

ANOVA

	df	SS	MS	F	Significance F
Regression	4	145,288.89	36,322.22	168.681	7.13E-54
Residual	147	31,653.64	215.33		
Total	151	176,942.53			

	Coefficients	Standard error	t Stat	p-value	$[1-(p/2)] \times 100$
Intercept	98.2523	11.5667	8.4944	0.0000	
ANEMIA	0.3928	0.1112	3.5322	0.0006	99.9725
ATTEND	−0.2883	0.0770	−3.7444	0.0003	99.9871
H_2O	−0.7085	0.1177	−6.0209	0.0000	100.0000
SSAFR	22.5792	3.3976	6.6456	0.0000	100.0000

When we compare the results of this model with those from equation (18.1), in light of our discussion of omitted variable bias, equation (18.1) is the statistically superior model.

9 We must also exclude any irrelevant independent variables, but this is generally less of an issue for econometric models. Since such models are grounded in economic theory, we are unlikely to include variables at odds with the model's theoretical underpinnings.

Bibliography

Allen, L. H. (2000) "Anemia and iron deficiency: effects on pregnancy outcome." *American Journal of Clinical Nutrition* **71**(5), 1280S–1284S. [Online] Available from <http://www.ajcn.org/cgi/reprint/71/5/1280S> [Accessed 1 November 2010].

Balnaves, M., and Caputi, P. (2001) *Introduction to Quantitative Research Methods: An Investigative Approach.* Sage Publications, Thousand Oaks, CA.

Beer Institute. (2010) *Brewers Almanac 2010.* [Online] Available from <http://www.beerinstitute.org/statistics.asp?bid=200> [Accessed 5 December 2010].

Bellù, L. G., and Liberati, P. (2006) *Inequality Analysis: The Gini Index.* Food and Agriculture Organization of the United Nations, Rome. [Online] Available from <http://www.fao.org/docs/up/easypol/329/gini_index_040en.pdf> [Accessed 28 July 2010].

Bernanke, B. S. (2010) *Monetary Policy and the Housing Bubble.* Speech to the American Economic Association. [Online] Available from <http://www.federalreserve.gov/newsevents/speech/bernanke20100103a.pdf> [Accessed 2 July 2010].

Bernat, Jr, G. A. (2001) "Convergence in state per capita personal income, 1950–99." *Survey of Current Business* June, 36–48. [Online] Available from <http://www.bea.gov/scb/pdf/2001/06june/0601cspi.pdf> [Accessed 8 January 2008].

Bucks, B. K., Kennickell, A. B., Mach, T. L., and Moore, K. B. (2009) "Changes in U.S. family finances from 2004 to 2007: evidence from the Survey of Consumer Finances." *Federal Reserve Bulletin* **95**, A1–A55. [Online] Available from <http://www.federalreserve.gov/pubs/bulletin/2009/pdf/scf09.pdf> [Accessed 13 September 2010].

Cliff, N. (1996) *Ordinal Methods for Behavioral Data Analysis.* Lawrence Erlbaum Associates, Mahwah, NJ.

Conference Board. (2001) *Business Cycle Indicators Handbook.* [Online] Available from <http://www.conference-board.org/pdf_free/economics/bci/BCI-Handbook.pdf> [Accessed 16 August 2010].

DeNavas-Walt, C., Proctor, B. D., and Smith, J. C. (2009) *Income, Poverty, and Health Insurance Coverage in the United States: 2008.* [Online] Available from <http://www.census.gov/prod/2009pubs/p60-236.pdf> [Accessed 5 October 2009].

DeNavas-Walt, C., Proctor, B. D., and Smith, J. C. (2010) *Income, Poverty, and Health Insurance Coverage in the United States: 2009.* [Online] Available from <http://www.census.gov/prod/2010pubs/p60-238.pdf> [Accessed 29 November 2010].

Diewert, W. E. (1976) "Exact and superlative index numbers." *Journal of Econometrics* **4**(2), 115–145.

Dismukes, R., and Durst, R. L. (2006) *Whole-Farm Approaches to a Safety Net.* Economic Research Service, U.S. Department of Agriculture, Washington, DC. [Online] Available from <http://www.ers.usda.gov/publications/EIB15/eib15a.pdf> [Accessed 17 August 2009].

Duly, A. (2003) "Consumer spending for necessities." *Monthly Labor Review* **126**(5), 5. [Online] Available from <http://www.bls.gov/ opub/mlr/2003/05/art1full.pdf> [Accessed 25 May 2010].

Economic Report of the President. (1947) United States Government Printing Office, Washington, DC. [Online] Available from <http://fraser.stlouisfed.org/publications/ERP/issue/1630/download/7538/ERP_1947_January.pdf> [Accessed 9 July 2010].

Economic Report of the President. (1983) United States Government Printing Office, Washington, DC. [Online] Available from <http://fraser.stlouisfed.org/publications/ERP/issue/1386/download/5917/ERP_1983.pdf> [Accessed 13 July 2010].

Economic Report of the President. (1994) United States Government Printing Office, Washington, DC. [Online] Available from <http://fraser.stlouisfed.org/publications/ERP/issue/1600/download/6050/ERP_1994.pdf> [Accessed 13 July 2010].

Economic Report of the President. (2004) United States Government Printing Office, Washington, DC. [Online] Available from <http://www.gpo.gov/fdsys/pkg/ERP-2004/pdf/ERP-2004-table31.pdf> [Accessed 15 March 2011].

Economic Report of the President. (2009a) United States Government Printing Office, Washington, DC. [Online] Available from <http://www.gpoaccess.gov/eop/2009/2009_erp.pdf> [Accessed 13 July 2010].

Economic Report of the President. (2009b) United States Government Printing Office, Washington, DC. Table B-100, Farm input use, selected inputs, 1948–2007. [Online] Available from <http://www.gpoaccess.gov/eop/tables09.html#erp8> [Accessed 30 August 2010].

Economic Research Service, U.S. Department of Agriculture. (2007) *2007 Vegetables and Melons Yearbook.* [Online] Available from <http://www.ers.usda.gov/Publications/vgs/2007/07JulYearbook/VGS2007.pdf> [Accessed 16 February 2009].

Economic Research Service, U.S. Department of Agriculture. (2009) *Agricultural Productivity in the United States.* [Online] Available from <http://www.ers.usda.gov/Data/AgProductivity/> [Accessed 15 July 2009].

Economic Research Service, U.S. Department of Agriculture. (2010a) *Food Availability (Per Capita) Data System.* [Online] Available from <http://www.ers.usda.gov/Data/FoodConsumption/FoodAvailSpreadsheets.htm> [Accessed 11 July 2010].

Economic Research Service, U.S. Department of Agriculture. (2010b) *State Fact Sheets: United States.* [Online] Available from <http://www.ers.usda.gov/StateFacts/US.htm> [Accessed 22 July 2010].

Elliott, D. J., and Baily, M. N. (2009) *Telling the Narrative of the Financial Crisis: Not Just a Housing Bubble.* The Brookings Institution, Washington, DC. [Online] Available from <http://www.brookings.edu/~/media/Files/rc/papers/2009/1123_narrative_elliott_baily/1123_narrative_elliott_baily.pdf> [Accessed 2 July 2010].

FRED, Federal Reserve Economic Data, Federal Reserve Bank of St. Louis (2010a) *Real Gross Domestic Product, 3 Decimal (GDPC96).* [Online] Available from <http://research.stlouisfed.org/fred2/series/GDPC96?cid=106> [Accessed 3 December 2010].

FRED, Federal Reserve Economic Data, Federal Reserve Bank of St. Louis (2010b) *New One Family Houses Sold (HSN1F).* [Online] Available from <http://research.stlouisfed.org/fred2/series/HSN1F> [Accessed 15 May 2009].

FRED, Federal Reserve Economic Data, Federal Reserve Bank of St. Louis (2010c) *30-Year Conventional Mortgage Rate (MORTG).* [Online] Available from <http://research.stlouisfed.org/fred2/series/MORTG> [Accessed 15 May 2009].

FRED, Federal Reserve Economic Data, Federal Reserve Bank of St. Louis (2010d) *Real Personal Consumption Expenditures, 3 Decimal (PCEC96).* [Online] Available from <http://research.stlouisfed.org/fred2/series/PCEC96> [Accessed 3 December 2010].

FRED, Federal Reserve Economic Data, Federal Reserve Bank of St. Louis (2010e) *Initial Claims for Unemployment Insurance, Seasonally Adjusted (ICSA).* [Online] Available from <http://research.stlouisfed.org/fred2/series/ICSA> [Accessed 5 December 2010].

FRED, Federal Reserve Economic Data, Federal Reserve Bank of St. Louis (2010f) *Consumer Price Index for All Urban Consumers: Education and Communication (CPIEDUSL).* [Online] Available from <http://research.stlouisfed.org/fred2/series/CPIEDUSL> [Accessed 5 December 2010].

Freddie Mac (2010) *Home Possible® Mortgages Frequently Asked Questions*. [Online] Available from <http://www.freddiemac.com/homepossible/> [Accessed 13 December 2010].

Friendly, M. (2010) *Gallery of Data Visualization: The Best and Worst of Statistical Graphics*. [Online] Available from <http://www.math.yorku.ca/SCS/Gallery/> [Accessed 16 November 2010].

Fuglie, K. O., MacDonald, J. M., and Ball, E. (2007) *Productivity Growth in U.S. Agriculture*. EB-9. Economic Research Service, U.S. Department of Agriculture, Washington, DC. [Online] Available from <http://www.ers.usda.gov/publications/EB9/eb9.pdf> [Accessed 5 December 2006].

Guthrie, H. W. (1966) *Statistical Methods in Economics*. Richard D. Irwin, Homewood, IL.

Hirsch, B. T., and Schumacher, E. J. (1998) "Union wages, rents, and skills in health care labor markets." *Journal of Labor Research* **19**(1), 125–147. [Online] Available from EBSCOhost Business Source Premier database <http://www.ebscohost.com/academic/business-source-premier> [Accessed 15 November 2010].

Just, R. E., and Weninger, Q. (1999) "Are crop yields normally distributed?" *American Journal of Agricultural Economics* **81**, 287–304. [Online] Available from JSTOR <www.jstor.org> [Accessed 7 December 2010].

Kaufman, B. E., and Hotchkiss, J. L. (2006) *The Economics of Labor Markets*, 7th edn. Thomson/South-Western, Mason, OH.

Lindsay, J. J., Wan, Y., and Gossin-Wilson, W. (2009) *Methodologies Used by Midwest Region States for Studying Teacher Supply and Demand*. U.S. Department of Education, Institute of Education Sciences, National Center for Education Evaluation and Regional Assistance, Regional Educational Laboratory Midwest, Washington, DC. [Online] Available from <http://ies.ed.gov/ncee/edlabs/regions/midwest/pdf/REL_2009080.pdf> [Accessed 27 August 2010].

Mroz, T. A. (1976) *Panel Study of Income Dynamics Data*. [Online] Available from <http://pages.stern.nyu.edu/~wgreene/Text/tables/TableF4-1.txt> [Accessed 25 October 2010].

National Bureau of Economic Research. (2010) *U.S. Business Cycle Expansions and Contractions*. [Online] Available from <http://www.nber.org/cycles/cyclesmain.html> [Accessed 16 November 2010].

National Center for Education Statistics, U.S. Department of Education (2008) *Digest of Education Statistics: 2007*. Table 320, Average undergraduate tuition and fees and room and board rates charged for full-time students in degree-granting institutions, by type and control of institution: 1964–65 through 2006–07. [Online] Available from <http://nces.ed.gov/programs/digest/d07>.

National Center for Education Statistics, U.S. Department of Education (2010) *Digest of Education Statistics: 2009*. [Online] Available from <http://www.nces.ed.gov/> [Accessed 6 August 2010].

OECD. Stat Extracts. (2010) [Online] Available from <http://stats.oecd.org/wbos/> [Accessed 20 August 2009].

Stevens, S. S. (1946) "On the theory of scales of measurement." *Science* **103**(2684), 677–680.

Tufte, E. R. (2007) *The Visual Display of Quantitative Information*, 2nd ed,. Graphics Press, Cheshire, CT.

United Nations (2011) *Knowledge Base on Economic Statistics*. [Online] Available from <http://unstats.un.org/unsd/EconStatKB/Knowledgebase.aspx>.

United Nations Children's Fund (UNICEF) (2010) *Millennium Development Goals*. UNICEF, New York. [Online] Available from <http://www.unicef.org/mdg/childmortality.html> [Accessed 1 November 2010].

United Nations Development Programme. (2007) *Measuring Human Development: A Primer*. [Online] Available from <http://hdr.undp.org/en/media/Primer_complete.pdf> [Accessed 9 August 2010].

United Nations Development Programme. (2010a) *Calculating the Human Development Indices*. [Online] Available from <http://hdr.undp.org/en/statistics/data/> [Accessed 1 December 2010].

United Nations Development Programme. (2010b) *Composite Indices — HDI and Beyond*. [Online] Available from <http://hdr.undp.org/en/statistics/indices/> [Accessed 9 August 2010].

United Nations Development Programme. (2010c) *The Human Development Index (HDI): Technical Notes*. [Online] Available from <http://hdr.undp.org/en/media/HDR_2010_EN_TechNotes_reprint.pdf> [Accessed 1 December 2010].

U.S. Bureau of Economic Analysis. (2010a) *Regional Economic Accounts*. [Online] Available from <http://www.bea.gov/regional/gsp/> [Accessed 24 June 2010].

U.S. Bureau of Economic Analysis. (2010b) "Why does BEA publish percent changes in quarterly series at annual rates?" *Frequently Asked Questions*. [Online] Available from <www.faq.bea.gov/> [Accessed 13 August 2010].

U.S. Bureau of Economic Analysis. (2010c) *National Economic Accounts*. [Online] Available from <http://www.bea.gov/national/index.htm> [Accessed 23 June 2010].

U.S. Bureau of Economic Analysis. (2010d) *Gross Domestic Product by State*. [Online] Available from <http://www.bea.gov/regional/gsp/> [Accessed 19 August 2010].

U.S. Bureau of Economic Analysis. (2010e) *National Economic Accounts, Interactive Access to National Income and Product Accounts Tables*, Table 7.2.6B, Real Motor Vehicle Output, Chained Dollars. [Online] Available from <http://www.bea.gov/national/nipaweb/>.

U.S. Bureau of Labor Economics. (2010) *Major Sector Productivity and Costs Index*. [Online] Available from <http://www.bls.gov/lpc/> [Accessed 2 April 2010].

U.S. Bureau of Labor Statistics. (2008a) *2008 Consumer Expenditure Survey. Standard Error Tables*. [Online]. Available from <http://www.bls.gov/cex/csxstnderror.htm#2008> [Accessed 14 September 2010].

U.S. Bureau of Labor Statistics. (2008b) *Consumer Expenditure Survey, 2006–2007*. [Online] Available from <ftp://ftp.bls.gov/pub/special.requests/ce/msa/y0607/midwest.txt> [Accessed 18 July 2010].

U.S. Bureau of Labor Statistics. (2008c) *Consumer Expenditure Survey, 2007*. [Online] Available from <http://www.bls.gov/cex/home.htm#tables> [Accessed 18 July 2009].

U.S. Bureau of Labor Statistics. (2008d) "The unemployment rate and beyond: alternative measures of labor underutilization." *Issues in Labor Statistics*. U.S. Department of Labor, Washington, DC. [Online] Available from <http://www.bls.gov/opub/ils/pdf/opbils67.pdf> [Accessed 9 July 2010].

U.S. Bureau of Labor Statistics. (2008e) *National Occupational Employment and Wage Estimates*. [Online] Available from <http://www.bls.gov/oes/current/oes_nat.htm> [Accessed 15 August 2009].

U.S. Bureau of Labor Statistics. (2009) *Current Population Survey, March 2009*. [Online] Available from DataFerrett <http://dataferrett.census.gov> [Accessed 18 May 2010].

U.S. Bureau of Labor Statistics. (2010a) *Consumer Expenditure Survey Glossary*. [Online] Available from <http://www.bls.gov/cex/csxgloss.htm#housing> [Accessed 18 July 2010].

U.S. Bureau of Labor Statistics. (2010b) *Consumer Price Index*. [Online] Available from <www.bls.gov/cpi> [Accessed 25 June 2010].

U.S. Bureau of Labor Statistics. (2010c) *Current Employment Survey*. [Online] Available from <www.bls.gov/ces/> [Accessed 25 June 2010].

U.S. Bureau of Labor Statistics. (2010d) *Current Population Survey*. [Online] Available from <www.bls.gov/cps> [Accessed 29 June 2010].

U.S. Bureau of Labor Statistics. (2010e) *Data Retrieval: Labor Force Statistics (CPS)*. [Online] Available from <http://www.bls.gov/webapps/legacy/cpsatab12.htm> [Accessed 19 November 2010].

U.S. Bureau of Labor Statistics. (2010f) *Labor Force Statistics from the Current Population Survey*. [Online] Available from <http://www.bls.gov/cps/lfcharacteristics.htm#laborforce> [Accessed 9 July 2010].

U.S. Bureau of Labor Statistics. (2010g) *Local Area Unemployment Statistics*. [Online] Available from <http://www.bls.gov/lau/> [Accessed 24 June 2010].

U.S. Bureau of Labor Statistics. (2010h) *State and Area Employment, Hours, and Earnings*. [Online] Available from <http://data.bls.gov:8080/PDQ/outside.jsp?survey=sm> [Accessed 29 July 2010].

U.S. Bureau of Labor Statistics. (2011) *Consumer Price Index, Frequently Asked Questions (FAQs)*. [Online] Available from <http://www.bls.gov/cpi/cpifaq.htm>.

U.S. Census Bureau. (2000a) *Census 2000. Summary File 1 (SF 1) 100-Percent Data*. [Online] Available from <http://factfinder.census.gov/> [Accessed 5 January 2009].

U.S. Census Bureau. (2000b) *Census 2000. Summary File 3 (SF 3) Sample Data*. Value, mortgage status, and selected conditions: 2000 (QT-H14). [Online] Available from <http://factfinder.census.gov/> [Accessed 5 January 2009].

U.S. Census Bureau. (2004a) *Statistical Abstract of the United States: 2003*. [Online] Available from <http://www.census.gov/prod/2004pubs/03statab/intlstat.pdf> [Accessed 28 April 2005].

U.S. Census Bureau. (2004b) *American Housing Survey for the United States: 2003*. [Online] Available from DataFerrett at <http://dataferrett.census.gov/index.html> [Accessed 23 August 2009].

U.S. Census Bureau. (2005) *American Housing Survey, 2004 Metropolitan Sample*. [Online] Available from DataFerrett at <http://dataferrett.census.gov/index.html> [Accessed 23 August 2009].

U.S. Census Bureau. (2006) *American Housing Survey for the United States: 2005*. [Online] Available from DataFerrett at <http://dataferrett.census.gov/index.html> [Accessed 23 August 2009].

U.S. Census Bureau. (2007a) *2007 Statistical Abstract*. [Online] Available from <http://www.census.gov/compendia/statab/2007/> [Accessed 3 August 2010].

U.S. Census Bureau. (2007b) *American FactFinder Glossary*. [Online] Available from <http://factfinder.census.gov/home/en/epss/glossary_o.html> [Accessed 16 October 2010].

U.S. Census Bureau. (2007c) *Statistical Abstract of the United States: 2007*. [Online] Available from <http://www.census.gov/compendia/statab/2007/labor_force_employment_earnings/labor_force_status.html> [Accessed 3 August 2010].

U.S. Census Bureau. (2007d) *American Community Survey, 2007*. [Online] Available from the American FactFinder <http://factfinder.census.gov> [Accessed 15 December 2010].

U.S. Census Bureau. (2008a) *American Community Survey, 2008*. [Online] Available from DataFerrett at <http://dataferrett.census.gov/index.html> [Accessed 26 August 2010].

U.S. Census Bureau. (2008b) *Current Population Survey, March 2008*. [Online] Available from <http://www.census.gov/population/www/socdemo/hh-fam/cps2008.html> [Accessed 16 August 2009].

U.S. Census Bureau. (2009a) *Current Population Survey, 2009 Annual Social and Economic (ASEC) Supplement*. [Online] Available from <http://www.census.gov/apsd/techdoc/cps/cpsmar09.pdf> [Accessed 9 July 2010].

U.S. Census Bureau. (2009b)*Current Population Survey, June 2009*. [Online] Available from DataFerrett at <http://dataferrett.census.gov/index.html> [Accessed 26 August 2009].

U.S. Census Bureau. (2009c) *Income*. [Online] Available from <http://www.census.gov/hhes/www/cpstables/032009/hhinc/new06_000.htm> [Accessed 19 November 2010].

U.S. Census Bureau. (2010a) *American FactFinder Glossary*. [Online] Available from <http://factfinder.census.gov/home/en/epss/glossary_m.html> [Accessed 5 January 2009].

U.S. Census Bureau. (2010b) *Current Population Survey Glossary*. [Online] Available from <http://www.census.gov/population/www/cps/cpsdef.html> [Accessed 31 July 2010].

U.S. Census Bureau. (2010c) *Income*. [Online] Available from <http://www.census.gov/hhes/www/income/data/historical/household/index.html> [Accessed 22 July 2010].

U.S. Census Bureau. (2010d) *Income Inequality*. [Online] Available from <http://www.census.gov/hhes/www/income/data/historical/inequality/index.html> [Accessed 29 November 2010].

U.S. Census Bureau. (2010e) *North American Industrial Classification System, Frequently Asked Questions*. [Online] Available from <http://www.census.gov/eos/www/naics/faqs/faqs.html#q5> [Accessed 16 November 2010].

U.S. Census Bureau. (2010f) *Statistical Abstract of the United States: 2009*, 129th edn. [Online] Available from <http://www.census.gov/compendia/statab/overview.html> [Accessed 5 July 2010].

U.S. Census Bureau. (2011) *Statistical Abstract of the United States: 2010*. [Online] Available from <http://www.census.gov/compendia/statab/2011/tables/11s0226.pdf> [Accessed 15 March 2011].

U.S. Department of Health and Human Services, Health Resources and Services Administration. (2010) *2004 National Sample Survey of Registered Nurses*. [Online] Available from <http://datawarehouse.hrsa.gov/nursingsurvey.aspx> [Accessed 17 April 2010].

U.S. Department of Housing and Urban Development and U.S. Census Bureau. (2005a) *American Housing Survey for the Sacramento Metropolitan Area: 2004*. [Online] Available from <http://www.census.gov/prod/2005pubs/h170-04-58.pdf> [Accessed 16 August 2010].

U.S. Department of Housing and Urban Development and U.S. Census Bureau. (2005b) *American Housing Survey for the Seattle–Everett Metropolitan Area: 2004*. [Online] Available from <http://www.census.gov/prod/2005pubs/h170-04-60.pdf> [Accessed 16 August 2010].

U.S. Government Accountability Office. (2005) *College Textbooks: Enhanced Offerings Appear to Drive Recent Price Increases*. [Online] Available from <www.gao.gov/cgi-bin/getrpt?GAO-05-806> [Accessed 20 September 2009].

Wooldridge, J. M. (2009) *Introductory Econometrics; A Modern Approach*, 4th edn. South-Western Cengage Learning, Mason, OH.

World Almanac and Book of Facts (2009) [Online] Available from EBSCOhost Academic Source Premier database <http://www.ebscohost.com/academic/academic-search-premier> [Accessed 31 August 2010].

World Bank. (1999) *Global Development Finance*, vol. 1. [Online] Available from <http://www-wds.worldbank.org/external/default/WDSContentServer/WDSP/IB/2000/12/20/000094946_99052707174626/Rendered/PDF/multi0page.pdf> [Accessed 28 November 2010].

World Bank. (2010a) *EdStats Database*. [Online] Available from <http://data.worldbank.org/data-catalog/ed-stats> [Accessed 20 October 2010].

World Bank. (2010b) *World Development Indicators*. [Online] Available from <http://data.worldbank.org/data-catalog> [Accessed 3 August 2010].

World Bank Institute. (2005) *Introduction to Poverty Analysis*. [Online] Available from <http://siteresources.worldbank.org/PGLP/Resources/PovertyManual.pdf> [Accessed 27 July 2010].

World Health Organization. (2010a) "Newborns: reducing mortality." [Online] Available from <http://www.who.int/mediacentre/factsheets/fs333/en/index.html> [Accessed 1 November 2010].

World Health Organization. (2010b) *Global Health Observatory*. [Online] Available from <http://apps.who.int/ghodata/> [Accessed 1 November 2010].

World Health Organization Statistical Information System. (2010) *World Health Statistics Indicator Compendium*. [Online] Available from <http://www.who.int/whosis/indicators/WHS10_IndicatorCompendium_20100513.pdf> [Accessed 25 June 2010].

Yahoo! Finance. (2010) [Online] Available from <http://finance.yahoo.com> [Accessed 25 July 2010].

Index

Page numbers in **bold** indicate where key concepts are introduced.